Comparative Island Archaeologies

Edited by

James Conolly
Matthew Campbell

BAR International Series 1829
2008

This title published by

Archaeopress
Publishers of British Archaeological Reports
Gordon House
276 Banbury Road
Oxford OX2 7ED
England
bar@archaeopress.com
www.archaeopress.com

BAR S1829

Comparative Island Archaeologies

ISBN 978 1 4073 0313 0

Printed in England by Alden HenDi, Oxfordshire

All BAR titles are available from:

Hadrian Books Ltd
122 Banbury Road
Oxford
OX2 7BP
England
bar@hadrianbooks.co.uk

The current BAR catalogue with details of all titles in print, prices and means of payment is available free from Hadrian Books or may be downloaded from www.archaeopress.com

Contents

Preface: comparative island archaeologies

Matthew Campbell, University of Auckland
James Conolly, Trent University

Islands hold a particular place in Western consciousness, and the long-term history of islands and their peoples continue to hold archaeologists' interest, despite recent attempts to deconstruct, or at least destabilize, notions of insularity (e.g., Rainbird 1999; 2004; 2007; Terrell 2004; Lape 2004). However, even critics who justifiably condemn outdated concepts like the 'primitive isolate' (Terrell et al. 1997), and perhaps even of the whole concept of 'island archaeology' as a distinctive enterprize (Rainbird 1999; 2007), still maintain their own island-based research foci and—critically—continue to support students who wish to engage with island cultures. Actively decentralizing islandness whilst continuing to promote island archaeology is not as incompatible as it seems, and our sense is that the reason for the continued strength of island-centered research agendas is that island archaeology continues to make, as it has in the past, a useful and, more importantly, distinctive, contribution to the study of long-term human ecology, culture and social change—the sum of which reinforces the value of island-based studies in the social sciences.

It is true, however, that islands are neither the "laboratories for the study of culture process" (Evans 1973) that they were once optimistically claimed to be (cf. Irwin 1992: 206; Rainbird 2007: 23–33), nor are they privileged places for building understanding of the human past. Islands nevertheless are still good places to 'do' archaeology, and there are likely as many reasons for this continued interest in islands as there are archaeologists who work in insular settings. At the most banal, islands demand our attention as archaeologists because of our mandate to document and understand the workings of culture history. Whilst banal, it is this task that requires us to be sensitive to the differences between terrestrial and oceanic islandscapes, and the interactions between them (cf. Broodbank 2000: 21–35). It is difficult and quite possibly futile to seek unifying generalizations about the nature of life on islands to help us in this regard, but yet the study of island history requires us to take on board a distinctive set of interwoven topics of investigation that cross-cut any specific geographical or temporal context.

Four obvious examples of clearly interlinked concerns with which island archaeologists typically engage include, first, *seafaring and voyaging*—the weaving together of the knowledge and know-how of both sea and weather conditions with maritime technologies in a way that is without obvious parallel in terrestrial cultures. Second, voyaging is linked to the ongoing concern with the processes and consequences of *colonization and abandonment*, a topic with a long and distinguished contextual history in Oceania and the Mediterranean, but a topic that is usually only addressed at much coarser, continental, scales elsewhere. Then, third, are the detailed studies of island *human ecology*, especially colonists' impact on island biodiversity—which is of course of much wider concern, but islands offer a unique perspective on human ecology because of their (usually) geographically circumscribed character. Fourth, there is a long-standing concern with the *social interaction* between maritime communities and the ways that islanders construct and maintain their on- and off-island social relationships. This too is a central theme in non-insular cultural geographies, but again it is the circumstances of island life that make the specific workings of this qualitatively different for those who live in island contexts.

To reiterate: none of these four basic but foundational topics of study are solely the concern of island archaeology, nor do islands necessarily offer a privileged perspective for addressing them; it remains the case, however, that for those who study with island archaeology these issues are not only important but are often central concerns—even to those who are not convinced of the heuristic value of a distinctly island archaeology. This is not to deny the diversity of other topics that archaeologists concerned with island peoples engage with—all that is emphasized here is that there are recurrent themes that help define why there remains a distinctive 'island archaeology'.

These four themes are explored in detail in the papers in this volume using data from the Pacific, the Caribbean, the North Sea and the Mediterranean. That these themes are reflected in this collection of essays is not coincidental. When we came up with the proposal (sitting in a pub in Auckland, in July 2003) of hosting a conference to explore the comparative archeology of islands, we anticipated that some unifying issues would emerge without direction simply because they are foundational. The aim thus was to bring together archaeologists who work on islands from as broad a geographical spread as possible, to provide a wide comparative perspective. The value of a global comparative approach is to provide insight into the underlying reasons for observed similarities and differences in vastly different social and environmental contexts. The benefit of such cross fertilization is exemplified by Broodbank's (2000) study of Cycladic island history, which draws heavily on the productive theoretical frameworks explored in Oceania. It is in this spirit—that there is enormous value to be gained from comparative study of the workings of cultural and environmental history—that the Auckland conference was organized.

The papers in this volume are organized in such a way as to promote a comparative island archaeology. The first group of papers (Chapters 1 to 4) is concerned with human ecology in a broad sense (i.e., inclusive of subsistence technologies), and includes contributions by David Addison (Marquesas Islands), Jacqueline Craig (Cook Islands), Dominique Guillaud, Hubert Forestier and Harry Truman Simanjuntak (Indonesian islands), and Paul Memmott, Richard Robins and Errol Stock (Wellesley Islands). While the coverage here is is weighted towards Australasia, each paper provides a distinctive approach and different contribution to the issue of island human ecology, including the dynamic interaction of technology and subsistence. The second group of papers (Chapters 5 to 9) deals with the topic of colonization, mobility and interaction amongst islands, and contrasts the nature of settlement dynamics in the context of inter-island and island-mainland connections in the Torres Strait (Melissa Carter and Ian Lilley), Australia (Michael Rowland) and the Mediterranean (Helen Dawson). Doug Sutton's paper provides a provocative review of approaches to Polynesian island colonization, whilst raising issues that are of broader relevance to other geographical contexts. Kimberley Owens also contributes an insightful case study, which combines textual and material evidence to build understanding of historic colonizational processes. Finally, the collection of papers in the third broad theme of 'community' (Chapters 10 to 14) provide a means of assessing the construction and maintenance of social relations in an island setting. Here, papers by Malcolm Lillie (England), Peter Sheppard and Richard Walker (western Solomons), Stephen Dyson (Sardinia), and Sharon Greene (Ireland) provide detailed contextual information about seafaring and the historical contingency of the specific social formations that arise at varying scales of islanders' community interactions. We can thus contrast, for example, Cooper's concerns with "questioning the validity of the marine island boundary" in terms of cultural identities (pg. 179) with Dyson's study of how the "complex dialogue between island initiative and external contact" (pg. 193) played out from Neolithic obsidian trade, through the 'Sea Peoples' and into the international struggles of the classical period on Sardinia. Similarly, Sheppard and Walter's statement—"the question was not whether there was interaction, but how much" (pg. 167)—can help us reflect on Greene's critical questioning of the perceived remoteness and isolation of islands of the west coast of Ireland during the early medieval period.

It is hoped that these papers, both individually and collectively, demonstrate why island archaeology remains a vibrant and relevant part of archaeological discourse. Clearly, islands are neither peripheral nor isolates in the context of their diverse histories, nor are they peripheral in the context of their contribution to archaeological thought.

References cited

Broodbank, C. 2000. *An Island Archaeology of the Early Cyclades.* Cambridge University Press: Cambridge.

Evans, J.E. 1973. Islands as laboratories for the study of cultural process. In C. Renfrew (ed.) *The Explanation of Culture Change: Models in Prehistory.* Duckworth: London. pp. 517–520.

Irwin, G. 1992. *The Prehistoric Exploration and Colonisation of the Pacific.* Cambridge University Press: Cambridge.

Lape, P.V. 2004. The isolation metaphor in island archaeology. In S. Fitzpatrick (ed.)

Voyages of Discovery: The Archaeology of Islands. Praeger: Westport, Conn. pp. 223–232.

Rainbird, P. 2007. *The Archaeology of Islands.* Cambridge University Press: Cambridge.

Rainbird, P. 1999. Islands out of time: towards a critique of island archaeology. *Journal of Mediterranean Archaeology* 12: 216–224.

Rainbird, P. 2004. *The Archaeology of Micronesia.* Cambridge University Press. Cambridge.

Terrell, J.E. 2004. Island models of reticulate evolution: the 'ancient lagoons' hypothesis. In S. Fitzpatrick (ed.) *Voyages of Discovery: The Archaeology of Islands.* Praeger: Westport, Conn. pp. 203–222.

Chapter 1

Risk management and variability in irrigation and agricultural production on Nuku Hiva, Marquesas Islands

David J. Addison, American Samoa Community College

Introduction

Islands vary greatly in size, proximity to other land, and the scope of their inhabitants' off-island interactions. This paper concerns a small, remote, and drought-prone island in Polynesia whose populations inhabited small and tightly confined territories. I examine environmental variability on the island and the role irrigation played in agricultural subsistence and risk management.

Physical setting

The Marquesas Islands are located at the eastern margin of Polynesia just south of the equator (8–10°S and 138–141°W). With a combined land area of about 1050 km^2, the archipelago is among the largest in tropical Polynesia. Nuku Hiva Island, at 339.5 km^2, is one of the two largest islands in the group. Each of the currently inhabited islands approaches 1000 m in elevation, with Nuku Hiva's highest point at 1224 m.

Millions of years of eruption and erosion have left Nuku Hiva with complex hydrology and topography. The initial deposition of Nuku Hiva's volcanic substrata and subsequent aging and weathering have left three legacies important in the current study: variability in aquifer location, size, and accessibility; topography affecting precipitation and hydrology; and rich volcanic soils of the valley floors and slopes that are fertile edaphic media.[1] The following brief summary (following Brousse 1993) outlines the origins of Nuku Hiva's geological features. The island's emerged part formed in four stages 4.3–2.9 million years ago in the following stages:

1. Building of the shield volcano (which survives mostly in the western planezes). The central eruptions were accompanied by explosive and effusive parasitic events.

2. The collapse of the calderas produced an elliptical depression whose main axis measures 20 km and is partly filled by basaltic lavas and lahars on To'ovi'i Plateau.

3. A large internal volcano formed in the bottom of the depression (roughly 15 km in diameter) caused by a second collapse episode. This volcano was formed initially of basaltic material then increasingly lighter coloured and differentiated rocks reaching trachytes. Explosive eruptions produced beds of ejecta covering To'ovi'i Plateau.

4. A 5-km wide collapse was accompanied by large domes of differentiated rocks. Basalts and hawaiites were ejected to the west before a massive slide of the southern part of the island along an east-west line of fragility.

Post-eruptive erosion has left Nuku Hiva with vistas of dramatic forms of dissection (figure 1.1). The volcanic relief of interlocking calderas and plateaus that dominates to the west and in the island's center grows in complexity to the north and east, where secondary ridges form promontories between amphitheater-shaped valleys at the head of often-deep bays (Anaho, Hatihe'u, A'akapa, etc.). To the south, the bays

[1] A few of Nuku Hiva's ridge tops have pyrophytic vegetation associations (dominated by *Miscanthus floridulus* and *Dicronopteris linearis*) that suggest depleted edaphic substrates or regular burning. Some of them are periodically burned (in contravention of local laws). It is unclear whether the depleted edaphic substrates (where they exist) are the result of erosion caused by burning, former cultivation (although most appear to be unsuited for agriculture), or the effects of recently introduced ungulates.

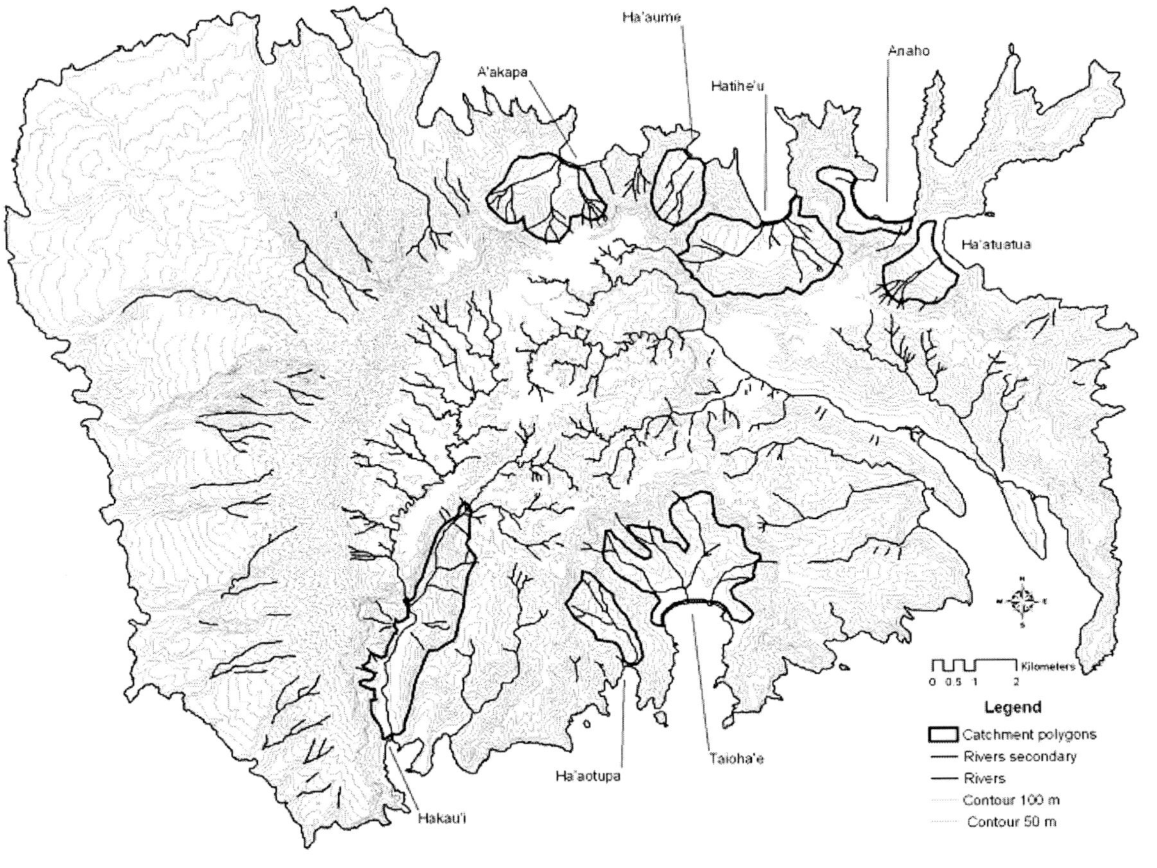

Figure 1.1: Nuku Hiva Island showing survey areas. The bounded area in each valley indicates the area used for calculating valley areas.

(e.g., Taioha'e and Taipivai) correspond to the low points of successive volcanic craters. Between bays, segments of rocky coast—often dominated by cliffs—encircle the island. In the interior, extreme slopes are a regular feature of the inner and outer edges of the calderas as well as the secondary ridges dividing valleys. Most habitation and cultivation areas are located in the valley floors or on the adjacent slopes, making the amount of usable land less than the raw figures for area indicate. Coral growth is generally confined to small isolated patches, however there are small fringing reefs in a few protected places (e.g., Hakatea, Anaho).

Normally, the Marquesan climate is a relatively benign one. Cauchard and Inchauspe (1978) have summarized the dominant characteristics of the Marquesan climate as follows: slight variation in temperature throughout the year with daytime temperature at sea level rarely below 25°C or above 30°C; rarity of stormy weather and absence of cyclones; stable easterly winds; little seasonality; mean annual precipitation between 700–1400 mm; and great interannual variability in precipitation with periods of drought. Although periodic droughts are a recurrent feature of the Marquesan climate, they do not occur at regular intervals. There is no annual dry season, and there is no regular interannual periodicity to the droughts. They happen on an unpredictable schedule, but that they will occur periodically is highly predictable. The droughts can last from a few months to more than a year. Rainfall patterns on Nuku Hiva are affected by general climatic conditions in the Marquesas and by the orographic effects of local topography.

ENSO and rainfall patterns

Agricultural potential on Nuku Hiva is affected by patterns of rainfall and hydrology. Amount of rainfall and dependability of groundwater resources vary among valleys. Some have streams and springs that flow permanently, while streams in others are affected by drought. During periods of high rainfall, lack of water is not a limiting factor for agricultural production; under normal conditions, most valleys have adequate water; with the potential for irrigated cultivation, valleys with perennial streams have an advantage during drought.

Climatic variables affecting rainfall amounts and periodicity in the Marquesas are likely to be complexly interrelated. One of the variables that may have important effects on the Marquesan climate is the South Pacific Anticyclone located off Rapa Nui (Easter Island). This high-pressure cell revolves counter-clockwise and generates winds, and resultant currents (Humboldt Current), that flow north along the coast of South America and westward toward the Marquesas. These winds are relatively cool and dry, and the currents bring cool southern water across the South Equatorial East Pacific. This produces what may be termed the normal state of the Marquesan climate, with adequate rainfall in most locations. In La Niña episodes, the cool-dry pattern is intensified and may result in drought conditions in the Marquesas. El Niño conditions pertain when the South Pacific Anticyclone weakens and the warm surface water normally kept at bay in the western Pacific is allowed to move eastward. During El Niño events strong enough to allow this warm surface water to reach the Marquesas, the weather may become warmer and wetter. On Nuku Hiva, during El Niño events, abundant rain coming from all quarters of the compass could be predicted. Windward/leeward location, elevation, and topography would have little effect and there would be abundant rainfall everywhere. This climatic patterning associated with La Niña and El Niño may be called the ENSO (El Niño Southern Oscillation) hypothesis.

To examine the ENSO hypothesis I turn to rainfall records for Taioha'e. The Taioha'e measurements are the longest continuous record available for the Marquesas and span the period 1950–present. Figure 1.2 depicts annual rainfall totals for this period. The ENSO hypothesis is not supported by these data. High rain levels occur in years with all ENSO states, likewise, drought years occur in years with all ENSO states (La Niña, El Niño, and ENSO-neutral). Total annual rainfall is not a sensitive enough measure to reflect the pattern suggested by the ENSO hypothesis. This may be because of "the briskness with which El Niño events emerge and the abrupt transitions into a La Niña state, or vice versa" (Caviedes 2001:153). These transitions often occur *within* a calendar year; hence one calendar year may encompass both El Niño and La Niña conditions, obscuring the pattern suggested by the ENSO hypothesis.

Figure 1.3 to 1.6 depict monthly rainfall for Taioha'e from 1950–1989[2] with one decade on each figure. There are fourteen periods of low rainfall lasting five months or more in this

[2]The "surge in the frequency of ENSO events" (Caviedes 2001:9) in recent decades has changed the pattern of rainfall on Nuku Hiva. This trend is shown in the dominance of "wet" years since 1979 (see figure 1.6), and I have not included measurements after 1989.

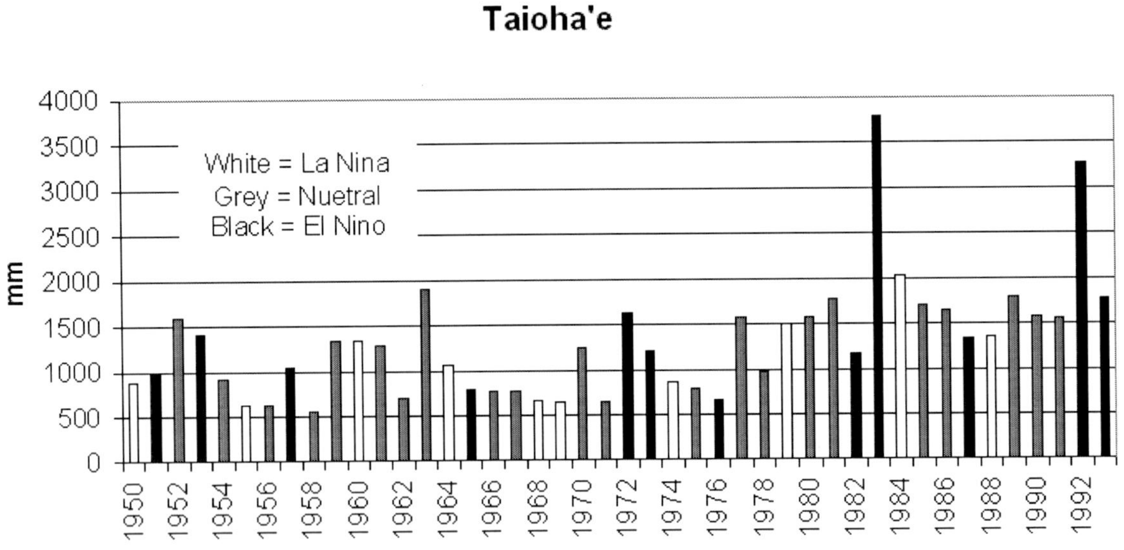

Figure 1.2: Annual rainfall totals at Taioha'e in mm. La Niña years are white, El Niño years black, and ENSO-neutral years grey.

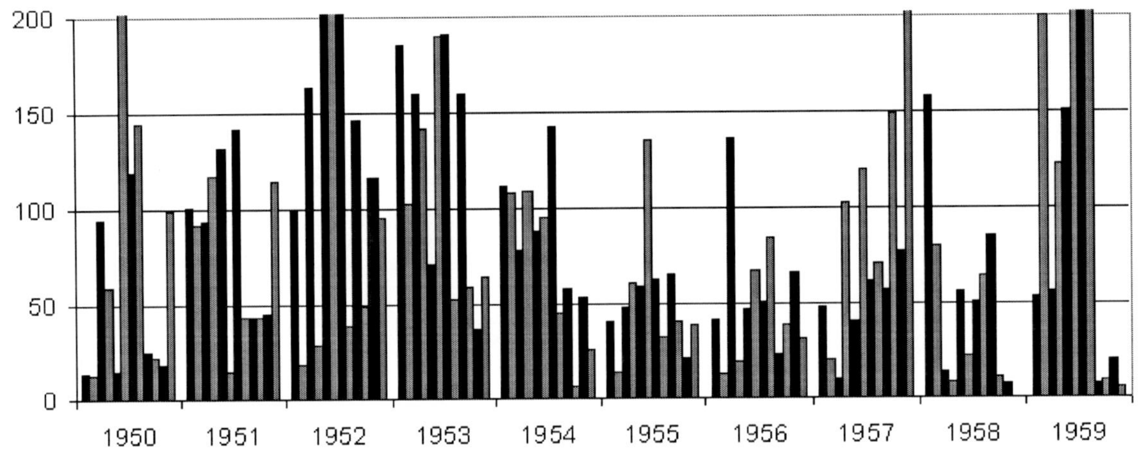

Figure 1.3: Monthly rainfall (in mm) at Taioha'e 1950–1959.

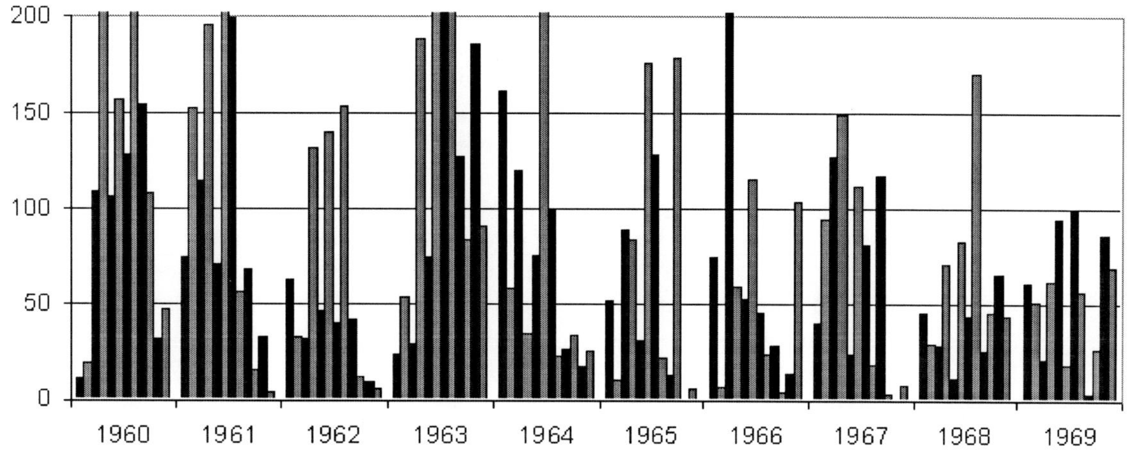

Figure 1.4: Monthly rainfall (in mm) at Taioha'e 1960–1969.

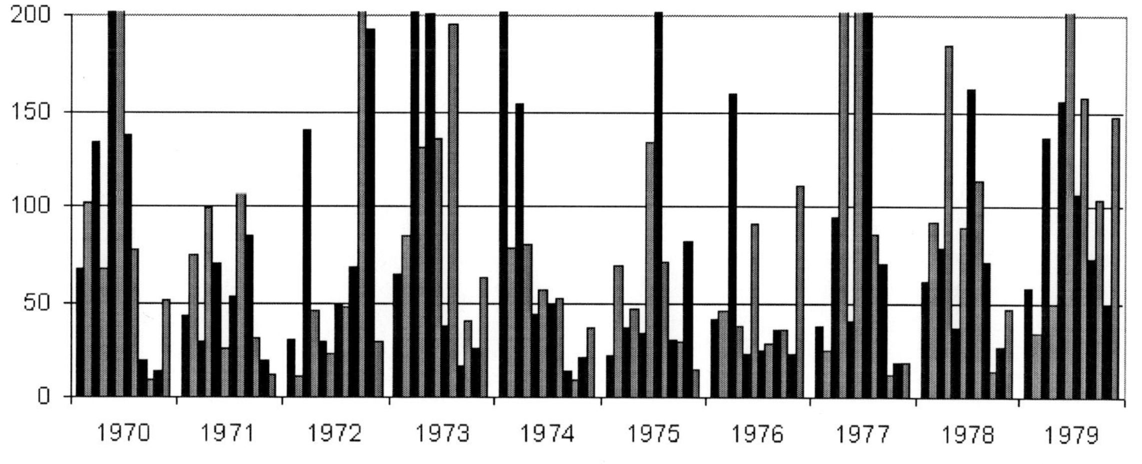

Figure 1.5: Monthly rainfall (in mm) at Taioha'e 1970–1979.

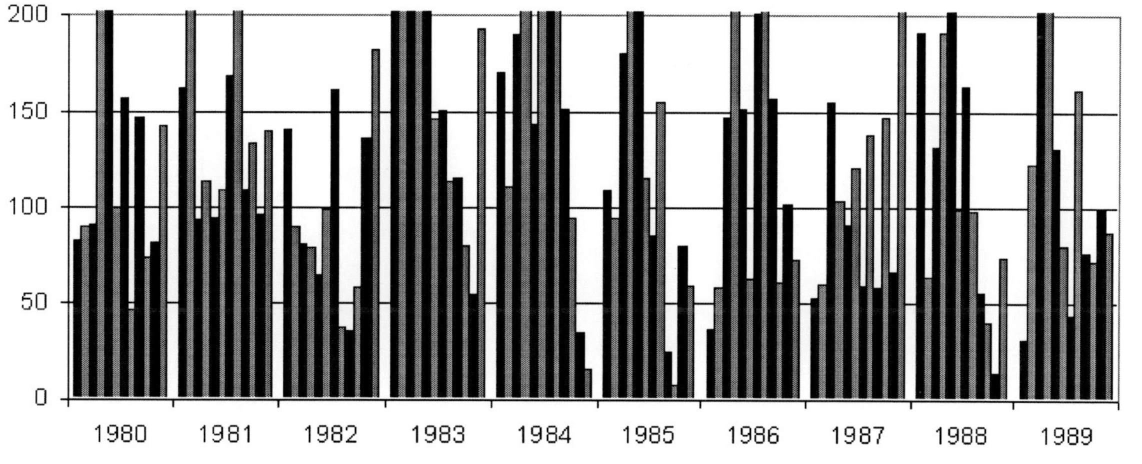

Figure 1.6: Monthly rainfall (in mm) at Taioha'e 1980–1989.

forty-year span (see table 1.1). All have average monthly rainfall below 50 mm. Five of these droughts lasted a year or more, and two approached three years in duration. Half of the droughts, including the two longest coincide with La Niña events.[3] This suggests support for the ENSO hypothesis.

Table 1.2 lists nineteen La Niña and El Niño events for the period 1950–89. Only three of these events—the La Niñas of 1950 and 1960 and the El Niño 1976—do not fit the expectations of the ENSO hypothesis. Climatic factors affecting rainfall in the Marquesas are surely multiple and complexly interrelated, but this short analysis of Taioha'e rainfall suggests that fluctuations in the ENSO state have an effect on Marquesan rainfall.

Spatial variability in rainfall on Nuku Hiva: effects of ENSO and topography

Having reviewed the evidence for ENSO on the Marquesan climate, I now turn to examining two valleys on Nuku Hiva with contrasting rainfall regimes. Figure 1.7 depicts rainfall for a 30-year period from five stations on Nuku Hiva (see figure 1.1 for station locations). On figure 1.7, note the uniformly high rainfall on Nuku Hiva in 1963, 1977, 1982, and 1992 corresponding to El-Niño events. Conversely, in La-Niña episodes—when the dry easterlies are particularly strong (such as 1964–65, 1967, 1976, 1978 and 1988)—most parts of Nuku Hiva received less rainfall.[4] The inter-annual variation is striking, but also apparent is the variability between locations. A comparison of two valleys—Taioha'e and Hatihe'u—will illustrate the variability between valleys on Nuku Hiva (see figure 1.1 for location). Both valleys are about the same size (ca. 600 ha) but vary dramatically in their rainfall patterns.

Figure 1.8 focuses on rainfall for Taioha'e and Hatihe'u. Two climatic periods are represented in this figure. The first fifteen years, from 1962 to 1976 are relatively dry, with stations throughout the Marquesas rarely reporting more than 1500 mm/yr and some consistently well below the 1000 mm/year mark. Taioha'e is in this latter category. By comparison, during this overall drier period, Hatihe'u had not only more rain (mean 1542 mm/yr, see table 1.3) than Taioha'e (mean 1041 mm/yr), but more importantly, the interannual variability was less pronounced (Hatihe'u coefficient of variation (V)=26%, Taioha'e V=39%).

The second period, beginning in 1977, is wetter throughout the Marquesas and this affected the two valleys differently. While rainfall at Hatihe'u remained high and stable (maximum and minimums almost unchanged, mean 1628 mm/yr, V=24%), Taioha'e's annual rainfall increased, with both the average (1791 mm/yr) and the minimum (1166 mm/yr) almost doubling. Taioha'e rainfall remained more variable (V=40%) than Hatihe'u. The reasons for the wetter period since the late 1970s are unclear but may be related to global warming (Salinger et al. 1995). In this period, there has been a marked increase in rainfall throughout the Intertropic Convergence Zone related to an increase in the number of ENSO events and their magnitude (Salinger et al. 1995). Given the worldwide increase in temperature over the last several centuries, it is likely that the pre-1977 pattern is more representative of earlier, historic rainfall patterns in the Marquesas. Generally drier and cooler conditions in the past may have reduced both the abundance and regularity of Marquesan rainfall, especially at drier valleys like Taioha'e.

Orographic effects

In normal conditions, orographic rainfall is a factor in the local precipitation regime, and there is a general windward/leeward distinction. The series of high ridges (ca. 1000–1200 m elevation) to the west of the To'ovi'i Plateau forms a strong orographic barrier creating the dry conditions of the western part of the island known as Henua Ataha. On eastern Nuku Hiva, the orographic effects of the local topography presents a somewhat more complicated situation. The adjacent valleys of Hatihe'u, Anaho, and Ha'atuatua have strikingly different patterns. Ha'atuatua has adequate rainfall caught by the high ridges (ca. 700–800 m) to the southwest. Hatihe'u, with ridges of similar height to the south and west, receives copious rainfall. But Anaho is, under normal conditions, at the lower limits of rainfall required for most Marquesan cultigens.

During drought conditions due to La Niña, these effects should be even more pronounced. There may be some rain at high elevation and at locations in the best orographic position, but most may receive inconsequential amounts of rain.

[3] La Niña episodes tend to follow El Niño events, but can precede them as well (Caviedes 2001:150).

[4] As will be seen below, generally there is little correlation of rainfall amounts between valleys, but for ENSO-active years the correlation between valleys is strong (r=0.79, significant at 95% confidence).

La Niña[a]	No. of consecutive drought months (<~ 50 mm)	El Niño[b]	Intensity[c]	No. Rainy months (> 100/150 mm)	Fit to model
1950	3			3/1	−
	4	1951	M−	5/0	+
	1	1953	M+	7/5	+
1955	8[d]			1/0	+
	3[e]	1957	S	5/2	+
1960	6[f]			8/4	−
	1	1963		6/5	+
1964	7[g]			3/2	+
	2	1965	M+	3/2	+
1968–69	4			1/1	+
	2	1969	M−	0/0	+
	5[h]	1972-73	S	8/5	+
1974	9			2/2	+
	5	1976	M	2/1	−
1979	1			7/3	?[i]
	2	1982-83	VS	14/8	?[i]
1984	2			9/7	?[i]
	0	1987	M	6	?[i]
1988	2			5/3	+

Notes: [a]Source: Caviedes 2001. [b]Source Bradley and Jones 1992. [c]Intensity ratings from Quinn and Neal 1992: VS=very strong; S+, S=strong; M+, M-, M=moderate. [d]Includes last 5 months of 1954. August 1954 through March 1957 had generally low rainfall at Taioha'e with no months above 150 mm and only 2 above 100 mm. [e]First 3 months of 1957 continue a moderately dry 1956. [f]Includes last 4 months of 1959. [g]Includes first 2 months of 1965. [h]The drought months are early in the year; rainfall increases starting September. [i]After 1979 the increased frequency and intensity of El Niño has made Taioha'e wetter and changed the previous rainfall pattern.

Table 1.1: Comparison of wet and dry periods with ENSO state fluctuations.

Start	End	La Niña year	Months duration	Mean monthly rainfall in mm	Total rainfall for period in mm
1954 Aug	1957 Mar	1955	32	47	1505
1958 Mar	1959 Jan		11	34	374
1959 Sep	1960 Feb	1960	6	12	72
1961 Aug	1962 Mar		8	38	306
1962 Sep	1963 Mar		7	25	176
1964 Aug	1965 Feb	1964	7	27	188
1966 Jul	Nov		5	23	115
1967 Oct	1970 Jan[a]	1968-69	29	45	1220
1970 Sept	1971 Jan		5	27	137
1971 Oct	1972 Sep[a]		12	33	368
1974 May	1975 May	1974	13	38	490
1976 Jul	Nov		5	29	147
1978 Oct	1979 Feb	1979	6	42	250
1988 Sep	1989 Jan	1988	5	43	215

Note: [a]August 1968 with 171.1 mm and March 1972 with 140.2 mm were not calculated.

Table 1.2: Periods of drought at Taioha'e.

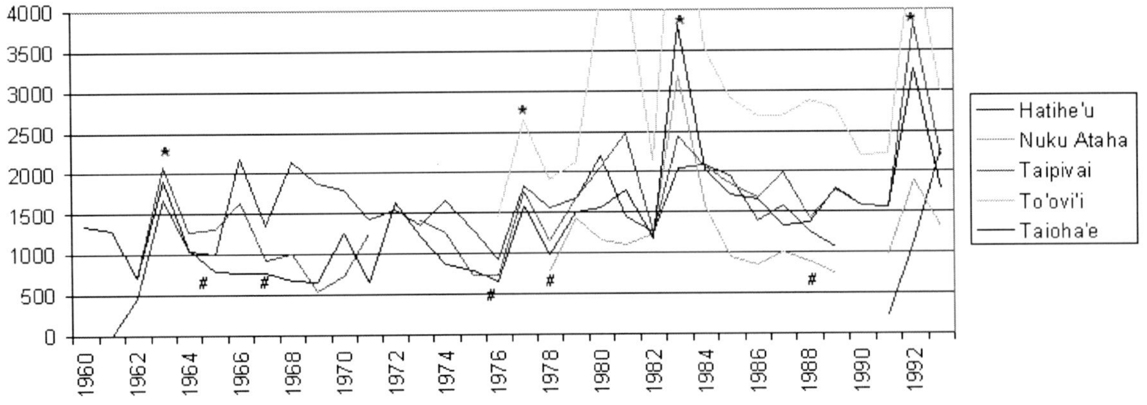

Figure 1.7: Rainfall (mm/yr) for five locations on Nuku Hiva. Breaks in lines indicate years with no data. El Niño events marked with *, La Niña with #. All stations are near sea level at the coast except To'ovi'i at ca. 800 m on the interior plateau.

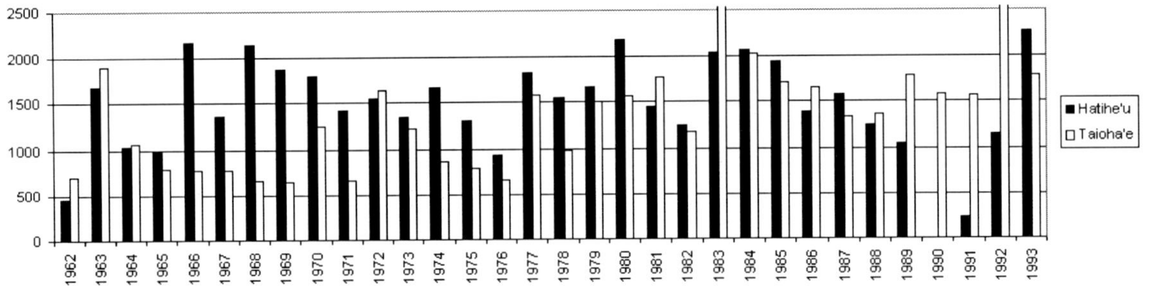

Figure 1.8: Rainfall (mm/yr) at Taioha'e and Hatihe'u. For the period 1962–1978 r=0.12; for 1979–89 r=0.45, indicating that during the wetter period since 1979 rainfall levels between the two valleys are about 14 times more correlated than for the 1979 period. Note that 500 mm/yr is the limit of sweet potato production and 1000 mm/yr the limit of breadfruit. Hatihe'u measurements for 1962, 1990 and 1991 are likely erroneous (Cliquet pers. comm.)

	Hat 1963–78	Hat 1979–89	Hat all years	Tai 1963–78	Tai 1979–89	Tai all years
Mean	1542	1628	1577	1014	1791	1330
Std Dev	379	390	378	396	710	660
Range	1252	1140	1268	1253	2637	3153
Minimum	929	1057	929	650	1166	650
Maximum	2181	2197	2197	1903	3804	3804
Co Var	26%	24%	24%	39%	40%	50%
Count	16	11	27	16	11	28

Table 1.3: Descriptive statistics for rainfall measurements at Hatihe'u (Hat) and Taioha'e (Tai).

Hydrology

Rainfall patterns, topography, geologic substrates and watersheds determine surface water occurrence (Stearns 1942). The same is not true of groundwater. An aquifer may occur under several topographic watersheds. Generally, on volcanic islands ash or tuff beds interstratified with basalt flows can produce perched aquifers (Stearns 1942). Nuku Hiva aquifers are poorly understood, but given the island's complex volcanic building sequence, and the appearance of complex geologic stratigraphy with a variety of volcanic materials, it is quite possible that this is an important factor in its hydrology. Tropical montane cloud forests may be important in replenishing some Marquesan aquifers during periods of low rainfall.

Streams

Regardless of rainfall, reliability of streams is an important aspect of irrigation-water management. Nuku Hiva's highly dissected landscape has innumerable steams and watercourses. Most of these only appear when it rains heavily enough that there is surface run off, although the inhabited valleys have reliable water sources. In only a few, people must rely solely on springs for water (e.g., Anaho, Hakatea), some have intermittent stream that may, during drought periods, have reduced flow or dry up for part or all of their course (e.g., Taioha'e, Ha'a'otupa). Still other valleys have streams that continue to flow even during droughts. This last category of streams appears to have dependable water flow for two reasons: (i) they are fed springs tapping drought-independent aquifers (e.g., Hatihe'u, Ha'aume, A'akapa) or; (ii) they drain large high-elevation catchments on the drought-free To'ovi'i Plateau (e.g., Hatihe'u, Taipivai, Hakau'i).

Cloud forests

Streams that drain large interior plateaus have substantially larger catchment areas than other streams on Nuku Hiva. This alone could account for their reliable flow. Another factor is moisture caught by cloud forests. Tropical montane cloud forests "receive substantial 'horizontal precipitation' through direct canopy interception of wind-driven cloud water" (Merlin and Juvik 1995:149). Even during periods of minimal rainfall, cloud forests can trap moisture and transmit it to the ground (Juvik et al. 1995; also Juvik and Nullet 1995). The degree to which this process feeds aquifers is undocumented for the Marquesas. Feral ungulates have had a devastating effect on Marquesan cloud forests (Merlin and Juvik 1995:158) and they were undoubtedly more extensive in pre-European times.

Springs

Through personal observation and discussions with Nuku Hiva people, I have noted that there is an uneven distribution of permanent springs on Nuku Hiva. The north coast appears to have springs whose flow is unaffected by drought. Thus this area has more reliable streams whereas Taioha'e has poor ones. This distribution suggests the possibility that the inner caldera that forms Taioha'e does not tap the aquifer(s) fed by high-elevation cloud forests on To'ovi'i Plateau and the ridges of the outer caldera.

Environmental summary

Variability of rainfall between valleys on Nuku Hiva is patterned by complex interactions of several factors. Windward/leeward location, orographic effects, local topography, and ENSO are all involved. Patterns of rainfall and hydrology are important environmental variables affecting agricultural potential on Nuku Hiva. Its valleys vary in the amount of rainfall they receive and in the dependability of groundwater resources. Some have streams and springs that flow permanently, regardless of drought, while streams in others dry up. During periods of high rainfall, lack of water would not have been a limiting factor for agricultural production; under normal conditions, most valleys have adequate water and there may be slight differences in productive potential; when drought strikes, valleys with perennial streams have a distinct advantage with the potential for irrigated cultivation. For example, Hatihe'u with both streams from the To'ovi'i Plateau and perennial springs; Hakau'i with its river draining more than half of To'ovi'i; or A'akapa and Ha'aume with perennial springs have adequate stream flow even during drought. Ha'a'otupa and Taioha'e, on the other hand have adequate rainfall in normal or El Niño years, but very poor water resources during drought. Anaho is agriculturally marginal in all but the wettest periods.

Marquesan agriculture

At the time of European contact on Nuku Hiva socio-political organization was based on ramified descent groups (*mata'eina'a*). *Mata'eina'a* boundaries could be coterminous with valley limits, but some valleys were the home of related but independent and potentially antagonistic *mata'eina'a* that could be united for varying periods. Proto-historic population levels for Nuku Hiva are unknown, but may have been near 10,000–20,000, with individual *mata'eina'a* ranging from hundreds to thousands of persons (also see discussion in Kirch 1991; Krusenstern 1813, 178; Porter 1970, 32). No polity extended beyond the confines of one valley, although temporary alliances were common.

A variant of the Polynesian chiefdom (*sensu* Kirch 1984), proto-historic Marquesan sociopolitical systems were relatively flexible and power relationships fluid (Dening 1971; Thomas 1990). Elites (chiefs/*haka'iki*/*ha'atepei'u*, war leaders/*toa*, prophets/*tau'a*, craft specialists/*tuhuka*) were the only landowners (*'akati'a*) and exercised a degree of individual authority over their personal holdings unusual in Polynesia. Landless persons (*po'i kikino*) served elites, providing domestic and agricultural labour. Traditional Marquesan agriculture was a variant of the oceanic agricultural complex with its heavy emphasis on vegetatively reproduced cultigens, root crops such as aroids and yams (*Dioscorea* spp.) and tree crops like breadfruit (*Artocarpus altilis*), banana (*Musa* spp.), and coconut (*Cocos nucifera*). Most historical and ethnographic sources describe traditional Marquesan agriculture as heavily dependent on breadfruit. Breadfruit is a crop that produces seasonally. In the Marquesas there is a major harvest in January and smaller crops in June and September. Aside from these periods of breadfruit abundance, minor amounts of the fruit ripen throughout the year. Vast amounts of breadfruit were stored as a pickled paste in silos dug into the ground and anaerobically sealed. Breadfruit stored in this way will last for decades. However, a closer reading of the early descriptions of Marquesan agriculture suggest that much of the arable land was devoted to relatively permanent multi-story gardens—vertically stratified gardens in which an upper layer of true tree crops composed mostly of breadfruit and coconut had a succession of crops such as aroids, bananas, and other food and economic cultigens grown underneath (Addison and Pearthree n.d.). The role of irrigation is not addressed in these sources but "taro" gardens are mentioned sporadically. Under adequate rainfall,[5] production differences between Nuku Hiva valleys are minimized with abundant production from multi-story gardens—a spatially homogenous caloric environment across the island. During periods of drought, however, the situation changes dramatically—predictability and abundance of food resources becomes more heterogeneous as drier valleys lose production from rain-fed gardens. Irrigated pondfield cultivation of taro (*Colocasia esculenta*) is among the most productive (kilojoules/m^2) traditional agricultural techniques in the world, providing a density of complex carbohydrates, rivaled among traditional production systems only by the homologous systems based on rice (*Oryza sativa*). Among the islands of Polynesia there is marked variability in the distribution and extent of pondfields. Islands such as Tongatapu and Rapanui/Easter Island—with no permanent surface streams—have no pondfields, whereas Hawaii, Rapa (Rapaiti), Mangaia, and Futuna have well-developed drainage networks feeding permanent streams with abundant pondfields. On Nuku Hiva, former pondfields are a more prominent feature of the constructed landscape than most archaeological summaries would suggest.

Irrigated terraces: variation among Nuku Hiva valleys

Each valley on Nuku Hiva has a particular configuration of environmental variables—its own microenvironment with differential potential for the cultivation of each of the traditional Marquesan cultigens. Key variables pertaining to agricultural production are related to the amount and reliability of water supply, and include aspects of geological substrates, and resultant hydrology, soils, and susceptibility to orographic rainfall.

Eight valleys were assessed for the degree of terracing for irrigated agricultural production. Valleys fell into three groups: relatively wet valleys with more terracing; dry valleys with substantially less terracing; and an intermediate group of valleys. Relationships between valley microclimate, reliability of steam flow, and degree of terracing were found to be more complex than might be predicted by simply noting their leeward/windward location.

Survey methodology

The eight valleys were: Ha'atuatua, Teonepoto (Anaho), Hatihe'u, Ha'aume, A'akapa, Hakau'i, Ha'a'otupa and Taioha'e (figure 1.1), note that some valleys have more than one stream). The

[5] Adequate is here defined as 1000 mm/year, at which level breadfruit can yield regularly Ragone (2004). Sweet potato, the Marquesan cultigen with the lowest water requirement, has a minimum of 500 mm/year.

survey was accomplished by walking up a single stream within each valley. Where streams branched only one branch was surveyed. Formerly irrigated terraces[6] adjacent to the streams on both sides were noted and approximate measurements of width and length were taken for terrace sets.

Because the valleys vary in size (120–610 ha, see table 1.4) comparing total terrace area does not give a realistic idea of the degree (not amount) of terracing between valleys.[7] Therefore, the degree of terracing is expressed as a ratio of the area of terraces documented on the survey to the distance along the stream surveyed. This measure provides a relative estimate of the amount of terracing in each valley making it possible to compare across valleys of different size, as well as for differences in stream length, another indicator of valley size.

Results

The following sections present the survey results for each valley. Meteorological measurements are available for only one of the surveyed valleys (Taioha'e). For this reason, at the beginning of each section I give a brief description of each valley climate based on my own observations of vegetation patterns, stream flow, and rainfall.

Ha'atuatua Ha'atuatua (120 ha) is a relatively wet valley. The interior of the valley benefits from orographic rainfall; the lower elevations are progressively drier. The stream surveyed here is the most reliable one in the valley. I have never observed it to be dry at the valley mouth. The springs at Vaikiki, one of the water sources for this stream, may be fed by an aquifer partially replenished during droughts by cloud forests in the mountains to the south. The large amphitheater-shaped valley rises slowly from the beach.

In the 1.23 km of the stream surveyed at Ha'atuatua, 6,280 m^2 of irrigated terraces were noted in 11 different sets. This gives an area/distance ratio of 5.14 (table 1.4). Most of the terraces at Ha'atuatua are likely near the stream due to the steep topography. It is likely that the terraces recorded here give a reasonably accurate sense of the degree of terracing in this valley.

Teonepoto (Anaho) Anaho (120 ha) is a relatively dry valley. There are no perennial streams within the small amphitheater-shaped valley, and there is little orographic rain. Teonepoto is an area at the eastern end of Anaho. The stream that flows through Teonepoto is fed partially by a spring that may draw on an aquifer fed by cloud forests in the mountains south of the valley. This stream is often dry at its mouth. When there is rainfall, the steep mountain, Tukemata,[8] catches a large amount of water, which forms a series of hanging waterfalls feeding the stream. This runoff drains quickly down to the coast. There is a narrow coastal plain at Teonepoto; beyond this the slope becomes very steep (40–80°).

The valley associated with Teonepoto stream is very constricted; all of the terraces in this portion of Anaho have been documented here. Only a single terrace set (measuring 600 m^2) was found across 820 m of survey. This provides an area/distance ratio of 0.73, the lowest among the eight valleys. That Anaho has the lowest ratio is unsurprising (table 1.4), for of the valleys surveyed it is the driest and has the least-dependable water resources.

Ha'aume Ha'aume Valley (180 ha) is relatively wet. One of the springs that feed the stream is said to never dry up. This spring, like that of Ha'atuatua, likely taps an aquifer partially fed by cloud forests in the mountains to the south.

In the surveyed 1.30 km of the stream at Ha'aume 3,350 m^2 of irrigated terraces were noted in 6 terrace sets. This gives an area/distance ratio of 2.58 (table 1.4). The steep topography at Ha'aume suggests that most of the terraces are near the stream. I suspect that the ratio estimate presented here is a fairly accurate representation of the degree of terracing at Ha'aume.

A'akapa A'akapa Valley (380 ha) is relatively wet. The portion of the stream surveyed is the most reliable in the valley. It likely taps an aquifer partially fed by cloud forests in the mountains to the south. This large amphitheater-shaped valley rises gently from the

[6]Terraces with level unpaved surfaces and arranged in step-like in relation to each other (terrace sets) were inferred as having been formerly irrigated to form pondfields. Taro was the only cultigen traditionally grown in ponded fields.

[7]This index gives an idea of the degree to which more intensified production could be achieved within the area of arable land available. Amount of terracing relates to total production—the more area terraced, the greater the potential production. A large valley that has more limited rainfall may still have more terraced area than a wet valley that is quite small but may also have had a larger population.

[8]Lit. eyebrow.

Valley	Area/ distance ratio	Total valley area (ha)	Surveyed terrace area (m^2)	Est. total valley area terraced (m^2)	Confidence[a]	Water
Teonepoto	0.73	120[b]	600	600	3	Dry
Ha'a'otupa	1.38	180	3,150	3,150	3	Dry
Taioha'e	1.40	610	3,050	12,000	1	Dry
Ha'aume	2.58	180	3,350	3,350	3	Intermediate
Ha'atuatua	5.14	120	6,280	12,000	2	Intermediate
Hatihe'u	10.72	600	11,260	60,000	1	Wet
A'akapa	12.11	380	23,850	48,000	2	Wet
Hakau'i	15.23	580	35,600	50,000	1	Wet

Notes: [a]1=low; 2=medium; 3=high. Confidence in the estimated total terrace area is based on the percentage of a valley's streams surveyed (higher percentage=higher confidence); visibility and topological likelihood of terraces away from stream edges; and my knowledge, based on previous survey, of the relative abundance or terracing in areas not included in this survey. [b]All of Anaho Valley.

Table 1.4: Degree of terracing, total land area, and water resources for eight Nuku Hiva valleys. Confidence in the estimated total terrace area is based on the percentage of a valley's streams surveyed (higher percentage=higher confidence); visibility and topological likelihood of terraces away from stream edges; and my knowledge, based on previous survey, of the relative abundance or terracing in areas not included in this survey.

beach. There is about 500 m of gently sloping land that is the site of the modern village. Most terraces at A'akapa are located near the stream because of the steep topography, and, I think that most terraces on this stream were recorded.

In the surveyed 1.97 km of the stream at A'akapa 23,850 m^2 of irrigated terraces were noted in 17 terrace sets. This gives an area/distance ratio 12.11 (table 1.4), among the highest of the surveyed valleys.

Hakau'i Hakau'i Valley (580 ha) is the most western and hence most leeward of the valleys surveyed. Because of this, one might expect it to be the driest. It is, however, relatively wet. The valley takes the form of a long canyon extending inland more than 4 km inland. The high ridges forming the west side of the valley catch orographic rainfall. In addition, the river at Hakau'i is one of the largest and most reliable on the island. Its catchment drains a large portion of the To'ovi'i Plateau in the center of the island.

The valley has a large flat bottom that extends about 2 km from the shore. People associated with Hakau'i say that much of the valley floor was once for growing irrigated taro. On Nuku Hiva, it is said that taro was the food of Hakau'i (*po'i kai ta'o to Hakau'i*). In the surveyed 2.33 km of the stream at Hakau'i 35,600 m^2 of irrigated terrace were noted in 5 terrace sets. This gives an area/distance ratio of 15.23 (table 1.4). The flat lower-valley floor is subject to alluvial deposition during flood events. During the regular floods the area is mostly knee-deep mud or standing water. Because the slope is so gentle, terrace retaining walls were probably quite low. Where they are visible on the surface, they are one course wide and high. Dirt bunds may have been extensively used in delineating the terraces in this area, but no trace of them remains. Because of the regular flooding, the terraces in this area have been extensively damaged and silted over. For two reasons, I think that the number of terraces at Hakau'i has been underestimated. The first is that discussed above; terraces in the lower valley are in poor condition and hard to identify. The second reason is that the survey followed the trail, which for the first 1.76 km gives fairly good visibility of terraced areas (though they might be difficult to identify). In the last section from 1.76 km to 2.33 km inland, the trail is not near the river and visibility of terraces is greatly reduced. I know from a previous reconnaissance survey that many terrace sets exist near the river in this section.

Ha'a'otupa This is a small (180 ha) amphitheater-shaped valley. It is relatively dry. In the surveyed 2.29 km of the stream at Ha'a'otupa 3,150 m^2 of irrigated terrace were noted in 4 terrace sets. This gives an area/distance ratio 1.38. This is remarkably similar to the ratio for neighbouring Taioha'e (table 1.4). Due to the presence of cattle, ground visibility from the stream was excellent. Because of the nature of the slope, most irrigated terraces are probably near the stream. I think that few escaped the survey, and that the numbers here are a good representation of the terraces at Ha'a'otupa.

Taioha'e The amphitheater-shaped valley of Taioha'e (610 ha)[9] rises gently from the beach. It is relatively dry.[10] The streams at Taioha'e are frequently dry at the beach. Apparently the springs at the back of the valley do not tap the same aquifers as perennial streams on the north coast.

In the 2.18 km of the stream at Taioha'e only 3,050 m^2 of irrigated terrace were noted in 4 terrace sets. This gives an area/distance ratio 1.40 (table 1.4). Due to the nature of the slope, most terraces at Taioha'e are likely located near the stream. Destruction of terraces from modern construction and development is likely, but it is mostly the lower stream would have been affected. Given that the stream flow is more unreliable on the lower part of the stream, I think that the numbers here are fairly representative of the degree of terracing at Taioha'e. Ha'a'otupa has had virtually no modern disturbance, and its microenvironment is similar to Taioha'e. The almost identical area/distance ratios of Taioha'e and Ha'a'otupa suggest that most terraces in the surveyed section of Taioha'e have been included here. The other sections of Taioha'e have environments similar to area surveyed and probably have a similar degree of terracing.

Implications of variability on Nuku Hiva

The valleys surveyed fall into three groups: the wet (with more terraces) and the dry (with few terraces), with Ha'atuatua and Ha'aume in an intermediate position (table 1.4). That the neighbouring valleys of Taioha'e and Ha'a'otupa have similar microenvironments and almost identical area/distance ratios, suggests that the methodology used here accurately reflects the relationship between environment and degree of irrigated terracing. Figure 1.9 graphically represents the data in table 1.4 and shows the lack of relationship between valley size and degree of terracing.

The fact that degree of terracing is closely correlated with reliability of water resources suggests that, to the extent environmental and climatic conditions allowed, pondfield production of taro was maximized. Total pondfield area is likely directly related to minimum stream flow. Each valley in the survey (with the possible exception of Hakau'i) contains large land areas topographically and edaphically suitable for construction of pondfields—yet not containing them. Irrigation water was likely the limiting factor. It is probable that the number of terraces in each valley is directly related to minimum stream flow, and that ancient Marquesans had built as many terraces as could be supported by the streams in each valley. The fact that Hakau'i appears to have had little land appropriate for pondfields that was not terraced, and has a reliable and high-volume river further reinforces the idea that minimal stream flow was the factor regulating total pondfield area in Nuku Hiva valleys.

As I have argued elsewhere (Addison 1996, 2001), the model of traditional Marquesan agriculture, drawn largely from historic accounts of Taioha'e and Vaitahu (on Tahuata Island) and emphasizing the role of breadfruit does not account for the variability in agricultural production strategies (e.g., Ferdon 1993; Kirch 1973; Rolett 1998; Suggs 1961; Thomas 1990, but see Kirch 1991). This model probably fits better the relatively dry Marquesan valleys such as Taioha'e and Ha'a'otupa or those with limited catchments such as Teonepoto. Wet valleys—those with either more rainfall or reliable perennial streams—show a different pattern, one with a greater emphasis on pondfield production. The differences are nontrivial; degree of terracing between dry and wet valleys can be on the scale of as much as an order of magnitude (figure 1.9, table 1.4).

These data have several implications for agricultural production on Nuku Hiva. People living in dry valleys would have had to rely more heavily on non-irrigated agriculture—techniques such as swidden and multi-storied arboricultural gardens. This would have meant less overall production per hectare, and the rate of calorie return would have been lower per unit of labour/time. Not only would more labour be required, but also production would have been less regular. Breadfruit is susceptible to fruit drop from wind or from soil water deficit throughout the period from fruit set to harvest. This means that breadfruit harvests could have var-

[9]The stream surveyed runs through a section of Taioha'e called Me'au. The other sections are Hoata, Ha'avau, and Pakiu. These subvalleys each occupy a separate drainage within the larger amphitheater of Taioha'e. Traditionally, separate *mata'eina'a* lived in each subvalley.

[10]The reasons for this are not clear to me. The high ridges surrounding it and its position facing the southeast trade winds should make it well watered. I have often noticed on arriving in Taioha'e from Hatihe'u or Taipivai that it is dry there while it had been raining in the place we departed from.

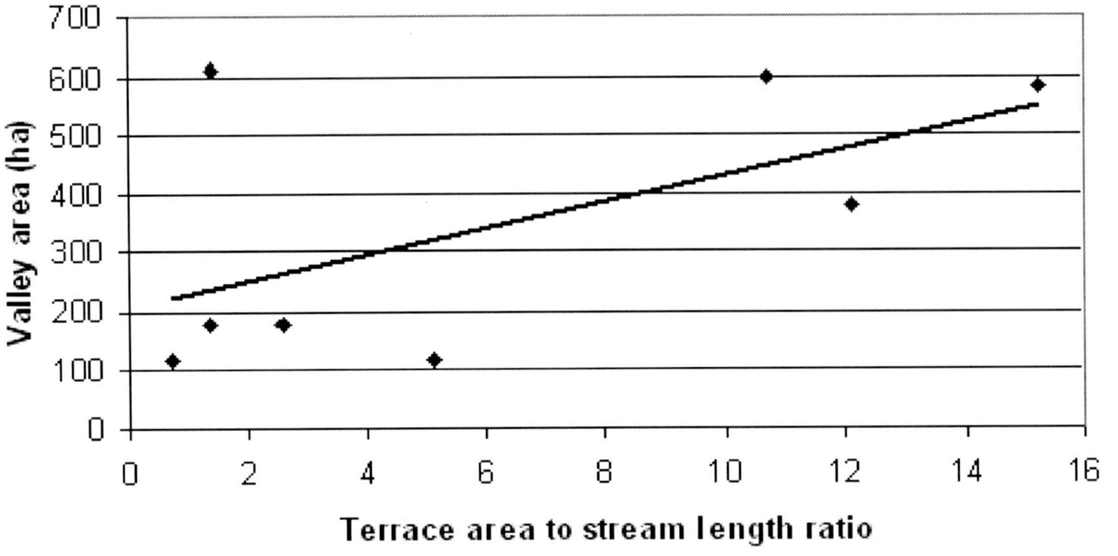

Figure 1.9: Relationship of terrace-area-to-stream-length ratio and total valley area (data from table 1.4; r=0.57, correlation not significant).

ied greatly from year to year. The advantage to breadfruit is that, in good years, it can produce an abundance of food well suited to storage in pit-silos. Such stored food would have been very important to people in the dry valleys because it could feed them through periods of bad harvest. Dry valleys would have had both lower overall production and more unpredictable production. Taioha'e and Hatihe'u have roughly similar total-production estimates (table 1.5). It should be remembered that these figures are for production from all agriculture and assume adequate rainfall. However, table 1.2 indicates that, because of low rainfall, in more than half the years in the 1950-1989 period, production at Taioha'e would not have been substantially reduced.

Much of the research on irrigation in Oceania has focused on the role of agricultural intensification in political development, specifically the evolution of chiefdoms (e.g., Allen 2001; Earle 1978; Kirch 1984, 1994; Kirch and Sahlins 1992; Spriggs 1981). Recently, other aspects of irrigation have received attention in the Pacific. Campbell (2001) in his work on Rarotonga noted the importance of irrigated taro fields in risk management; Field (2002, 2004a, 2004b) points out the value of the dense and reliable food resources produced from irrigated terraces in Fiji; Allen's (2001) Hawaii work, while not on ponded fields, focuses on the importance of constructed agricultural infrastructure in managing risk. This appears to be the primary role of irrigation on Nuku Hiva.

If surplus production for political purposes was the aim, we would expect more pondfields and more even distribution. It should be remembered that the dry and wet valleys described here are only such in periods of drought. In normal and El Niño years, all valleys are fairly equally well watered. In most years, each of these valleys could irrigate many more pond fields than are present. This would have resulted in a substantial surplus production. But such pond fields were not built. The degree of pondfield construction corresponds very well to each valley's water resources during drought. This pattern suggests that irrigation was especially important in these periods of scarcity. Marquesans had limited food options during drought. The bulk of traditional Marquesan agriculture depended directly on rainfall. During drought, once soil moisture reaches a certain level plant growth stops. For fruit or seed producing plants this mean that food production also stops. For some major Marquesan food crops the results are disastrous—immature breadfruit fall from the tree, coconuts don't form, bananas wither on the stalk. Some other cultigens have a degree of capacity for field storage (plant left in the ground to be harvested as needed). Yams (*Dioscorea* spp.), *Alocasia macrorrhizos*, and sweet potato (*Ipomoea batatas*) produce food in thickened stems or roots. These plants, al-

Valley	Total valley area (ha)	Multistory gardens[a,b]	Annual taro production[a,c]	Total production[a]	Dependability of production
Anaho	120	598.2	1.5	599.7	1
Ha'a'otupa	180	897.3	7.9	905.2	1
Ha'aume	180	897.3	8.4	905.7	2
Ha'atuatua	120	598.2	30.0	628.2	2
Taioha'e	610	3,040.9	30.0	3,070.9	1
Hatihe'u	600	2,991.0	250.0	3,241.0	3
A'akapa	380	1,894.3	120.0	2,014.3	3
Hakau'i	580	2,891.3	125.0	3,016.3	3

Notes: [a]In metric tons. [b]Based on Kirch's (1994:182) figures for *maumu* 'orchard gardens'; assumes half the valley's total area planted in multistory gardens. [c]Calculated at 25 t/ha/yr (Kirch 1994; Spriggs 1984) and not adjusted for fallow.

Table 1.5: Estimates of productive potential of Nuku Hiva valleys.

though they stop growing during drought, remain alive and hold substantial carbohydrate reserves. Growing such field-storage crops is one way to assure food during drought; storage is another. Certainly, the huge amount of breadfruit stored in silos as fermented paste was an important risk buffering strategy. Irrigation was a third such strategy. Plants in irrigated terraces would have produced longer during droughts. As stream flow was reduced some taro pondfields may have received less than optimal amounts of water, but could have had enough for continued growth and production. In valleys with reliable springs, drought may have had only minor impacts on taro production if total amount of ponded fields was related to minimum stream flow.

Another line of evidence suggests the importance of irrigated fields during drought. At Hatihe'u, intensive mapping and surveying was done (Addison 2001; Addison and Pearthree, n.d.). All former pond-fields in this area have ancient house foundations directly adjacent. In some cases houses were actually surrounded by ponded terraces. This indicates that proximity and surveillance were important aspects of Marquesan irrigation. In time of food scarcity ancient Marquesans needed to keep a close eye on one of the few food sources available.

Valleys with less irrigation would have had to rely on the other two agricultural risk reduction strategies, and probably on non-agricultural ones such as exchange, mobility, or aggressive appropriation. Kirch (1994) has noted that polities in less productive environments were more aggressive and expansive (e.g., Alo on Futuna and the leeward Hawai'i Island chiefdom). It is interesting to note two possible parallel examples from the Marquesas. Taioha'e—the driest of the large valleys on Nuku Hiva came under the rule of one chief near the time of European contact and was regularly in conflict with other Nuku Hiva valleys until the French colonization. At the time of European contact, the entire island of 'Ua Pou was ruled by a chiefly line from the dry northeastern side of that island.

Summary

Valleys on Nuku Hiva vary in the amount and regularity of rainfall. ENSO has important effects on Nuku Hiva rainfall, with La Niña related to an increased chance of droughts and El Niño associated with rainy periods. Not only are valleys like Taioha'e generally drier, they are more susceptible to drought. Wet valleys like Hatihe'u receive more rain, receive it with greater regularity, and are less susceptible to drought.

The analysis of terrace abundance, water availability, and valley size presented here demonstrates that there is variability in relative amount and total area of terraces in each valley; that the amount of terracing is correlated to the availability of irrigation water; that valleys would have had different productive potential both in the amount of food produced as well as the dependability of that production; and production does not follow a clear leeward/windward pattern nor is valley size a *simple* measure of productive potential. Because of differential susceptibility to drought, valleys of similar size vary dramatically in their productive potential. Irrigation was an important risk-reduction strategy employed by Marquesans in the past, and people living in valleys with unreliable streams would have had to rely largely on other risk-reduction strategies such as storage or aggressive appropriation.

References cited

Addison, D.J. (1996). Traditional agriculture of the Marquesas Islands (French Polynesia). In I. Glover and P. Bellwood (eds.), *Indo-Pacific Prehistory: The Chiang Mai Papers, vol 2*, Canberra: Australian National University.

Addison, D.J. (2001). Irrigation in traditional Marquesan agriculture: surface survey evidence from Hatihe'u Valley, Nuku Hiva. In C.M. Stevenson, G. Lee, and F.J. Morin (eds.), *Pacific 2000: Proceedings of the Fifth International Conference on Easter Island and the Pacific*, pp. 267–272. Los Osos: The Easter Island Foundation.

Addison, D.J., and E. Pearthree (n.d.). New Sites on Leeward Nuku Hiva, Marquesas Islands: An Archaeological Reconnaissance Survey of Henua Ataha. Manuscript report.

Allen, M.S. (2001). The Kona field system in spatial and temporal perspective. In M.S. Allen (ed.), *Gardens of Lono: Archaeological Investigations at the Amy B. H. Greenwell Ethnobotanical Garden, Kealakekua, Hawai'i*, pp. 137–155. Honolulu: Bishop Museum Press.

Brousse, R. (1993). La géologie des îles hautes. In *Atlas de la Polynésie Française*. Paris: Editions de l'ORSTOM.

Campbell, M. (2001). *Settlement and Landscape in Late Prehistoric Rarotonga, Southern Cook Islands*. Doctoral thesis, University of Sydney, Sydney.

Cauchard, G., and J. Inchauspe (1978). Climatologie de l'archipel des Marquises. *Cahiers du Pacifique* 21:75–106.

Caviedes, C.N. (2001). *El Niño in History*. Gainesville: University Press of Florida.

Cliquet, M. (1994, pers. comm.). Conversation with M. Cliquet, Director of the French Polynesia Meteorological Service.

Dening, G. (1971). *Tapu and haka'iki in the Marquesas, 1774–1813*. Doctoral thesis. Cambridge: Harvard University.

Earle, T.K. (1978). *Economic and social organization of a complex chiefdom: the Halelea District, Kaua'i, Hawaii*. Anthropological Papers no. 63. Ann Arbor: University of Michigan, Museum of Anthropology.

Ferdon, E.N. (1993). *Early observations of Marquesan culture, 1595–1813*. Tucson: University of Arizona Press.

Field, J.S. (2002). GIS-based analyses of agricultural production and habitation in the Sigatoka Valley, Fiji. In T.N. Ladefoged and M.W. Graves (eds.), *Pacific Landscapes: Archaeological Approaches*, pp. 97–124. Los Osos: The Easter Island Foundation.

Field, J.S. (2004a). Environmental and climatic considerations: a hypothesis for conflict and the emergence of social complexity in Fijian prehistory. *Anthropological Archaeology* 23:79–99.

Field, J.S. (2004b). *The Evolution of Competition and Cooperation in Fijian Prehistory: Archaeological Research in the Sigatoka Valley*. Doctoral thesis. Honolulu: University of Hawai'i.

Juvik, J.O., C. dos Anjos, and D. Nullet (1995). Direct cloud water recovery by inertial impaction: implications for large scale water supply in the Cape Verde Islands. *Theoretical and Applied Climatology* 51:89–96.

Juvik, J.O., and D. Nullet (1995). Comments on "A proposed standard fog collector for use in high-elevation regions". *Journal Of Applied Meteorology* 34:2108–2110.

Kirch, P.V. (1984). *The Evolution of Polynesian Chiefdoms*. Cambridge: Cambridge University Press.

Kirch, P.V. (1991). Chiefship and competitive involution: the Marquesas Islands of eastern Polynesia. In *Chiefdoms, Power, Economy, and Ideology*, edited by T. Earle, pp. 119–45. Cambridge: Cambridge University Press.

Kirch, P.V. (1994). *The Wet and the Dry*. Chicago: University of Chicago Press.

Kirch, P.V. (1973). Prehistoric subsistence patterns in the northern Marquesas Islands, French Polynesia. *Archaeology and Physical Anthropology in Oceania* 8:24–40.

Kirch, P.V., and M. Sahlins (1992). *Anahulu: The Anthropology of History in the Kingdom of Hawai'i*. Chicago: University of Chicago Press.

Krusenstern, A.J. (1813). *Voyage Round the World in the Years 1803, 1804, 1805, and 1806*. London: John Murray.

Merlin, M., and J. Juvik (1995). Montane cloud forest in the tropical Pacific: some aspects of their floristics, biogeography, ecology, and conservation. In L. Hamilton, J. Juvik, and F. Scatena (eds.), *Tropical Montane Cloud Forests*, pp. 149–162. New York: Springer-Verlag.

Porter, D. (1970). *Journal of a Cruise Made to the Pacific Ocean, Vol 2.* Upper Saddle River, N.J.: Gregg Press.

Ragone, D. (2004). *Artocarpus altilis* (breadfruit). In C.R. Elevitch (ed.). Species Profiles for Pacific Island Agroforestry. `www.traditionaltree.org`. Holualoa, Hawai'i: Permanent Agriculture Resources (PAR).

Rolett, B.V. (1998). *Hanamiai: Prehistoric Colonization and Cultural Change in the Marquesas Islands, East Polynesia.* New Haven: Dept. of Anthropology and The Peabody Museum Yale University.

Salinger, M.J., R. Basher, B. Fiztharris, J. Hay, P. Jones, J. McVeigh, and I. Schmidely-Leleu (1995). Climate trends in the southwest Pacific. *International Journal of Climatology* 15:285–302.

Spriggs, M.J.T. (1981). *Vegetable Kingdoms: Taro Irrigation and Pacific Prehistory.* Doctoral thesis, Canberra: Australian National University.

Stearns, H.T. (1942). Hydrology of volcanic terranes. In O.E. Meinzer (ed.), *Hydrology.* pp. 678–703. New York: Dover Publications.

Suggs, R.C. (1961). *The Archeology of Nuku Hiva, Marquesas Islands, French Polynesia.* New York: American Museum of Natural History.

Thomas, N. (1990). *Marquesan Societies: Inequality and Political Transformations in Eastern Polynesia.* Oxford: Clarendon Press.

Chapter 2

Resource competition between pigs and humans: isotopic evidence from Aitutaki, Cook Islands

Jacqueline A. Craig, University of Auckland

Introduction

As the ancient Polynesians settled the islands of West and East Polynesia, they carried with them a suite of four commensal animals: pigs, dogs, chickens and rats. Some islands received the full complement, other islands only one or two species; for example, there is no evidence for the introduction of the pig or chicken to New Zealand. Of these four, pigs were far more important culturally than the other species, and were considered to be a high status food on most islands. The consumption of pork was limited to important feasts and ritual occasions, and usually to men—women and children on many islands would not have had access to pork at all (Kirch 2000:428, Oliver 2002:81).

Despite this cultural importance, pigs disappeared during the prehistoric period from many islands, among them Tikopia, Mangaia and Rotuma (Bay-Peterson 1984:125, Kirch 2000:431–2). As Bay-Peterson (1984:126) points out, on very small islands such as Kapingamarangi which in 1810 kept only eight pigs, accidental extirpation would be a distinct possibility. However, the great majority of the islands would have been large enough to support a reasonable pig population, and were in any event most likely in periodic contact with other islands that still kept pigs and could supply new breeding stock, so it is possible to rule out chance events that might have wiped out the indigenous pig population permanently. In most cases where extirpation occurred it must have been a conscious decision on the part of the islanders.

Both Bay-Peterson (1984) and Kirch (2000) have examined this problem and come to very similar conclusions about the factors involved in the disappearance of pigs. Kirch breaks the process down into four causal factors:

Small island size and relative isolation: Bay-Peterson (1984:125) shows that the majority of pig extirpations took place on islands with an area of approximately 50 km^2 or less; Kirch's three examples of Tikopia (4.6 km^2), Mangareva (15 km^2) and Mangaia (52 km^2) fit well within these parameters (Kirch 2000:433–4). These islands are also relatively isolated, with at least an overnight canoe voyage and in some cases voyages of several days required to reach the nearest neighbour.

High human population density: Kirch (2000:434) estimates that the population density of the three islands in question was most likely more than 200 persons per km^2 of productive land.

Intensive resource competition: A small area available for cultivation combined with a high population density necessarily results in stress on agricultural systems. In such a tight system pigs may place a disproportionate strain on resources. Rapaport's (1968, 1971) classic study of the Tsembaga Mareng of the New Guinea Highlands and their use of pigs in ritual feasting gives a good idea of the resources that are required to maintain a viable pig population. One pig consumes almost as much food as one human does, in this case largely garden produce. On a small island with limited resources, where there are already pressures on the human population to support themselves, pigs will place an added burden on production systems. As second trophic level[1] feeders, they have a very low

[1] Although communities can have any number of trophic levels, most seem to have three or four. The first of the four basic levels is represented by plants, the second by herbivores, the third by primary carnivores (feeding off herbivores) and secondary carnivores (eating both herbivores and primary carnivores) (Begon *et al.* 1996:833,848). Energy transfer from one trophic level to the next is low, averaging about 10% (Begon *et al.* 1996:735).

return for the amount of energy they consume. Bay-Peterson (1984:124) states "each adult pig cost the amount of food required to maintain an adult human being, and returned only a percentage of this to the economy as meat." Of course having a high value placed upon them culturally would ensure that their presence was not dependent on a strict cost/benefit basis.

Internal strife and warfare: On an island where the previous three conditions are met, it is likely that conflict will arise as a consequence of competition for the limited food and land resources.

Kirch targets resource competition, specifically trophic level competition between humans and pigs, as the prime mover behind pig extirpation on islands that meet the above requirements. He places pigs firmly in the second trophic level (although acknowledging that they are potentially omnivorous), thus coming into direct conflict with humans, who as omnivores occupy both the second and the third trophic levels (Kirch 2000:437).

However, much of this is conjecture given that we know very little about Polynesian pig husbandry in prehistory. We know pigs were present because their bones are often plentiful in archaeological middens, sometimes there are the remains of stone-walled pens, and we have the ethnographic records of early explorers such as Captain Cook, and others on his voyages, as to pig population numbers at European contact (Cook 1955, Ledyard 1963). However, it is generally a mistake to project the ethnographic present wholesale into the past; pig bones and pens alone do not tell the whole story; and of course, each island will have its own unique response to the problem of feeding its pigs. In order to test fully the hypothesis that trophic level competition between humans and pigs led to pig extirpation it is necessary to know what their prehistoric diet was in more detail.

One method that allows us to do this is stable isotope analysis. This technique allows archaeologists to determine the diet of an individual through examination of chemical traces left in the skeleton. Because it is centred on the individual it escapes many of the drawbacks of relying on archaeological methods for determination of past diets. Although the technique has its own inherent problems, most can be controlled for and the results can provide independent confirmation (or refutation) of trends seen in the archaeological record. Individual results can be extrapolated to describe the population as a whole. There are a number of isotopes commonly used in this research, each with their individual strengths and weaknesses. For examining problems centred around trophic levels in which both marine and terrestrial protein is available, carbon (^{13}C) and nitrogen (^{15}N) stable isotopes are particularly useful. δ^{13}C values differentiate between marine and terrestrial sources of protein as marine environments are often enriched in δ^{13}C relative to terrestrial environments. δ^{15}N values increase the higher up the food chain an animal is, so ^{15}N is well suited to determining an individual's trophic level.

Aitutaki in the Southern Cook Islands provides an ideal opportunity for testing the utility of stable isotope analysis on another potential case of pig extirpation in prehistory.

Prehistoric Aitutaki

Aitutaki, an almost atoll, is the most northern island of the Southern Cook Group, approximately 225 km north of Rarotonga. The term 'almost atoll' is used to describe morphology in which the main island is small enough that the surrounding reefs 'almost' form an atoll (Dickinson 1998:1051). The main island of Aitutaki is approximately 20 km^2 in area (Milne 1991:5), and is surrounded by a large, roughly triangular reef enclosing a 50 km^2 lagoon (Allen 1996:103). Within the lagoon there are a number of small islands (*motu*), two of which, Rapota and Moturakau, are volcanic in origin, comprised mostly of basalt. The other *motu*, concentrated along the eastern edge of the reef, are coralline (Allen 1996, Milne 1991). While the population levels in prehistory are unknown, the missionary John Williams (1838:5) reported it as "about 2000 people" in 1821. The first missionary to reside permanently on Aitutaki, Henry Royle, also placed the population at around 2000 in 1839, at the same time recording high levels of disease (Royle 1840 in McArthur 1967:179). The prehistoric population density is likely, therefore, to have been at least 100 people/ km^2, which is below the 200/ km^2 cited in Kirch (2000:434). However, Kirch's figure is based on "productive land", not the actual size of the island in question and this qualification would raise the population density on Aitutaki somewhat. That there was internal strife and competition is clear however, as in Williams' accounts there are several references to endemic warfare on Aitutaki: "[in 1821] they were constantly killing, and even eating each other, for they were cannibals" (1838:17) and "subsequently [to the missionaries' arrival] war had thrice broken out"(1838:18).

Once settled, Aitutaki appears to have maintained regular contact with the other islands in the southern Cook Islands, trading raw materials and cultural information. In the group as a whole during this early period (550 BP and before) there is evidence for trading further afield as well, with basalt from Samoa found on Aitutaki and other islands in the Southern Cooks (Allen and Johnson 1997, Walter and Dickinson 1989, Walter 1996, Walter and Sheppard 1996, Walter 1998:98, Weisler *et al.* 1994). However, these contacts with other island groups would not have been as important as the inter-group contacts within the Cook Islands, and the system can be characterized as one in which regular voyaging took place between the islands of the Southern Cooks and there was trade in a variety of raw materials such as basalt and pearlshell (Walter and Dickinson 1989:98). The 'late' period of exchange (post 550 BP) is characterized by the contraction of spheres of interaction. Goods from other archipelagos become rarer, and exchange within the southern Cook Islands also seems to tail off. This is particularly evident in the decline in use of pearlshell (an imported raw material for most islands though not necessarily Aitutaki) and basalt from Rarotongan sources (Allen 1992a, Kirch *et al.* 1995:52, Walter and Campbell 1996, Walter 1998). During this period Aitutaki would have become more and more reliant on its own lithic, agricultural and marine resources.

There is archaeological evidence for the presence of all four commensals dating from initial settlement; however there is conflicting evidence regarding the presence of pigs (and dogs) throughout the sequence. Although the archaeological evidence points to the continuous presence of pigs from settlement until early historic times (Allen 1992b:385), it stands in contrast to European reports that there were no pigs or dogs present at contact. In 1789 Captain Bligh reported "They said they had no Hogs, Dogs or Goats upon the Island... Notwithstanding they said there were no Hogs, Yams or Tarro, they called them by Name, and I am rather inclined to beleive [sic] they were imposing upon me" (Bligh 1937:95). He did not, however, land on the island and therefore was not in a position to verify either his opinions or what the Aitutakians had told him. John Williams states that on Aitutaki in 1821 "there being no quadrupeds on the island, but a few millions of rats, we sent from Raiatea a number of pigs and goats" (Williams 1838:18); and the same on Mangaia "there were no animals except rats until I visited it, these formed a common article of food" (Williams 1838:64). Pigs were present on Rarotonga however: "they had no other than a breed of small native pigs, of which there were but few, as they were particularly tender and difficult to rear" (Williams 1838:40). Two possible conclusions arise from this: firstly that the early Europeans were mistaken when they said that pigs were not present on Aitutaki and there was a small population hidden away; or that pigs had indeed been eliminated from the island, but for such a short time that the inhabitants still remembered them and the small gap in their presence on the island is not reflected in the archaeological assemblage. Given that Williams resided on Aitutaki for an extended period, and specifically mentioned the fact that rats were the only "quadrupeds" on Aitutaki while recording the presence of pigs on Rarotonga, the latter explanation is more reasonable.

On the face of it, therefore, Aitutaki seems to present a typical example of pig extirpation: it is a small, relatively isolated island that was intensively cultivated in prehistory; most likely had a fairly high population density, though perhaps not as high as other islands that experienced extirpation; and experienced internal conflict and warfare. As with most other islands in Polynesia however, we know very little about their methods of pig husbandry in prehistory, or indeed even into the early historic period.

Isotopic analysis of Aitutaki fauna

Sites on Aitutaki range in date from cal. 1052–804 BP at 1 sigma (Allen 1994) through to contact and historic times. The large faunal collections from two main phases of excavation on Aitutaki (Allen 1992b, Bellwood 1978) contained both human and pig bone from a variety of stratigraphic zones and sites, and thus offered an ideal opportunity to examine the diet of each species and evaluate trophic level competition, as well as any changes in diet through time that may have resulted from this competition.

While some stable isotope studies of animal diet in relation to human diet have concentrated largely on the animals and used human results from other work in the area for comparison (Cannon *et al.* 1999, Clutton-Brock and Noe-Nygaard 1990, Day 1996, Schulting and Richards 2002, van der Merwe *et al.* 2000), this was not possible in the case of Aitutaki, which has had no prior stable isotope work done on humans, nor, for that matter, have the Cook Islands as a whole. It was also necessary to establish a database of isotope values of potential dietary sources on Aitutaki. This was done by collecting a wide range of plants identified as being important in prehistoric times, as well as a range of lagoon and outer-reef fish species.

Oliver (2002:81) states that in early historic times in Polynesia pigs were generally fed the same foods as their owners, usually vegetables, although he qualifies this by saying the same "kind" of diet, not necessarily the same quality of diet since the pigs were often fed leftovers or substandard fruits or tubers. This similarity of diet would be even more likely on smaller islands with a more limited range of available food types. Even if allowed to forage at will during the day, the pigs will be largely limited to foods that are being grown for human consumption—on a small island like Aitutaki there is little else available. There are published reports (Spennemann 1990, Swadling and Chowning 1981) of pigs foraging for shellfish on the tidal flats, and anecdotal stories of attempts at fishing in the same areas, which may have contributed some extra protein to their diet although the success rates are unknown. The other source of protein available to free-roaming pigs is human faeces. However, consumption of this will simply reflect human access to resources and human diet in general, although it will raise the δ^{15}N value of the pigs (Cannon *et al.* 1999:404).

In general though, humans and pigs eating a similar diet will have very similar stable isotope values. If, however, humans are occupying a higher trophic level than pigs, then their δ^{15}N values will be enriched relative to the pig δ^{15}N values. In such a scenario the δ^{13}C values may be very similar if the ratios of marine and terrestrial foods being consumed by humans and pigs are similar; or the δ^{13}C values of one may be enriched if there is a relatively greater marine component in the diet.

If humans and pigs are eating a similar diet in an environment with limited resources they are likely to be in competition with each other. According to the principle of competitive exclusion this may lead to a reduction in the survivorship, growth and/or the reproduction of at least some of the individuals concerned. The more similar the diet, the greater the competition will be (Begon *et al.* 1996:278–9, Oosting 1956:207). In theory, in such a situation one species will achieve competitive advantage over the other, forcing the less well adapted species to either shift its diet or become extinct in that ecological niche. In a situation such as Aitutaki, where humans control the diet of pigs almost entirely, it is most likely that the pigs will have their diets altered so they are in less direct competition with the humans, or they will be extirpated altogether.

The amount of competition will depend on the composition of the human and pig diet, and therefore it is crucial to know the composition and therefore the trophic level of the diets of the two species. If both pigs and humans were predominantly second trophic level then the competition would be greater than if the humans are occupying both second and third trophic levels. If both were occupying the third trophic level the competition would be even more intense, given the loss of energy between the levels. The greater the competition, the faster we might expect to see pig diet moving away from human diet.

Analysis and results

A sample of 24 human bones and 27 pig bones were selected for analysis from the archaeological assemblages. The majority of the human bone (13 samples) was excavated at Ureia by Bellwood in 1970, and most likely dates from ca. 350 BP to the period shortly after European contact (Zone C), although there is some uncertainty about its provenance (Allen 1992b, Bellwood 1978, pers. comm). Three samples from Allen's 1987–9 Ureia excavations (Allen 1992b:107–9) have also been included and are from earlier stratigraphic zones: E/G (1050–750 BP); and J (pre-1050 BP). The eight Moturakau samples are later in date, ranging from approximately 600 BP to European contact (Allen 1994:63). The pig bone was sourced from four sites, Ureia, Moturakau, Aretai and Hosea, and covers the entire range of stratigraphic zones. In addition to the human and pig bone, 24 modern samples were taken from potential dietary sources on Aitutaki: twelve C3 plants (all prehistoric crop plants), one C4 plant (a weed that may have been eaten by pigs), seven herbivorous or omnivorous fish, and four carnivorous fish. Along with an already published shellfish value (Leach *et al.* 2000:153), these form the background against which the dietary reconstructions take place.

All samples were sent to the Rafter Isotope Laboratory at the Institute of Geological and Nuclear Sciences (IGNS) for analysis. Bones were brushed to remove burial soil, washed and sonically cleaned in deionised water, and dried in a vacuum oven. Each sample was pulverized in mortar and pestle, and demineralized in 0.5M HCl. Collagen was filtered from the solution and gelatinized with 0.01M HCl in a nitrogen atmosphere at 100°C for 16 hours. The gelatin was double-filtered, and lyophilized to determine yields (Beavan, pers. comm.).

Carbon and nitrogen in the samples were analysed by an ANCA-SL elemental analyser. The CO_2 and nitrogen gases were analysed for δ^{13}C, δ^{15}N, %C and %N, and C/N ratios. Measurement error values are ± 0.1‰ for carbon and

±0.3‰ or better for nitrogen (Beavan, pers. comm.).

The IsoSource program (Phillips and Gregg 2003) has been used to quantitatively analyse the stable isotope results. In contrast to linear mixing models which are only able to deal with proportional contributions of n+1 different sources, IsoSource allows for up to eight potential sources. All possible combinations of sources (0–100%) are examined. Combinations that fall within specified tolerances are considered feasible solutions from which the frequency and range of potential source contributions can be determined. As each feasible solution must equal 100% of the diet, a higher contribution of one source will necessarily mean a lower contribution of other sources. Since each of these feasible source combinations is constrained to sum to 100%, there are tradeoffs among the sources within their feasible ranges. For example, if one source had the maximum feasible contribution consistent with the isotopic data, then some of the other food sources must have contributions closer to the lower end of their range. Results should report the distribution of feasible solutions rather than just the mean, in this study the 1–99 percentile values are presented. Histograms generated from the results show the distribution of feasible contributions from each source to the human and pig diet. The isotopic signatures of each food source were corrected to account for fractionation during digestion and assimilation; −5‰ for δ^{13}C and −3‰ for δ^{15}N to account for trophic fractionation (Ambrose and Norr 1993:101, Keegan and Deniro 1988, Phillips and Gregg 2003, van der Merwe 1982); −3.7‰ for δ^{13}C to convert bone values to flesh values. The pig diet was calculated without including pig and dog protein based on the assumption that, as a high status food, pork would probably not have been commonly eaten by humans, and therefore even less often by pigs themselves. Dog was excluded on the same basis.

The results show that the average human and pig δ^{13}C and δ^{15}N values (figure 2.1) are indeed very similar. Humans are at a slightly higher trophic level (reflected in δ^{15}N), but not a whole trophic level higher. When compared to the average background values it is clear that both humans and pigs are consuming significant quantities of protein in general, which is reflected in their high δ^{15}N values; and they have a tendency towards marine protein in particular which is reflected in their enriched δ^{13}C values. Both are approaching the carnivorous fish in δ^{15}N levels, indicating that they are occupying both the second and third trophic levels.

The histograms in figure 2.2 were generated using IsoSource and show the probability distributions of the contribution of the various potential dietary sources to the average diet of both species. For example, the histogram for carnivorous fish in the pig diet shows that the range of possible contribution to the entire diet is approximately 15–30%, with the mean (and highest bar which represents the category with the most viable solutions) at 22%. On the other hand, the herbivorous/omnivorous contribution ranges from 0–17% of the diet, with a mean of 3.5%—the height of the bar in the 0–2% of diet category also indicates that that majority of the viable solutions fell towards the bottom of the range. The two species histograms show that while there are definite similarities between the human and pig diets, they are not identical. Both are quite specific and there is not a wide range in the possible amount that each dietary source could contribute to the diet. They are taking smaller amounts of protein from a wide range of sources, and both species also show a heavy reliance on the C3 plants. Pigs have a slightly more specialized diet with a heavier reliance on carnivorous fish than other marine protein, but the C3 plants contribute the bulk of the diet.

Based on these results there is a high potential for trophic level competition and therefore competitive exclusion; far more so in the case of Aitutaki than envisioned by Kirch who characterized the pigs as essentially being second trophic level consumers and therefore not in such direct competition with humans. In this situation there are two species, both high on the food chain, potentially eating very similar diets, on a small island with limited resources.

Change over time

Given that there is direct trophic level competition occurring between humans and pigs on Aitutaki in prehistory, examining pig diet over time may reveal the timing and tempo of any changes that occurred. There were significant dietary differences between pig samples taken from the 1050–750 BP and 550–350 BP periods, and the 350 BP to contact period.

Table 2.1 shows the change in pig diet over these three time periods, compared to human diet from the site of Ureia for the same time periods. It shows a shift from an early diet that relied heavily on carnivorous fish (mean 21%) and C3 plants (59.8%), to one that utilized the available resources more evenly and emphasized C3 plants more (mean 70.3%). The human diet remains essentially the same, although showing

	Carnivorous fish %	Omnivorous/ herbivorous fish%	Shellfish %	C3 plant %	C4 plant %	Pig %	Dog %	Total mean protein
Pig								
1050–750 B.P	11–35 (21)	0–23 (6.3)	0–22 (6.1)	54–69 (59.8)	0–24 (6.7)			33.4%
550–350 BP	28–35 (30.7)	0–6 (1.6)	0–6 (1.5)	63–67 (64.5)	0–7 (1.7)			33.8%
350 BP–contact	0–24 (6.7)	0–22 (6.2)	0–31 (8.9)	60–88 (70.3)	0–28 (7.8)			21.8%
Human								
Pre–1050 BP	0–26 (7)	0–24 (6.5)	0–32 (9.2)	24–60 (42.6)	0–30 (8.2)	0–46 (15.4)	0–38 (11.1)	49.2%
1050–750 BP	0–30 (7.8)	0–26 (7)	0–34 (9.8)	16–54 (36.6)	0–32 (8.8)	0–50 (17.5)	0–40 (12.6)	54.7%
350 BP–contact	0–28 (7.5)	0–22 (5.8)	0–28 (7.8)	12–52 (33.5)	0–26 (7.1)	0–60 (23.1)	0–48 (15.3)	59.5%

Table 2.1: IsoSource results for human and pig diet change over time (percent contribution of dietary sources). 1–99 percentile, mean in brackets.

	Carnivorous fish %	Omnivorous/herbivorous fish %	Shellfish %	Total mean %
Pig				
1050–750 B.P	11–35 (21)	0–23 (6.3)	0–22 (6.1)	33.4
550–350 BP	28–35 (30.7)	0–6 (1.6)	0–6 (1.5)	33.8
350 BP–contact	0–24 (6.7)	0–22 (6.2)	0–31 (8.9)	21.8
Human				
Pre–1050 BP	0–26 (7)	0–24 (6.5)	0–32 (9.2)	22.7
1050–750 BP	0–30 (7.8)	0–26 (7)	0–34 (9.8)	24.6
350 BP–contact	0–28 (7.5)	0–22 (5.8)	0–28 (7.8)	21.1

Table 2.2: Percentage of marine protein in pig and human diet over time. 0–99 percentile, mean in brackets.

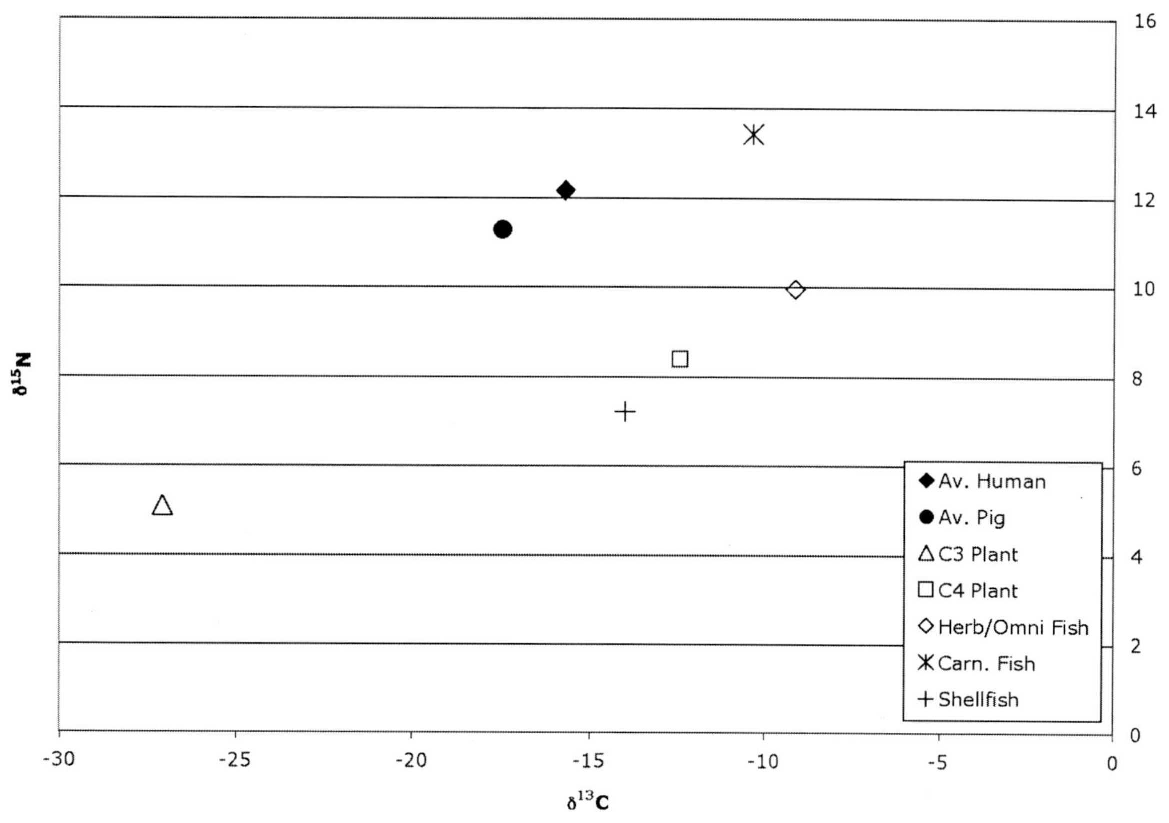

Figure 2.1: Average δ^{13}C and δ^{15}N values for human, pig and dietary sources.

an increase in over-all protein towards the end of the prehistoric sequence. Table 2.2 shows that clear directional changes have occurred in the amount of marine protein available to pigs over time, whereas the amounts of marine protein in the human diet have remained relatively stable over time.

Again, competitive exclusion seems to be the key to the change over time. Pig protein consumption is initially relatively similar to humans (means of 33.4% and 49.2% respectively) and they are eating significant amounts of off-shore carnivorous fish; after an increase in the amount of protein in the diet 550–350 BP, there is a decrease in the amounts of the carnivorous fish which results in a more generalized diet with less protein overall by the end of the prehistoric period (mean 21.8% of diet). Humans remain steady in the amounts of marine protein over time and see a rise in the overall protein component of their diet. Although IsoSource shows that the increase in protein 350 BP—contact is due to more pig and dog in the diet, it must be remembered that these are probability distributions and if we were to specify, based on what we know ethnographically about the absence of pigs and dogs at European contact, that the pig and dog contributions must be towards the bottom of the possible range of contributions, the amount of marine protein would rise accordingly to compensate (Phillips and Gregg n.d.). However, this does not change the fact that *overall*, protein consumption increased at this time.

The first significant change in pig diet occurs during the same period that archaeologists speculate inter-island contacts began to decline, around 550 BP (Walter 1998:98). This decline would have resulted in Aitutakians becoming more reliant on local resources and would have placed more stress on the agricultural and material resources available, thus intensifying the competition between humans and pigs for increasingly limited resources. The second significant shift in diet occurs at some point towards the end of the prehistoric sequence, 350 BP to contact—at this point the pig diet has changed from its original composition similar to the human diet, to a diet that is more generalized, lower in protein, and one that presumably offered less competition to humans. However, if extirpation did take place at the end of this period, the shift in diet was clearly not enough to remove them as potential competitors for the more limited resources on the island.

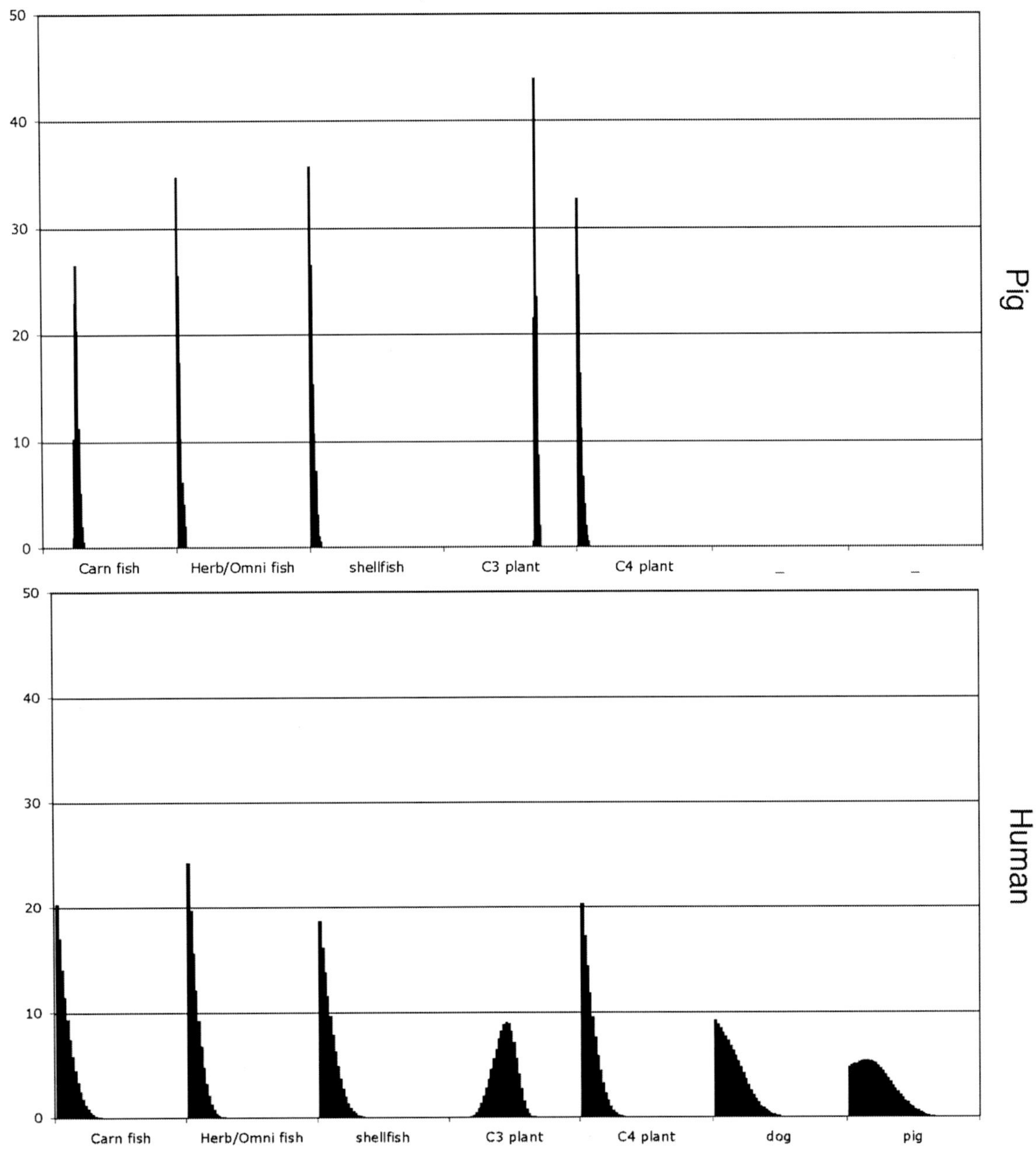

Figure 2.2: IsoSource histograms for average pig and human diet.

Discussion

Given the similarity between early pig and human isotopic values, human diet did not change as radically over time as pig did. This is likely due to trophic level competition eventually leading to competitive exclusion of pigs on Aitutaki in the late prehistoric or very early contact periods.

On Aitutaki pigs are occupying a higher trophic level than predicted by Kirch; they are definitely not solely occupying the second trophic level, but extending well into the third trophic level as well, as do the humans. This illustrates the danger of assuming that the situation immediately post-contact (and possibly post-reintroduction) for pig husbandry can be extended back into the past. While it is quite likely that on many Polynesian islands the pigs were largely second trophic level consumers, it is clear that on Aitutaki, and probably on many other islands, they ate a similar diet to humans, and this included a large marine protein component.

The fact that Aitutaki is an island also plays an important part in this scenario. The concept of islands as laboratories (Evans 1973) has been a popular one, their presumed isolation allowing archaeologists to study processes that may be more difficult to discern in mainland settings where there are many more outside influences operating (Anderson 2004). Not only may these processes be easier to observe on an island, the isolation may also serve to intensify outcomes. Despite close cultural contacts with the other islands in the Cooks, and further afield early in prehistory, Aitutakians would have been largely reliant on their own natural resources, particularly when it came to food procurement. If this was the case for humans, then it was doubly so for pigs—on a small island the opportunities for wild foraging would have been very limited both in terms of what was available, and also what access pigs had to those resources. A relatively closed small island system is far less likely to be able to support natural foraging by a pig population than a continental location, meaning that the feeding of pigs is likely to be at the discretion of the owners rather than a result of pig preferences and/or access. Thus the pig diet will reflect the access that humans have to particular resources and also what resources they felt were appropriate or available to feed the pigs with. Earlier in the sequence there was enough high quality marine protein available that feeding some to the pigs was a sustainable drain on resources; as that easy access decreased over time the pig consumption of seafood would have constituted competition with the human population, and proved to be a drain on the economy of the island.

The response of humans to this competition was not immediate extirpation but a gradual change in pig diet over almost a millennium. This clearly shows that pigs formed an important part of the subsistence system and were a culturally important item—people were reluctant to give them up and experimented with different diets over time in order to maintain the population for as long as possible. It was only towards the very end of the prehistoric sequence that the decision to finally rid the island of pigs was made.

Acknowledgements

The samples were taken from the Aitutaki excavation collections of Melinda Allen and Peter Bellwood, held in the Department of Anthropology, University of Auckland. Permission to use the human bone was kindly granted by the people of Aitutaki, and in particular the Bishop, Makimare, Turu, Tetevano, Paerau, Mouroa, and Mapu families who own the land upon which the Ureia and Moturakau sites were located. Special thanks to Mayor Tai Herman, Ina Solomona and Teau Maki for help in obtaining those permissions. Permission was gained for the collection of background dietary samples from the Government of the Cook Islands, thanks to Fred Charlie and Tiraa Arere of the Ministry of Agriculture, and to Metu Korosela and Richard of the Ministry of Fisheries for help in the collecting.

Thanks to Nancy Beavan of the Rafter Radiocarbon Laboratory, IGNS, for the sample processing, and so much help and advice on this project. Melinda Allen, Judith Littleton and Matthew Campbell have all read, commented on, and made valuable suggestions about the paper—many thanks.

Funding for the work has come from the Wenner Gren Foundation (Individual Research Grant 6844), the Postgraduate Research Fund, University of Auckland, and the Department of Anthropology, University of Auckland.

References cited

Allen, M.S., 1992a. Temporal variation in Polynesian fishing strategies: the Southern Cook Islands in regional perspective. *Asian Perspectives*, 31(2):183–204.

Allen, M.S., 1992b. *Dynamic Landscapes and Human Subsistence: Archaeological Investigations on Aitutaki Island, Southern Cook Islands.* Unpublished Ph.D. thesis, University of Washington.

Allen, M.S., 1994. The chronology of coastal morphogenesis and human settlement on Aitutaki, Southern Cook Islands, Polynesia. *Radiocarbon*, 36(1):59–71.

Allen, M.S., 1996. Style and function in East Polynesian fish-hooks. *Antiquity*, 70:97–116.

Allen, M.S. and K.T.M. Johnson, 1997. Tracking ancient patterns of interaction: recent geochemical studies in the Southern Cook Islands. In, M.I. Weisler (ed.), *Prehistoric Long–Distance Interaction in Oceania: An Interdisciplinary Approach.* New Zealand Archaeological Association Monograph, No. 21, pp. 111–133.

Ambrose, S.H. and L. Norr, 1993. Experimental evidence for the relationship of the carbon isotope rations of whole diet and dietary protein to those of bone collagen and carbonate. In, J.B. Lambert and G. Grupe (eds.), *Prehistoric Human Bone—Archaeology at the Molecular Level.* Berlin:Springer-Verlag, pp. 1–38.

Anderson, A.J., 2004. Islands of ambivalence. In, S.M. Fitzpatrick (ed.), *Voyages of Discovery. The Archaeology of Islands.* Westport, Connecticut: Praeger, pp. 251–273.

Bay–Peterson, J., 1984. Competition for resources: The role of pig and dog in the Polynesian agricultural economy. *Journal de la Société des Océanistes*, 77:121–129.

Begon, M., J.L. Harper and C.R. Townsend, 1996. *Ecology: Individuals, Populations, and Communities.* Oxford: Blackwell Science.

Bellwood, P.S., 1978. Archaeological research in the Cook Islands. *Pacific Anthropological Records*, 27.

Bligh, W., 1937. *The Log of the Bounty.* Volume II. London: Golden Cockerel Press.

Cannon, A., H.P. Schwarcz and M. Knyf, 1999. Marine–based subsistence trends and the stable isotope analysis of dog bones from Namu, British Columbia. *Journal of Archaeological Science*, 26:399–407.

Clutton–Brock, J. and N. Noe–Nygaard, 1990. New osteological and C-isotope evidence on Mesolithic dogs: companions to hunters and fishers at Star Carr, Seamer Carr and Kongemose. *Journal of Archaeological Science*, 26:643–653.

Cook, J., 1955. *The Journals of Captain James Cook on His Voyages of Discovery: Edited From His Original Manuscripts by J.C. Beaglehole.* [J.C. Beaglehole (ed.)]. Cambridge: University Press.

Day, S.P., 1996. Dogs, deer and diet at Star Carr: a reconsideration of C-isotope evidence from early Mesolithic dog remains from the Vale of Pickering, Yorkshire, England. *Journal of Archaeological Science*, 23:783–787.

Dickinson, W.R., 1998. Geomorphology and geodynamics of the Cook-Austral Island-Seamount Chain in the South Pacific Ocean: Implications for hotspots and plumes. *International Geology Review*, 40:1039–75.

Evans, J.D., 1973. Islands as laboratories for the study of culture process. In, C. Renfrew (ed.), *The Explanation of Culture Change: Models in Prehistory.* London: Duckworth, pp. 517–520.

Keegan, W.F. and M.J. DeNiro, 1988. Stable carbon- and nitrogen-isotope ratios of bone collagen used to study coral-reef and terrestrial components of prehistoric Bahamian diet. *American Antiquity*, 53(2): 320–336.

Kirch, P.V., D.W. Steadman, V.L. Butler, J. Hather and M.I. Weisler, 1995. Prehistory and human ecology in Eastern Polynesia: excavations at Tangatatau Rockshelter, Mangaia, Cook Islands. *Archaeology in Oceania* 30: 47-65.

Kirch, P.V. 1996. Late Holocene human-induced modifications to a central Polynesian island ecosystem. *PNAS* 93(11): 5296–5300.

Kirch, P.V., 2000. Pigs, humans and trophic competition on small Oceanic islands. In, A. Anderson and T. Murray (eds.), *Australian Archaeologist: Collected Papers in Honour of Jim Allen.* Canberra: Coombs Academic Publishing, pp. 427–39.

Leach, B.F., C.J. Quinn, G.L. Lyon, A. Haystead and D.B. Myers, 2000. Evidence of prehistoric Lapita diet at Watom Island, Papua New Guinea, using stable isotopes. *New Zealand Journal of Archaeology*, 20: 149–159.

Ledyard, J., 1963. *John Ledyard's Journal of Captain Cook's Last Voyage*. J.K. Munford (ed.) Corvallis: Oregon State University Press.

McArthur, N., 1967. *Island Populations of the Pacific*. Canberra: Australian National University Press.

Milne, J.D.G., 1991. *Soils of Aitutaki, Cook Islands*. Lower Hutt: DSIR Land Resources, Department of Scientific and Industrial Research.

Oliver, D., 2002. *Polynesia In Early Historic Times*. Honolulu: The Bess Press.

Oosting, H.J., 1956. *The Study of Plant Communities: An Introduction to Plant Ecology*. San Francisco: W.H. Freeman and Company.

Phillips, D.L. and J.W. Gregg, 2003. Source partitioning using stable isotopes: coping with too many sources. *Oecologia* 136: 261–269.

Phillips, D.L. and J.W. Gregg, n.d. Post-processing IsoSource results to incorporate additional non–isotopic constraints. Unpublished manuscript.

Rapaport, R.A., 1968. *Pigs for the Ancestors: Ritual in the Ecology of a New Guinea People*. New Haven: Yale University Press.

Rapaport, R.A., 1971. The flow of energy in an agricultural society. *Scientific American*, 22:116–132.

Schulting, R.J. and M.P. Richards, 2002. Dogs, ducks, deer and diet: new stable isotope evidence on early Mesolithic dogs from the Vale of Pickering, North-East England. *Journal of Archaeological Science*, 29:327–333.

Spennemann, D.H.R., 1990. The role of pigs and dogs in the taphonomy of archaeological assemblages from Tonga. In, S. Solomon, I. Davidson and D. Watson (eds.), *Tempus. Archaeology and Material Culture Studies in Anthropology, Volume 2*.

Swadling, P. and A. Chowning, 1981. Shellfish gathering at Nukakau Island, West New Britain Province, Papua New Guinea. *Journal de la Société des Océanistes*, 37:159–167.

van der Merwe, N.J., 1982. Carbon isotopes, photosynthesis, and archaeology. *American Scientist*, 70:596–606.

van der Merwe, N.J., R.H. Tykot, N. Hammond and K. Oakberg, 2000. Diet and animal husbandry of the preclassic Maya at Cuello, Belize: Isotopic and zooarchaeological evidence. In, S.H. Ambrose and M.A. Katzenberg (eds.), *Biogeochemical Approaches to Paleodietary Analysis*. New York: Kluwer Academic/Plenum Publishers, pp. 23–38.

Walter, R. and W.R. Dickinson, 1989. A ceramic sherd from Ma'uke in the Southern Cook Islands. *Journal of the Polynesian Society*, 98:465–470.

Walter, R., 1996. Settlement pattern archaeology in the Southern Cook Islands: a review. *Journal of the Polynesian Society*, 105:63–99.

Walter, R. and M. Campbell, 1996. The Paraoa site: fishing, and fishhooks in 16th century Mititaro, Southern Cook Islands. *Man and Culture in Oceania*, 12:47–60.

Walter, R. and P.S. Sheppard, 1996. The Ngati Tiare adze cache: further evidence of prehistoric contact between West Polynesia and the Southern Cook Islands. *Archaeology in Oceania*, 12:33–39.

Walter, R., 1998. *Anai'o: The Archaeology of a Fourteenth Century Polynesian Community in the Cook Islands*. New Zealand Archaeological Association Monograph 22. Auckland: New Zealand Archaeological Association.

Weisler, M.I., P.V. Kirch and J.M. Endicott, 1994. The Mata'are basalt souce: implications for prehistoric interaction studies in the Cook Islands. *Journal of the Polynesian Society*, 103:203–216.

Williams, J., 1838. *A Narrative of Missionary Enterprises in the South Sea Islands: with remarks upon the natural history of the Islands, origin, languages, traditions, and usages of the inhabitants*. London: J. Snow.

Chapter 3

Insular models of technical change: Sumatra, Nias and Siberut (Indonesia)

Dominique Guillaud, Institut de recherche pour le développement
Hubert Forestier, Institut de recherche pour le développement
Harry Truman Simanjuntak, University of Jakarta

Introduction

The following text is based upon research undertaken since 2000 in Sumatra, which aims to understand human adaptation to tropical environments from the onset of the Holocene period to the present. Technology is the central theme of our approach; in this, our work is partially influenced by the theories of Leroi-Gourhan (1973), who demonstrated that technology accounts for the relations between humans and their environment. To epitomize the diversity of this relationship, its evolution, and the complexity of the factors involved, we will not restrain ourselves to one period only, but we will consider the dynamics of lithic and metallic industries during prehistoric and protohistoric times.

The nature, the quantity and the quality of the raw materials (stone and metal) available for the making of tools, implements and weapons, influences the technical system of a society and its impact on the environment. This constraint of raw materials is of great significance in the islands, where the resources are generally more limited than in continental areas. This is why, for a better understanding of the importance of resources for the techniques, we chose to compare different insular areas in the western part of the Indonesian archipelago. A portion of a large island, the South Sumatra province, which presents a certain continental character and offers locally some useful resources in stone and metal, will be compared to two much smaller islands, Nias and Siberut, located about 150 km off the western coast of Sumatra, which offer very little raw material resources (figure 3.1).[1]

We will only deal here with the results of our studies that are useful for our demonstration, and refer to other publications for more details. The results will concern two connected fields: (1) archaeological data, which contribute to the knowledge of this insular region of Southeast Asia; in particular a cultural chronology of the past 12,000 years will show the main technical changes; (2) methodological considerations, which indicate how the comparison between islands can be a tool for the study of technical evolution. This will include a brief analysis of 'islandness' itself.

In the first place, we will deal with some landmarks of the main episodes of human settlement that were acknowledged when we started our survey in Sumatra. The most ancient record of modern humans is dated to about 10 to 12,000 years and related to the 'Australoïd' expansion in the region. Those dates were obtained in the shell middens of northeast Sumatra, and were related to the so-called 'Hoabinhian technocomplex' (containing 'sumatraliths') (McKinnon 1975 and 1991; Bellwood 1997; Glover 1978).

Although the age of the Neolithic period in the island was unknown until now, we could date it in the south of Sumatra to about 3–4000 years BP. Whether these Neolithic populations were of Austronesian or Austroasiatic[2] origin is a question that the present article will not address. Considering present-day linguistic and cultural

[1] In 2005, we will also include the island of Bangka, on the eastern coast of Sumatra, amongst our fields.

[2] The Austronesians and Austroasiatics are originally Neolithic populations from the centre-east of China (Yellow River basin). In their migrations, the Austronesians would have chosen a sea-route, settling first in Taiwan about 5000 years ago, and expanding later in the whole of insular Southeast Asia and in Oceania. The Austroasiatic would have taken a land-route and settled in the Indochinese peninsula.

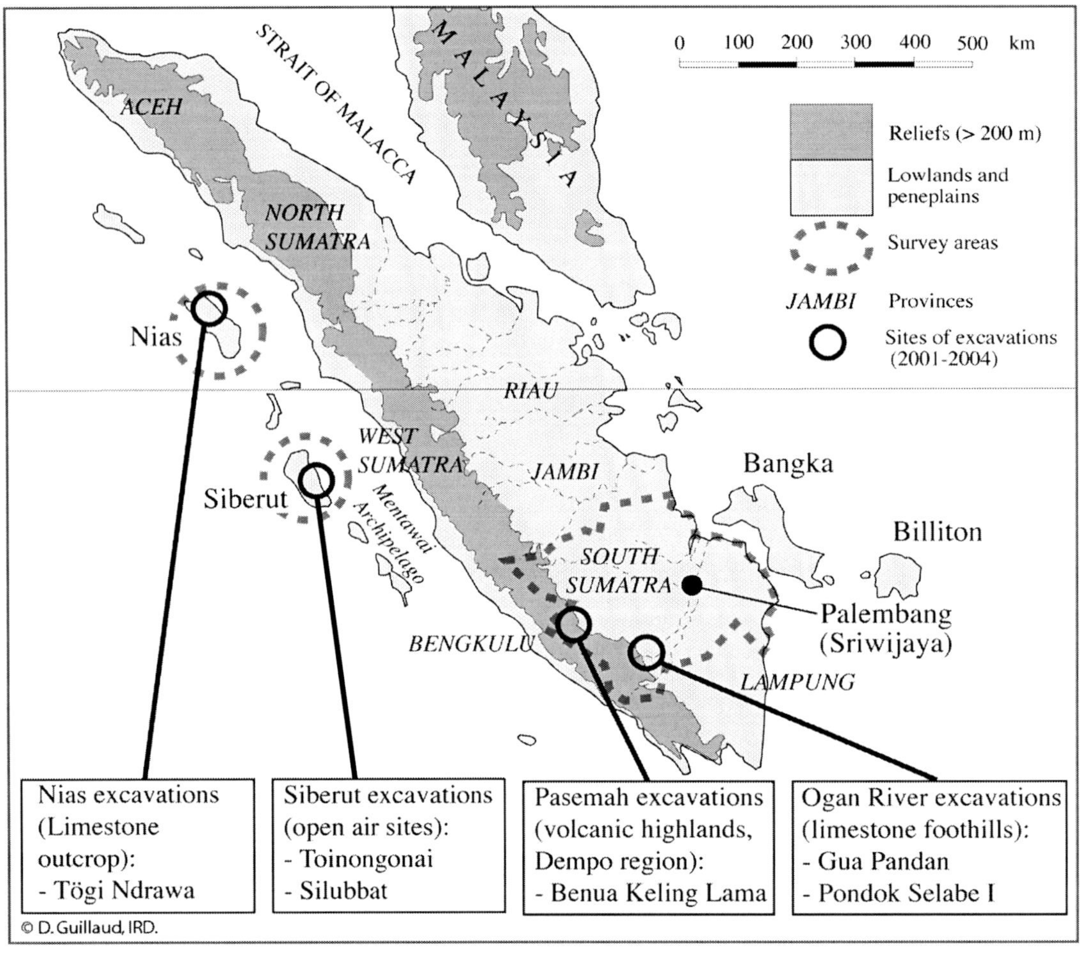

Figure 3.1: The islands of Sumatra, Nias and Siberut and the main excavations undertaken by the IRD/Puslit Arkenas team (2001–2004).

evidence, one tends to favour the Austronesian hypothesis, but this is still to be confirmed. The arrival of the Austroasiatics in the Indonesian archipelago is not very well documented, although some authors consider that the non-Austronesian *asli* languages of some indigenous groups of hunters-gatherers would indicate that the Austroasiatics settled at uncertain times in Kalimantan and maybe in Sumatra. At about 2500–2000 BP, an important change occurred with the arrival of metal, and this know-how, which shows more clearly an Austroasiatic influence, seems to have been imported from continental Southeast Asia. Metal is related to large-scale trade networks, involving intricate schemes of transmission, and arrived at very different periods depending on the regions.

South Sumatra: the sequence of human settlement in the context of a large island

During the first stage of our project, only the province of South Sumatra was surveyed. The province is characterized by a succession of contrasting ecological zones, from west to east, the volcanic highlands, the foothills with limestone formations (karsts), the long peneplain with its podzolic soils, cut off by large rivers, and finally the swampy coastline (figure 3.2). This combination of ecosystems is found throughout most of Sumatra. The ecological diversity of the region is ideal for building understanding of human settlement choices and adaptations to certain environments. Our study has shown that some zones seem to have been favoured by the populations during certain periods, because they offered the resources necessary to their technical needs.

The prehistoric appeal of karstic zones

During 'lithic' periods (i.e., the end of the Pleistocene until the mid-Holocene), the karstic zones have proved essential for human settlement. Our research (Forestier et al. 2005) has revealed for the first time the presence of Acheulean hand-axes and classical cleavers ('biface' and bifacial pieces) in the Ogan river bed. Incidentally, this finding fills a gap concerning the distribution in the world of this technical model, ascribed to *Homo erectus*.

During the Holocene period, a cave site dating from 8–9000 years BP onward has been located in the same area (Gua Pandan cave). The most ancient levels (figure 3.3) comprise a classical Hoabinhian tool-kit: massive pebble-tools such as choppers, unifacial pebble-tools (figure 3.4), tools with a thick end-scraper front, and big flake tools.

Following this first modern occupation, the neighbouring cave of Pondok Selabe I reveals the comprehensive sequence of occupation in the karstic zone (figure 3.5; Simanjuntak et al. 2004, 2005).

A very particular 'pre-Neolithic' level at 4500 BP, showing no ceramics nor polished stone, can be characterized by a macro-industry of chert, flint nodules and river pebbles, found together with some remains of forest fauna (figure 3.6). This occupation seems to be linked to hunting activities.

At 2700 BP, another specific 'Neolithic' level consists of thin smooth ceramics, incised or corded, but it lacks polished stone; instead it shows very small knapped tools of obsidian, flint and andesite. The knapping method is very simple (an algorithm method defined as 'a basic system of alternating platform'—see Forestier 2000) and implies a direct percussion with a hard hammer (figure 3.7). The level, including a lot of faunal remains (deer, wild boar, monkeys, civet-cats, etc.) suggests that hunting activities were still predominant, but coexisted with the very beginning of agriculture on the foothills. The same cave exhibits in its higher levels a continuous occupation until the metal age, including some intrusive burials.

The useful raw materials for lithic activities (chert and flint) originated from the limestone formations of the foothills, which therefore attracted human settlement. This occupation was obviously not confined to these karstic zones, for there have always been movement of people, and therefore exchange networks, allowing a measure of independence from the sites of raw material. The obsidian found in the Neolithic level clearly indicates the existence of networks with the nearby volcanic highlands. We will also see below that, owing to the vegetal industry, this dependence has probably been less constrained than believed.

The highlands: the continuity from the Neolithic to the Metal age

Let us now consider the South Sumatra highlands, in the surroundings of the Dempo Volcano: the excavations in the spectacular open-air site of Benua Keling Lama, a complex of more than fifty earth mounds, have revealed the following sequences of occupation (figure 3.8).

First, a 'classical' Neolithic period with a radiocarbon age of 3600 years, which revealed a polished stone adze and numerous ceramic pot-

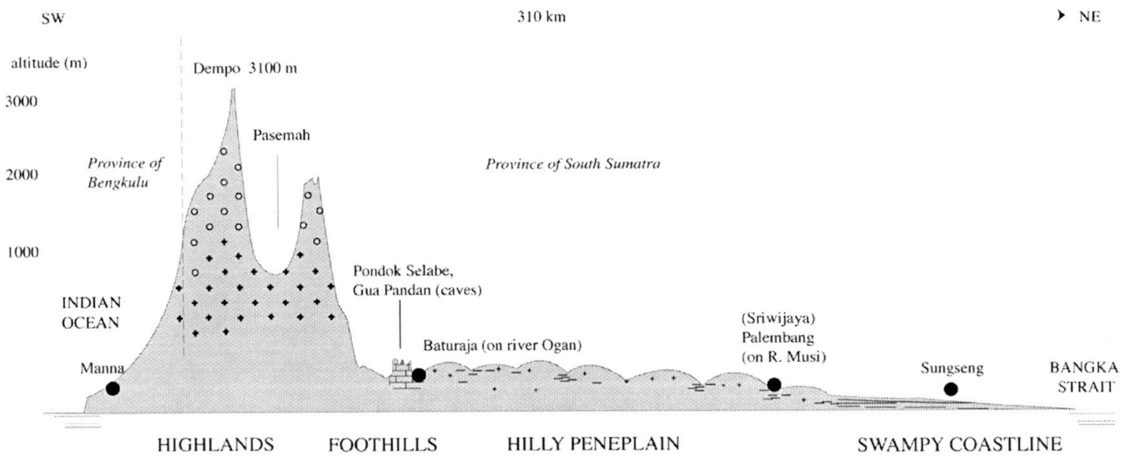

Environment	volcanic soils	limestone formations	podzolic soils, alluvium	alluvium, peat soils
Archaeological finds	Neolithic occupations, Stone monuments, "Dong son" society, fortified sites	Paleolithic, pre-Neolithic and Neolithic occupations	hindo-bouddhist temples, harbor-city of Sriwijaya	Pre-Sriwijaya?

Figure 3.2: Section of the island of Sumatra, showing the succession of environments and the archaeological characteristics of each zone.

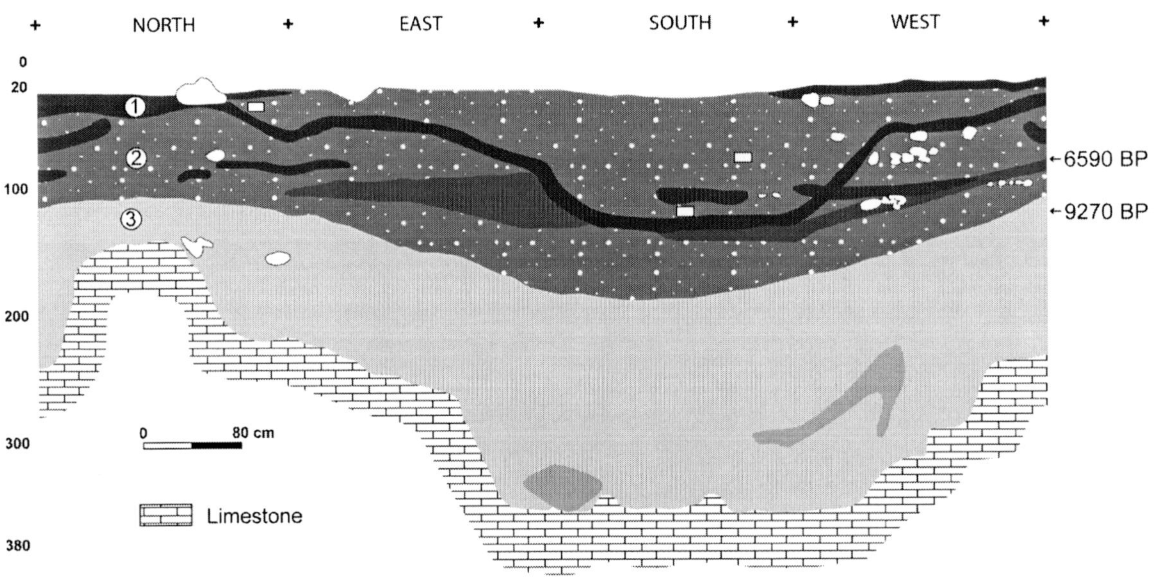

Figure 3.3: Stratigraphy of Gua Pandan cave, South Sumatra. 1. brown sandy clay; 2. brown sandy clay with a lot of stone flakes, bones, molluscs, obsidian and stone; 3. brown-yellow clayey sand.

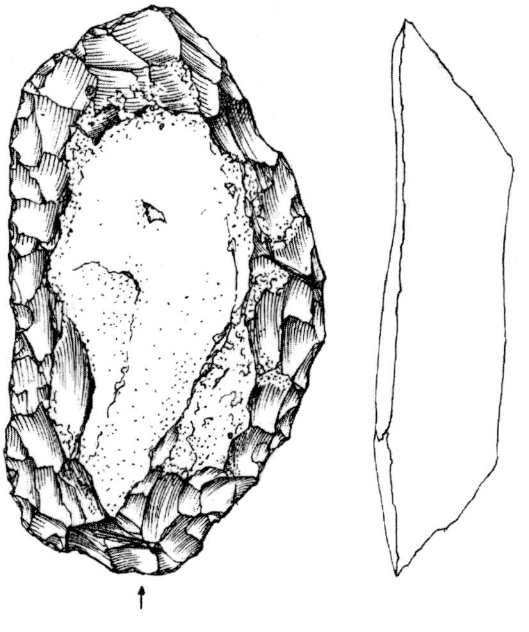

Figure 3.4: Unifacial pebble-tool from Gua Pandan cave, South Sumatra (Z=120).

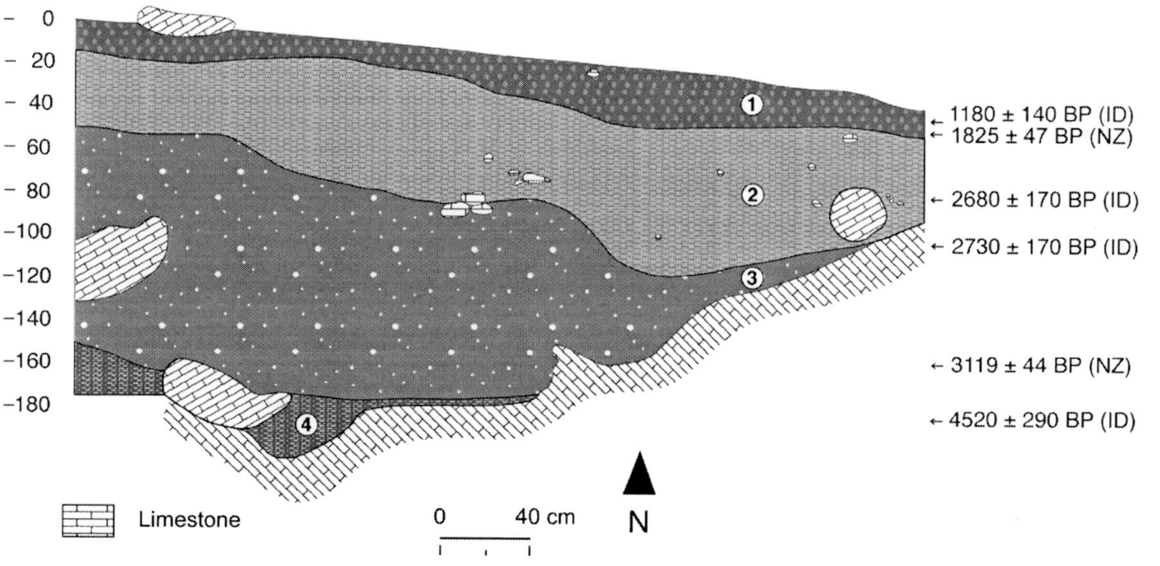

Figure 3.5: Stratigraphy of Pondol Selabe I cave, South Sumatra. 1. disturbed layer (metal age); 2. neolithic layer, reddish-brown; 3. pre-Neolithic layer, dark brown, with limestone rocks; 4. pre-Neolithic layer in contact with the bedrock.

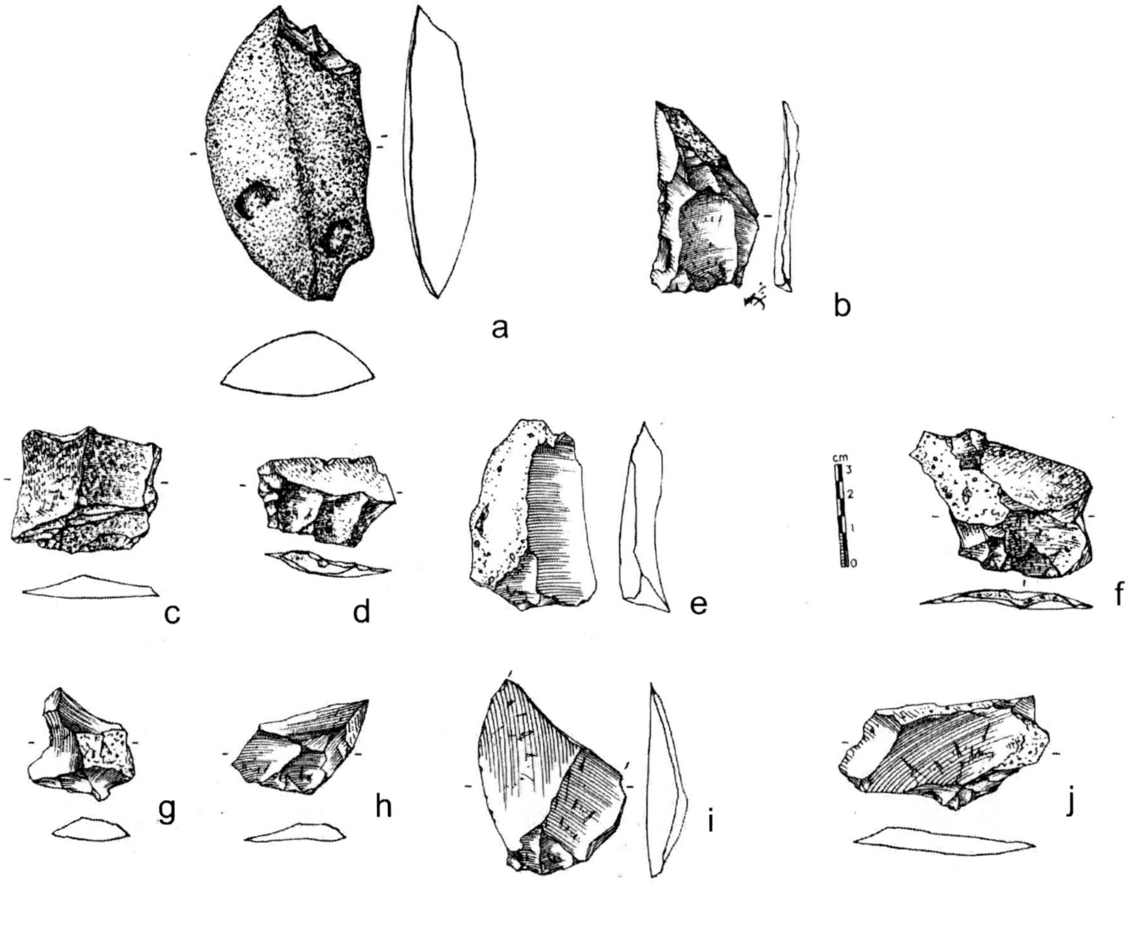

Figure 3.6: Some artefacts from the pre-Neolithic level of SLB1, South Sumatra. (a) pebble tool; (b) pseudo-Levallois flake; (c) to (j) flakes on different raw materials.

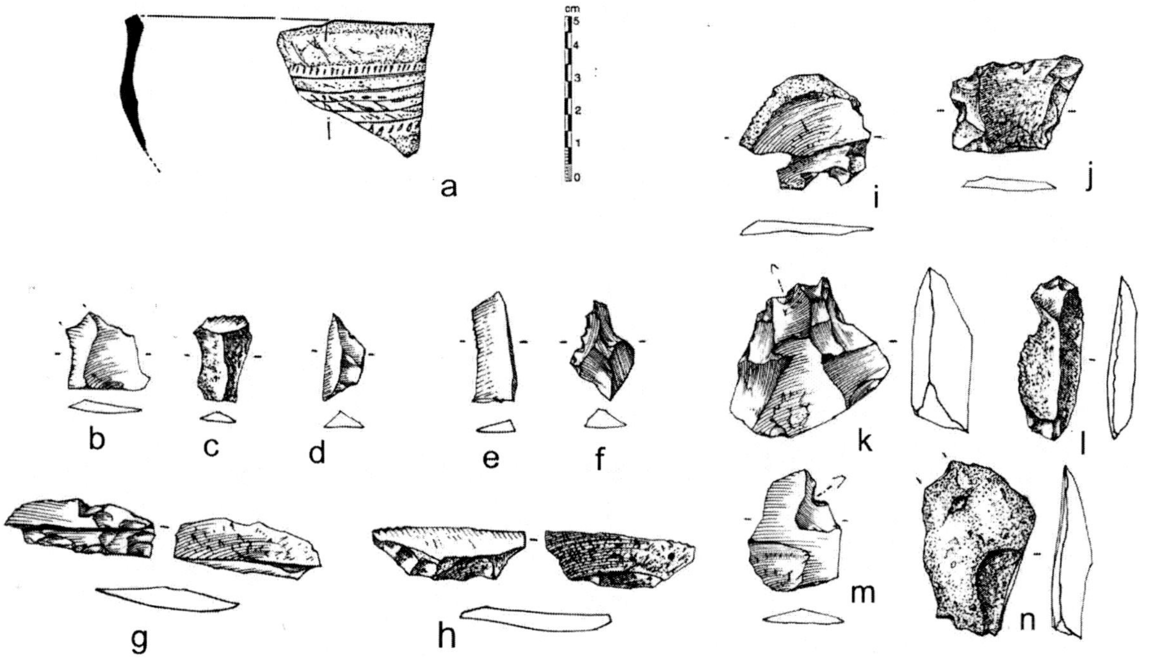

Figure 3.7: Some artefacts from the Neolithic level of SLB1, South Sumatra. (a) incised ceramic; (b) to (h) micro flake tools; (i) to (n) macro flake tools.

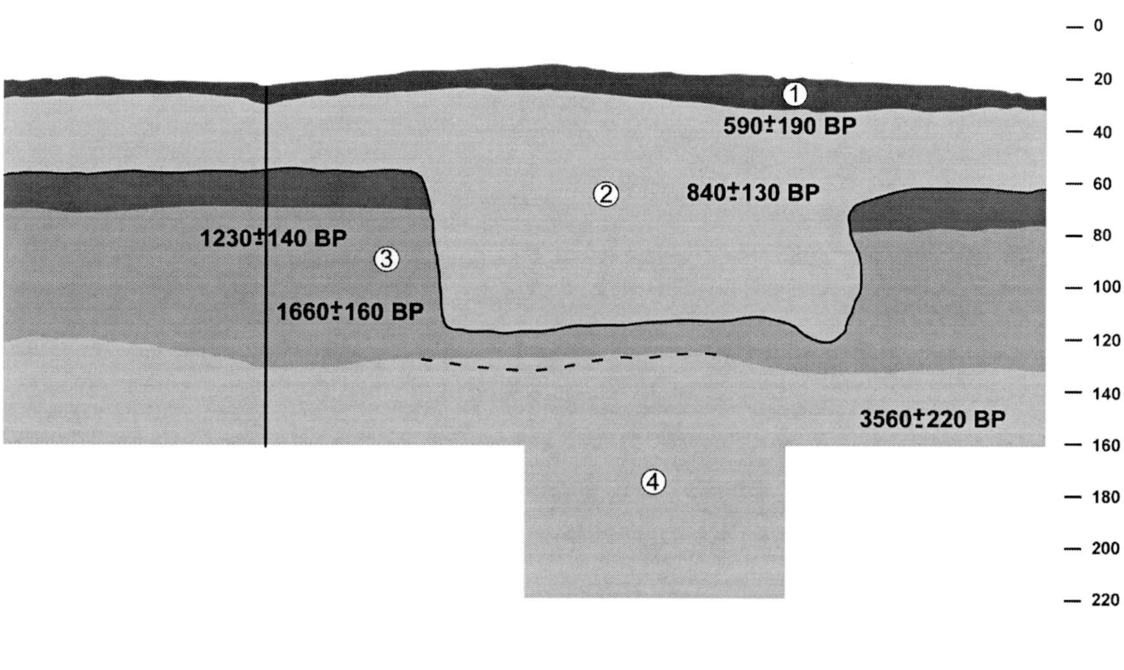

Figure 3.8: Western section of the earth mound at Benua Keling Lama, Pasemah, South Sumatra. 1. humus, black sand layer (disturbed); 2. brown-red sand layer; 3. brown-grey sandy clay (undisturbed); 4. brown-red sandy clay (undisturbed).

sherds. This occupation could document the arrival of a population who developed cultivation on the fertile volcanic soils. It is interesting to note that the dates of this Neolithic level in the highlands are older than the dates on the foothills: this could indicate that the 'Neolithization' of the region occurred first in the highlands, and then went down to the lowlands. Second, the metal age, which in the region is related to the megalithic tradition found around the Dempo. This level is radiocarbon to about 1700 BP. The occupation of the site comes to an abrupt end in the 14th century, when it is turned into a burial site. This episode is well documented by oral tradition, which describes the arrival of the present-day population, the Pasemah. They reject and partly absorb the previous settlement, the so-called Rejang, who in their tradition refer to the megaliths.

As we have seen, societies tend to favour specific areas according to their technical skills and systems and their material production. In the earliest periods the limestone formations offered raw material for the making of tools and weapons, as well as water supply, shelter in the caves, and hunting opportunities. Later, along the riverbeds, some acceptable soils for the very beginning of horti/agriculture were used for cultivation. The fertile volcanic lands of the highlands were also occupied and exploited and, with the Neolithic period and the setting of the first structured networks, populations became less and less dependent on the limestone formations.

Metal, the arrival of which corresponds to a strengthening of political and trade networks, gives insular Southeast Asia its true 'Asiatic' identity. Continuity in the choice of settlement sites is discernible (the site from the metal-age being found on the same spots as those from the Neolithic), but networks expand considerably (Bellwood 1997). The metal-age is in fact related to the numerous megaliths of the region, whose figures document a 'Dong Son' society. This name refers to the bronze kettle-drums crafted in the north of Vietnam, and scattered all over the Indonesian archipelago. Very little is known about this Dong Son society, whose presence might be connected to the gold and copper ore found in the highlands. Eventually, all centres of power rely on the control of the ore sources and of the sites of forge.

The lowlands and the remains of the Classic and Islamic period

While some metal can be found in the highlands, the lowlands offer almost none. It is only off the coast, on the islands of Bangka and Billiton that some iron and a large amount of tin are found. Tin in particular was important because it generated, from the first centuries AD onward, a profitable trade with China.

Some time after the settlement of a Dong Son society in the highlands, a centre of power merges and strengthens in the lowlands. The hindo-bouddhist harbour-city of Sriwijaya appears in the lowlands around 500–600 AD, and developed thanks to the long distance trade with China, whose routes pass through the strait of Malacca (Coedes 1918; Wolters 1970; Manguin 1991). The city controls and uses the metal ore from Bangka and Billiton, and regulates the exports from the hinterland of Sumatra (gold, products from the forest and probably slaves).

The conclusion of this first part of our study is somewhat unsatisfactory: the local history is documented but the mechanisms of the technical changes and their significance for the regional context remain unexplained. In return, comparisons with other areas enable us to understand the changes and to formulate hypothesis regarding their occurrence.

The islands: Nias and Siberut, two independent 'laboratories' of technical evolution

Human societies settling on islands seem to invest their limited space with a special value, stronger than on continental areas and different in its nature—the scarcity of land incited islanders to shape their world into a microcosm, which is useful for our purposes of archaeological comparison. Our method consists of using a small and less complex island to illuminate and help explain the organization and functioning of a larger and more complex island.

The small islands of Siberut and Nias present a stronger insular character than Sumatra. None of these islands is volcanic, which explains important differences with Sumatra's mainland, in particular in the availability of resources. Concerning the raw material sources, stone is of a very mediocre quality, and there is no metallic ore. We started our survey on these islands more recently than in South Sumatra, and our research here is still in progress.

Siberut, the island of the 'Flower-Men' displaying a vegetal technology

Siberut, part of the Mentawai archipelago, presents a difficult environment for human settlement. On the eastern coast, the mangrove

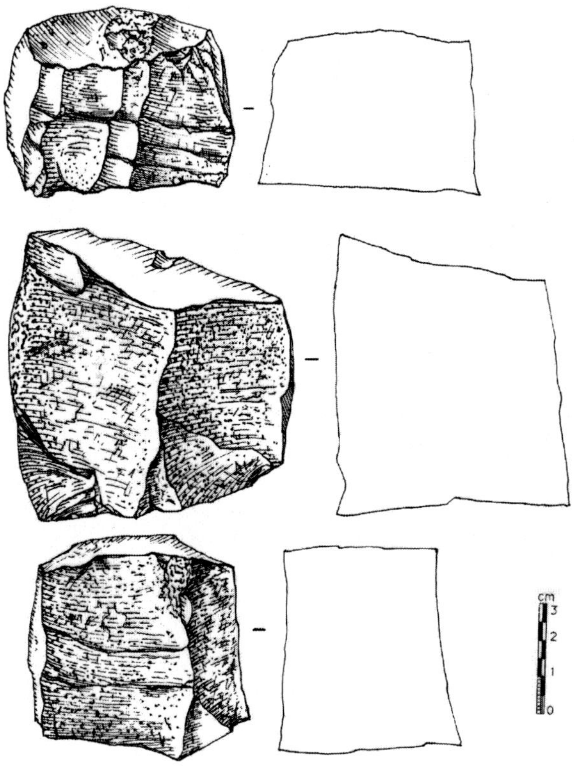

Figure 3.9: 'Horse-hoof' tools from Toinongonai, Siberut Island.

swamps stretch as far as two kilometres inland, and are followed by small sedimentary hills. During our survey[3] we located at Toinongonai an open-air site on an ancient marine terrace, 400 m inland, to the east of the island. The site shows numerous remains of knapping and a few tools on very desilicified material. This site is located near the sources of raw material themselves (sandstone, andesite, flint, fossilized wood and quartzite). The tools consist of end scrapers, notch and denticulated, core tools with an abrupt retouch (like a horse-hoof: figure 3.9) and, more specifically, a 'plane tool' with a thick back (a *rabot*). Only a small number of tools have been made from pebbles, which are rather used as cores for producing flakes ('core-tools'). This industry on volcanic stone (andesite, basalt) can be ascribed to a phase of settlement by hunter-gatherers, contemporaneous with the Hoabinhian culture that was widely spread in the north of Sumatra and in continental Southeast Asia.

Elsewhere on Siberut, polished stone tools (adzes in thin andesite) have also been collected (by Schefold 1991 and Forestier 2004). They document the Neolithic phase of the settlement on the island, which might be attributed to the arrival of an Austronesian population, bringing along the famous complex 'polished stone–bow–tattoos–long houses–yam–taro', more or less still in use today.

Another enigmatic site has been spotted in the southeast of Siberut, along a riverbank. It consists of a shell midden from the 13th century in a dwelling site with postholes. A few pieces of iron were found scattered there on the surface, but they do not necessarily imply that metal was already known at that period.

Present-day Siberut society is renown for its 'Flower-Men' who speak an Austronesian language and exhibit all the characteristics of a classical Austronesian society: tattoos, a strict clanic organization, pig farming, and a horticulture of taro, yam and sago (*Metroxylon* spp.). This society has maintained important hunting and gathering activities in a sparsely populated forest area. Two elements of this culture are true witnesses of the past, and help document archaeological situations: (1) horticulture, sustained by a certain kind of social system and (2) vegetal industry.

Horticulture Yam and taro tubers can be stored in the ground and there is no special season for harvesting: each cultivator, alone, may

[3] IRD—Pusat Arkeologi—Unesco, see Forestier 2004.

collect in his garden the tubers he needs. The sago palm is also very specific: each family owns plantations which have been installed with little effort (planting of the shoots) and which can be exploited after 10–15 years. Our inquiries have shown that each family owns around a thousand sago trees, and that each tree cut and processed in order to obtain starch (demanding hard work indeed) provides the family with food for at least 3 months (and usually more!). This horticulture appears all the more flexible as the starch can be stored damp during several months, and that each family independently can therefore plan its own activities. Moreover, the remains of the processing of sago, or even pieces of the trunks, are used to feed the pigs. We are dealing here with a highly decentralized social organization, based on the clan. The light horticultural work enables the population to invest more time in other activities such as hunting, which is rendered possible by the low density of population and by the importance of the forest on the island. The forest represents a real domesticated element, a part of the productive area of the villages.

Vegetal industry According to the sources, islanders have known metal since the 17th century but it was not widely used. Missionaries and the Dutch administration introduced it more extensively at the beginning of the 20th century. Before that, there was probably a system based on the use of stone, and, according to present-day observations, on the use of vegetal material. Iron implements today are mainly used to shape other tools made of vegetal material (such as bamboo), which are still very much in use; stone probably had the same function as metal in the past. These objects made of vegetal material play a considerable role in the islander's collection of tools and implements for cultivating or hunting and for fulfilling the needs of everyday life.

This observation has important consequences for the knowledge of 'lithic' periods. In the archaeological excavations, in Sumatra and elsewhere in Southeast Asia, the faunal remains that are found are of forest animals, suggesting hunting techniques that were adapted to the forest environment: bows, arrows, traps and nets, instead of spears or javelins. However the stone tools found in the excavations, and more specifically the Hoabinhian tools, look more like tools designed for scraping, grinding or cutting plants, than like implements requiring a haft or directly used as hunting weapons. The observation of the techniques used by the hunters and horticulturists of Siberut helps us to reconstitute the collection of objects made of vegetal material that apparently is absent in the excavations. In the technical systems adapted to the forest of insular and continental Southeast Asia, there were probably complementary uses between stone and vegetal materials, as stone was usually rare, and generally of a poor quality.

Nias: metal, megalithism and warrior societies

The island of Nias is partly sedimentary and partly metamorphic. It offers no metal resources and a very small amount of 'good stone' for knapping or polishing. The island presents a succession of small valleys, which limit the different socio-political units.

The most ancient settlement on Nias extends as far back as approximately 12,100 years, as shown in our excavations of the Tögi Ndrawa cave, to the northeast of the island (Driwantoro et al. 2004). These excavations have revealed the continuous stratigraphy (with a depth of 4 meters) of a shell midden *in situ* in the cave (figure 3.10). Very little cultural changes can be observed during this sequence going from 12,000 BP to 1500–1000 BP, as the whole profile produces pebble-tools of a Hoabinhian unifacial type ('Sumatraliths') or choppers, flakes tools (notches, scrapers, etc.), a great amount of sea- and freshwater shells, and a lot of remains of forest fauna. This period corresponds to a phase of important rise in the sea-level, which started with the last glaciation of 18,000 BP and ended around 6–5000 years BP, when the sea reached today's level, giving their present shape to Sumatra and the islands off its coasts.

There is still a gap in our knowledge of the island's history between these cave dwellings with shell middens, documented by archaeology, and the settlements known by the ethnological descriptions of the last centuries. Nothing is known of the period in-between the shell midden and the metal age, which in fact proves to be very recent. Our investigations tend to demonstrate that the introduction of metal and the beginnings of megalithism happened at the same time not that long ago, in the 17th century at the earliest.[4] Bonnatz (2002: 254) confirmed that 'the practice of erecting anthropomorphic stone monuments in northern Nias...coincides with the time when the first slave trade contracts were

[4] Test-pits have been made at the foot of some stone monuments and radiocarbon dating is currently expected. Interviews with the inhabitants seem to confirm that metal and megalithism happened only 6 to 10 generations ago, and the period varies from one area to another.

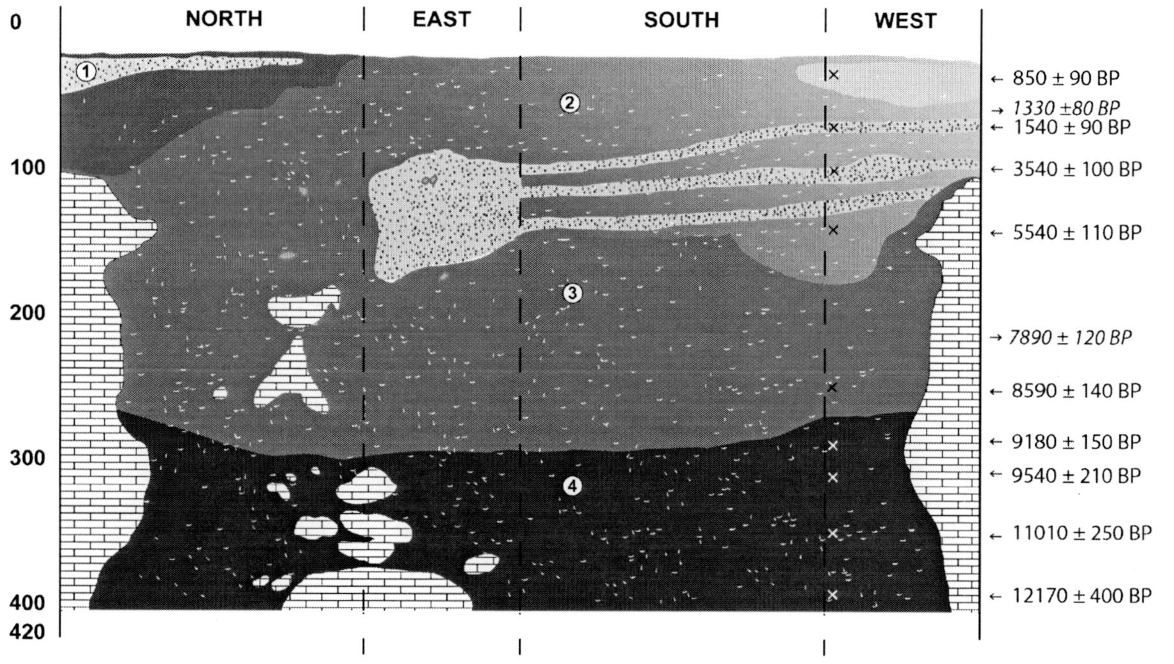

Figure 3.10: Stratigraphy of the Togi Ndrawa cave, Nias. 1. powdery clayish sand, brown-grey; 2. brown and white clayish sand, with a lot of ashes; 3. brown and white sandy clay; 4. compact brown sandy clay. (Italics: dates obtained by K. Wyrandnyana).

signed with Dutch traders'. Those related phenomena, metal and megalithism, are therefore particularly well documented by the oral tradition of Nias, which sheds a very useful light on the technical and social transition to the metal age.

According to Meyers (2003: 82), the introduction of metal, as witnessed in New Guinea, 'did not lead to a new production pattern, but reduced the labour-intensity and time spent on production, allowing people to spend more time on ceremonies, social activities and warfare'. The case of Nias emphasizes a notable shift of the society. Not only did metal, replacing stone, allow a saving of time, but it also induced important changes:

The introduction of metal went together with megalithism, which did not occur previously. Megalithism in Nias is not only related to funeral practices, but is also clearly linked to a social competition: local chiefs hoarded up a great amount of goods (pigs or food) for the holding of ostentatious feasts, crowned by the erection of a megalith showing their prestige. These practices clearly suggest a kind of 'Big Man' system. Even though the society prior to metal could have been characterized by a strong social competition, metal seems to have considerably sped up the social dynamics. The social stratification is much stronger in the south of Nias, as opposed to the north where megaliths are a little less frequent and less ostentatious. This difference might reflect a more recent change in the latter zone, as metal came first from the south. The result of its introduction seems to be an increased social stress, revealed in the considerable head hunting practices and in the frenzied slave trade with foreign merchants in order to obtain the precious metal, be it iron or copper (Marschall 2002).

Thus several phenomena occurred concomitantly: the introduction of metal, and the domination of some groups over others (megalithism being the individual expression of this domination). Ultimately, metal could be obtained in exchange for the island's only resource: the inhabitants. This situation rekindled social competition and hostilities.

Marschall (2002) ascribes the slave trade in the island to demographic pressure, which can be interpreted as a shortage of the required resources regarding the needs of the population. This imbalance can be attributed to an increase in the population and/or to a lack of resources. Several explanations could be evoked: a natural disaster, or an uneven distribution of wealth and goods, due to a sudden disruption of the social order. It is also possible that metal came along with a new wave of immigrants, which would have caused a sizeable population growth. All these factors may also have combined and other explanations still have to be ascertained.

Comparison

The neighbouring island of Siberut is located more or less on the same migration routes and seems to have experienced the same settlement episodes; therefore it can provide some landmarks for a better understanding of the island of Nias. Today, the societies of Siberut are more or less egalitarian. We are still conducting a survey on the past significance of metal on the island, but one of our hypotheses is that there has been a kind of rejection of metal and of the unequal social principles that it conveyed.[5] The upholding of horticulture was also a condition for the upholding of the egalitarian relationships between the clans, and vice-versa. Several examples from Oceania could be summoned up in support of this model. It is also possible that, for some reason or another, and in particular for ecological reasons, the demographic pressure has never reached in Siberut the level which made Nias go over to another technical system. Subsequently, the population in Siberut could have been too small or the society too consistent to sustain the trading of slaves. Also, following Leroi-Gourhan (1973: 424), 'we can establish as a general principle (but not an absolute one) that any need that is normally fulfilled tends to keep its usual ways', and that Siberut's society was not really in need of change.

How can we now extrapolate to the mainland of Sumatra, the observations made in Nias and Siberut?

The 'phantoms' of archaeology: the vegetal industry

Some similarities come to light by comparing the Hoabinhian industry from the cave on Nias, aged 12,000 years, the stone tools from Gua Pandan at the foothills in South Sumatra, dated back to 9000 years, and the industry 9–10,000 years old from the shell middens of northeast Sumatra. The technical system is the same, revealing a similar shaping of the pebbles: the 'sumatraliths' found in Nias and in North Sumatra's shell middens are identical to the other unifacial pebble-tools from South Sumatra. Although the sumatraliths have been linked for a long time to the shell middens and therefore to a coastal environment, the pebble-tools from Gua Pandan in the hinterland of Sumatra confirm that the same industry can be found in all sorts of environments, whether continental or coastal. The function of the tools is identical: they were very certainly meant for the processing of vegetal materials (Forestier 2003). Therefore what seems to determine the Hoabinhian technical system is the 'phantom' of vegetal tools—not the resources that they enabled people to gather. This technical system proved to be flexible and efficient enough to adapt to environments as different as the coast (providing sea products) and the interior (where hunting activities took place).

Moreover, the complementary nature of stone and vegetal materials has probably enhanced the geographical mobility of prehistoric human groups. The large collection of tools made of bamboo or of another inexhaustible material from the forest allowed people to move easily away from the stone sites. The observation of present-day foragers from the Sumatra peneplain (the Anak Dalam) strengthens this idea, and tends to prove that the association between 'lithic and lignic' is significant for all the periods of the stone industry.

Changes in techniques and society with the introduction of metal

Of course, the stone monuments on Nias are very specific, and by their symbolic expression and their codified and repetitive character differ from the megaliths from the South Sumatra highlands, indicating in a way a different society (figure 3.11). Some information can nevertheless be gathered from the insular example:

There is not a systematic link between megalithism and funeral practices, contrary to what has been usually asserted. Although some elements of megalithism (like slab graves) are clearly funerary, the stone carvings of humans and animals in the highlands of Sumatra could be reconsidered as being marks of social status, and/or as markers of time and territory (what we can call 'geosymbols'). The geographical distribution of some types of megaliths is to be considered from this viewpoint.

Megalithism is the sign of a strong social stratification; besides, slab graves indicate a hierarchy in the burial practices. This phenomenon is acknowledged for the metal age (Bellwood 1997), when the metal ore and the techniques of forge were controlled by a small group (an elite).

Domination by a small group is ensured by war (or by other means of violence, abduction and plundering). War is also a way to obtain

[5] Of course, this process of technical choices was not necessarily that explicit among the human groups.

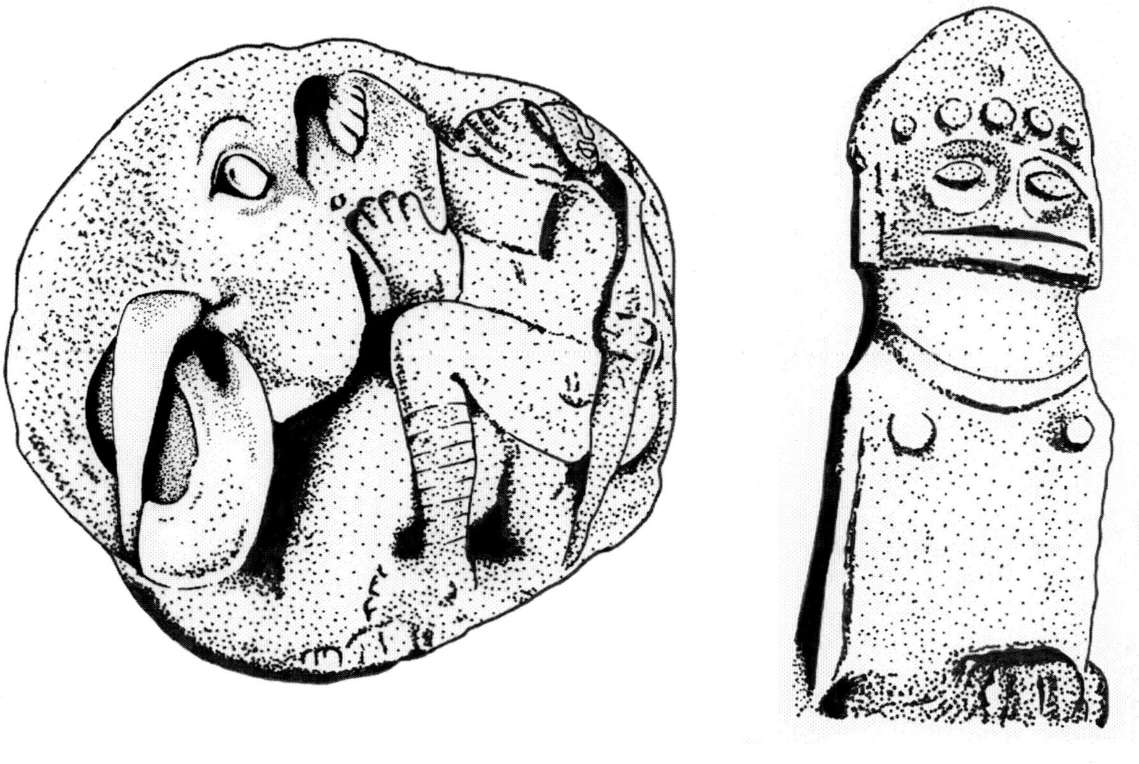

Figure 3.11: Two examples of megaliths: (a) Batu Gajah, Pasemah highlands (height approx. 1.8 m), held in the museum of Palembang; (b) Balugu from Nias (height approx. 1 m), Ds. Lawalawaluo, held by Polres, Gunung Sitoli.

other resources, such as metal ore[6], by the means of force, or more frequently by the means of trade, in exchange of slaves or goods provided by the dominated populations.

The relation between metal and demographic increase is still vague and has to be looked into more closely. Chronologically, metal opens up the way for the basic Asiatic crop, rice, allowing higher densities of population than horticulture. Rice, especially irrigated, is found in hierarchical societies: the ownership of land, the work in the fields with their constraining schedule, the control of irrigation, the storage of the harvest and its protection—all these elements require an organization at a relatively high level and in all cases, a strong social hierarchy.

Thus we deal in the highlands of South Sumatra with a stratified (rice-growing?) society, practising war, where the social competition is strong and the demographic pressure is rather high. Curiously enough, the writings of the first Dutch administrators describe in these very terms the society of Pasemah whom they encountered there in the 19th century. Gramberg (1865) evokes the paths of the Pasemah as 'their warpaths... mainly used for incursions and banditry' (1865: 16). 'A matter of great importance tied Pasemah to Palembang [the capital in the lowlands]: the considerable trade of slaves... in cases of disagreement, the people from Pasemah were swift to attack the territory of Palembang' (1865: 26). 'Real gangs here and there carry on attacks, ransack, kill, and take away cattle and people [from the neighbouring populations]... [the inhabitant of Pasemah] believes that it is a rightful tribute that the weakest, willy-nilly, owe to the strongest' (1865: 31). The society of Pasemah, as seen through this very colonial description, proves to be a truthful heir of the megalithic societies which it dethroned probably in the 14th century.

Besides, all this refers to a topic that would in itself require a longer analysis than can be provided in these pages: the technical change and notably the change to a totally different system, such as going over from stone to metal. In this regard, one should not let the above descriptions impose wrong ideas, such as the metal age being a sinister fatality that human groups rapidly embrace when they have no other choice left. In fact, 'the normal condition is, for the group, a constant effort to assimilate external inputs' (Leroi-Gourhan 1973: 422), and technical change only occurs when, and if, it has a meaning and a place in the evolution of the local society. Moreover, the spreading of techniques involves a 'receiving' society, but also a 'giving' one. This leads us to consider, beyond Nias or South Sumatra, the origins, motives and ways of the external groups that brought along metal. Diffusion often follows war processes and conquering policies, although commerce, intellectual culture and a policy of civilising expansion can also be summoned up (idem).

[6] This hypothesis might be very significant in the South Sumatra highlands, where the metal ore was available in limited quantities.

Islands and technical changes

Reaching the end of our journey through the islands and their history, can we pretend to have a more accurate idea of 'islandness'? The small islands which supplied facts for our analysis of technical changes represent cultural and geographical models that are relatively simple and easy to observe: there is more difference to be found between two islands than within the same island, where, due to the limited geographical scale, the same patterns of settlement will be repeated. The large island of Sumatra shows the opposite: the ecological contrasts within the island induce a succession of different patterns, each one adapted to a specific environment. 'Islandness' therefore finds its most remarkable characteristic in the individuality of each (small) island.

The extreme value of scarce insular land leads the societies to a stricter control of their social and spatial organizations, and strengthens their territorial settling. However, no island is a closed world: it is always linked to nearby lands by exchange routes, which bring along most of the changes. This situation induces a double dynamics, the slow internal maturation of insular societies, and the sometimes brutal changes brought from outside.

Transitions can be very quick in the islands, where, due to the reduced scale, the relation between resources and population can rapidly be exacerbated. The introduction of metal in Nias shows at what speed the local societies have changed. Two to three centuries were sufficient to set up the warrior society, whether of an external or internal origin, which has accounted for the fame of the island. This swiftness reminds of what biologists call the 'insular syndrome', a process where the slightest ecological change sweeps away the existing ecosystems, protected until then by their isolation. The classical dynamic of resource and population is at the heart of technology, and probably plays an important part in the changes observed. Islands, where this dynamic is exacerbated, provides us with much needed insight into this.

Acknowledgements

The drawings of stone tools are by Hubert Forestier. All other illustrations are by Laurence Billault, IRD.

References cited

Bellwood, P. (1997). *Prehistory of the Indo-Malaysian Archipelago.* Sydney: Academy Press.

Bonnatz, D. (2002). Megaliths on Nias: the retention of identity. *Indonesia and the Malay world* 30/88: 253–276.

Coedes, G. (1918). Le royaume de Çrivijaya. *Bulletin de l'Ecole Française d'Extrême-Orient* 18/6: 1–36.

Driwantoro, D., Forestier, H., Simanjuntak, H.T., Wyradnynana, K., and Siregar, D. (2004). Tögi Ndrawa cave site at Nias Island. New data on life during the Holocene Period based on dating. *Berkala Arkeologi 'Sangkhakala'* Balai Arkeologi Medan 13: 10–15.

Forestier, H. (2000). Chaînes opératoires lithiques en Asie du Sud-Est au Pléistocène supérieur final et au début de l'Holocène. *L'Anthropologie* 104: 531–548.

Forestier, H. (2003). Des outils nés de la forêt. L'importance du végétal en Asie du Sud-Est dans l'imagination et l'invention technique aux périodes préhistoriques. In A. Froment and J. Guffroy (eds.), *Peuplements anciens et actuels des forêts tropicales*, pp. 317–337. Paris: IRD-éditions.

Forestier, H. (2004). *Siberut Archaeological Aurvey, Mentawai Archipelago, West Sumatra.* Unesco/IRD Jakarta.

Forestier, H., Simanjuntak, H.T., Driwantoro, D. (2005). Les premiers indices d'un faciès acheuléen à Sumatra sud. In V. Zeitoun (ed.) *Dossiers d'archéologie*, n° spécial Asie du Sud-Est: 16–17.

Glover, I. (1978). Report on a visit to archaeological sites near Medan, Sumatra. *Bulletin of Indo Pacific Prehistoric Association* 1: 56–60.

Gramberg, J.S.G. (1865). *De inlijving van het landschap Pasoemah.* Batavia: Van Dorp.

Leroi-Gourhan, A. (1973) [1945]. *Milieux et techniques.* Paris: Albin Michel.

Manguin, P.-Y. (1991). The Merchant and the King. *Indonesia* 52: 41–54.

Marschall, W. (2002). Social stratification and slavery on Nias and its reflection in oral history. *Indonesia and the Malay world* 30/88: 309–318.

McKinnon, E.E. (1975). A brief note on the current state of certain of the Kitchen middens of east Sumatra. *Sumatra Research Bulletin* 4/2: 45–50.

McKinnon, E.E. (1991). The Hoabinhian in the Wampu, LauBiang of northeastern Sumatra: an update. *Bulletin of Indo Pacific Prehistoric Association* 10: 132–142.

Meyers, K. (2003). *The Changing Cultural and Ecological Roles of Siberut People in the Management and Conservation of their Natural Resources.* Jakarta: Unesco.

Schefold, R. (1991). *Mainan gagi Koh: kebudayaan Mentawai.* Balai Pustaka: Jakarta.

Simanjuntak, H.T., Forestier, H. (2004). Research progress on the Neolithic in Indonesia: special reference to the Pondok Silabe cave, South Sumatra. In V. Paz (ed.), *Southeast Asian Archaeology, Wilhem G. Solheim II, Festschrift*, pp. 104–118. Diliman, Quezon City: The University of the Philippines Press.

Simanjuntak, H.T., Forestier, H., Jatmiko, Bagyo Prasetyo. (2005). Gens des karsts au Néolithique à Sumatra. In V. Zeitoun (ed.), *Dossiers d'archéologie* n° spécial Asie du Sud-Est: 46–49.

Wolters O.W. (1970). *The Fall of Sriwijaya in Malay History.* Kuala Lumpur-Singapore : Oxford University Press.

Chapter 4

What exactly is a fish trap? Methodological issues for the study of aboriginal intertidal rock wall fish traps, Wellesley Islands region, Gulf of Carpentaria, Australia

Paul Memmott, University of Queensland

Richard Robins, Everick Heritage Consultants Pty Ltd (QLD)

Errol Stock, Griffith University

Introduction

The authors and their colleagues are currently re-activating a large-scale research project on the pre-history and cultural change of the Aboriginal people of the Wellesley Islands and adjacent mainland in the Southern Gulf of Carpentaria (figure 4.1). It includes a study of rock wall fish traps. In this paper we outline a range of methodological problems which impact on the way our fish trap research might be grounded and theorized. In this paper we address the question of what are the most useful conceptual units for traps and how that decision influences the way we measure densities of traps for inter-island and inter-group comparisons. Understanding the relative ages of proximate fish traps and whether they were used contemporaneously become further complicating issues. We explore these questions as essential components of a general methodological debate on fish trap research.

Why is the question 'What is a fish trap?' relevant to global perspectives on the archaeology of islands? Why is such a prosaic question of potential importance to issues of Australian Aboriginal offshore colonization? Few places in Australia have had such a significant input into archaeological considerations of Aboriginal offshore island colonization than the Wellesley Islands. The archaeological substantiation for this influence is however, minimal.

The largest of the 15 Wellesley Islands, Mornington, and various smaller surrounding islands, are home to the Lardil people. Mornington is linked to the mainland by a number of smaller 'stepping-stone' islands. These intervening islands, home to the Yangkaal people, approach the mainland at Bayley Point. The mainland was occupied by the Ganggalida. This mainland coastline and the North Wellesley Islands are inter-visible in all seasons, with easy crossings over no more than 3.5 kilometre of open sea, traditional transport being by timber rafts. The South Wellesley Islands lie east from Bayley Point, the largest being Bentinck, home to the Kaiadilt people. The open sea gap between the mainland and the South Wellesleys (10.5 km) is more substantial than in the North Wellesleys, and was sufficiently challenging on rafts that the voyage was not undertaken lightly, perhaps several occasions in one's lifetime (Evans 2005).

In the early 1960s Norman Tindale and colleagues of the South Australian Museum undertook socio-cultural and biological research amongst the Islanders, and particularly amongst the Kaiadilt (Tindale 1962a, 1962b). Tindale's work led to several significant findings that have influenced debates about the Aboriginal colonization of offshore islands. Tindale (1977, 1981) argued that:

- There were important differences in language, social systems, material culture and biological markers—most significantly the presence amongst the Kaiadilt of rare blood type for Aboriginal Australia;
- the Kaiadilt were an isolate of early Australian settlers who "bore the mark of Wajak" (referring to Pleistocene Java), and their social system, technology and material culture were impoverished due to lack of contact with other island and mainland groups; and

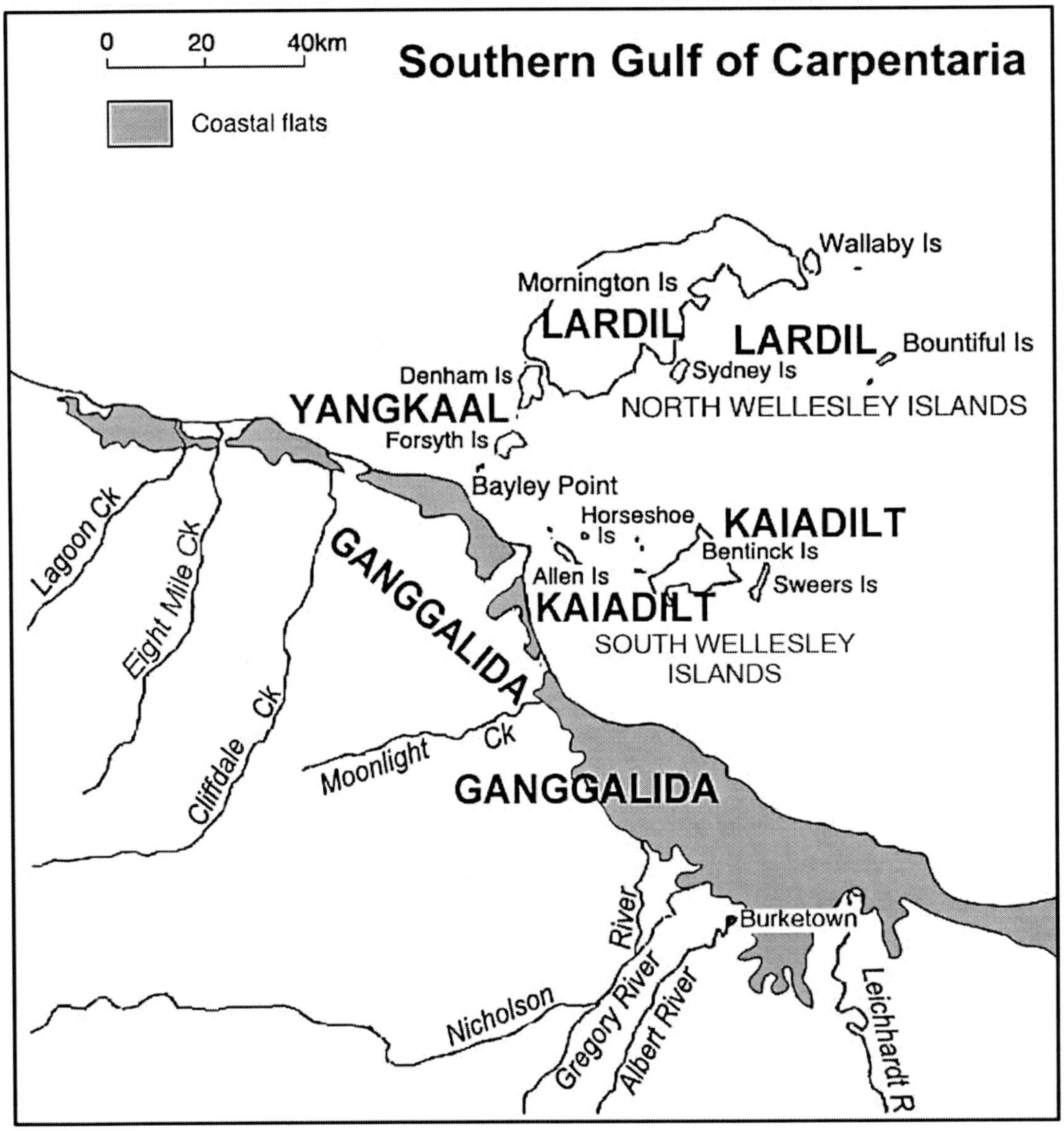

Figure 4.1: Wellesley Islands and the mainland coast showing the territories of the four Aboriginal study groups.

- the distance separating Bentinck from the mainland marked the limit for sure and successful movement by people using simple watercraft.

Recent linguistic, anthropological, archaeological and geomorphological research has challenged these assumptions (Memmott et al. 2004; Evans 2005). The upshot of this research indicates that the Kaiadilt are not a relict population—linguistic evidence for example, points to them having settled about 1500 years ago. The 10.5-kilometre distance is not the barrier to travel assumed by Tindale—Allen Island near the mainland coast is identified Kaiadilt country, and was periodically visited.[1] There were clearly cultural commonalities as well as obvious differences. These differences reflect a number of factors including the likely sequence of island settlement and the variation between the North and South Wellesley Islanders in the degree of permeability and acceptance of social transmission and socioeconomic adaptations.

Although the picture of isolation presented by Tindale was exaggerated, the debate about isolationism still remains. New data and arguments indicate a more complex set of relationships than Tindale envisaged, and the case of the Kaiadilt has become more important. Is there some archaeological substantiation for this picture? Can the archaeology provide some time depth? Can it point to changes in technologies? Can it assist in the identification of significant environmental changes that might have influenced events? Can archaeological markers that might culturally differentiate groups be identified?

The answers are 'maybe' and 'yes'. Archaeological reconnaissance indicates that traditional forms of archaeological evidence of the Southern Gulf are limited. There are no large rockshelters on the mainland or in the islands. There are no large shell mounds, although some small ones have been investigated. Much of the stone for manufacturing artefacts is imported from outside the region. Shell is quickly eroded in the warm, wet conditions. Sites with long stratigraphic sequences appear to be rare, if they exist at all. Three test excavations by Robins on the mainland have provided ages ranging from 1500 to 400 BP, whilst one site on Mornington has provided basal date of 1710 BP. Another potential source of archaeological information could come from human remains, however there are a number of cultural sensitivities that make this a difficult option to explore in the contemporary context.

The most prominent form of archaeological evidence is in the form of fish traps which have been created by building rock walls in strategically placed points of the intertidal zone (figure 4.2).[2] Fish, turtle, dugong, crabs, and shellfish can be obtained from them, although they have not been in regular usage since the early contact period.[3] Reconnaissance surveys of most of the Wellesley Islands and the mainland between Moonlight Creek and Bayley Point, as well as aerial photo interpretation, ground surveys[4] and ethnographies of their use from traditional owners, indicate that there are some 108 fish trap sites containing at least 334 traps.[5] This makes it the largest complex of stone walled fish traps in Aboriginal Australia. Their distribution is limited to the coastlines of the North and South Wellesley Islands and along the mainland coastline opposite those islands (a coastal stretch of

[1]See evidence submitted for Wellesley Sea Claim. However the Ganggalida and Yangkaal also claim rights in Allen Island, and successfully argued so during the Native Title claim in the late 1990s.

[2]The first written accounts of these fish traps appear in early exploration and colonial records (Boyd 1896:57, Roth 1901a)

[3]For the Lardil and the Yangkaal this period intensified in 1914 with the permanent establishment of a Mission on Mornington Island whilst for the Kaiadilt, traditional lifestyle was more or less maintained until the late 1940s. The Ganggalida however, being on the mainland experienced violent contact and degrees of local disruption as early as the 1880s (Memmott 1979:Ch. 6).

[4]Ground surveys coupled with ethnographic recording were undertaken on selected traps at Bayley Point, Point Parker, Mornington and Sweers Islands.

[5]Four techniques were used to identify the location of fish traps. The first was an aerial survey conducted in 1983. Low altitude flights in a small aeroplane were conducted over most of the islands and mainland. Fish traps were identified, counted, photographed obliquely and plotted on to topographic maps. The second technique involved ground truthing and documenting some of the identified or known traps on the mainland coast, Mornington and Sweers Island. Where possible plans were made and photos and measurements taken. The third method involved the identification of traps from aerial photographs and then plotting them onto master maps. The fourth method involved using local Aboriginal knowledge to elicit and identify traps, often during trips for other research purposes.

The quantity of ethnographic data elicited in preliminary interviewing with the members of the Aboriginal study groups varied inversely with the depth of their culture contact. Hence the least information was obtained from the Ganggalida whose cultural disruption had commenced in the 1880s, or possibly earlier. There is a medium corpus of data from the most elderly Lardil informants who are relying on their memory knowledge of the earliest years of the Mornington mission (1914-1925) and the period just prior to this (1905–1914). The most detailed information comes from the Kaiadilt who were using their fish traps at the time of their migration to Mornington mission (1947–48), and who then built a trap in the proximity of their new camp in the Mission and used it for several decades (Memmott 1979:243, 1982:38, 41A).

Figure 4.2: The typical fish trap of the Wellesley Islands with a single-wall and layout adjusted to substrate contour; located outside a fringe of mangroves, Allen Island. (Photo by Connah and Jones of the University of New England, 12/5/82.)

some 43 kilometres).[6] They represent an aspect of shared material culture throughout this area and demonstrate a regional cultural similarity in technology. (Robins 1983, 1998, 2000; Memmott et al. 1984; Memmott 1985, Robins et al. 1998; Trigger 1985, 1987; Memmott and Trigger 1998:114). The subtle differences in the construction, use and design of these traps between the four language groups remains the subject of ongoing investigation by the authors and their colleagues.[7]

How might the traps contribute to an understanding of cultural dynamics and diversity? The most striking characteristic of fish trap distribution is the variation in densities of traps between islands. The lowest densities are to be found in the Lardil country amongst the Lardil in the North Wellesley Islands, whereas the highest densities are to be found amongst the Kaiadilt in the South Wellesleys and the Yangkaal of the smaller North Wellesley Islands. Bentinck Island, in Kaiadilt country, is enormously rich with an average of one site every 0.9 kilometres and one trap per 0.4 kilometres. In contrast Mornington Island, the largest island of the Wellesleys and occupied by the Lardil, has traps at roughly every 20 kilometres of coastline.

Why this difference in density, especially between Lardil traps and Kaiadilt ones? Our preliminary investigations indicate that the variation in numbers is not directly related to access to rock sources or to particular types of marine or terrestrial environments. An alternative hypothesis to explain the variation is that whilst the insularity of the Lardil facilitated cultural independence, at the same time the closeness of the islands facilitated regular trade and exchange of technologies and beliefs. This interaction permitted participation in socio-ceremonial events which fostered the use of large fishnets rather than stone-wall traps. Whereas the Kaiadilt, due to their relative isolation, vigorously maintained a sense of social differentiation and underwent a process of local intensification through the specialized development of trap technology. However, the evidence is more intractable than it might first appear, and there are a number of issues that need to be addressed before we can make definitive statements about trap density and variation (let alone degrees of insularity).

Problems of selecting trap units

In trying to decide what units we should be defining and analysing in a study of fish traps, an initial problem comes from our access to different types of aerial photographs. The oblique aerial photos shot with hand-held cameras during several surveys (1981–2004) are in different formats[8] and at different stages of the tide. Similarly, the 1983 vertical aerial photos (1:25000) were taken on different days and stages of the tide. Objects like traps are recognized in all the aerial photos because of contrast (tone and/or colour) against their surroundings and the degree of linear coherence in the image. Recognition of trap walls in the oblique photos is relatively easy because they were visible targets for the airborne photographer. Locating traps in the vertical air photos is more problematical because of the scale factor: a 25-metre long wall is only one millimetre long in a contact print. Consequently, much of our work with the vertical aerial photos has required optical magnifications and enlarged prints. In addition, photography at high tide may have obscured many traps through submersion, and windy conditions may have created turbid water reducing the visibility of some traps. Despite our care and cross comparisons between oblique and aerial photographs, we expect additional work will add new traps to the 334 confirmed.

A current task of the authors is to map all fish traps and their surrounding coastal land systems at a scale of 1:25000. This will enable us to carefully re-count the numbers of traps and recompare relative densities of traps. However this raises the question of what are the most useful units to consider. For preliminary working pur-

[6]The rock wall fish traps on the mainland coast are on that southerly part of the Gulf of Carpentaria coastline that is oriented roughly in a north-west/south-east orientation. The most south-easterly of these traps (as located to date by the authors) is at approximate latitude 17°14.5' (aeroplane GPS reading: S17°14.451', E139°10.674'.) The nearest map landmark to this trap is the mouth of James Creek about 2.25 kilometres back up the coast (i.e., NNW of the site). The most north-westerly of the traps (as located to date by the authors) is at Bayley Point shown on topographical maps at latitude 16°55'. The distance between these two extreme locations tracing along the undulating mainland coastline is about 43 kilometres. Robins and Trigger have determined the extent of the traps along the mainland coast through a combination of (i) the knowledge of Ganggalida Elders (or ethno-geographers, e.g., Willie Doomadgee), (ii) observation by aerial passes, (iii) ground truthing. Checks were made as far as Eight Mile Creek, some 53 kilometres WNW of Bayley Point.

[7]The colleagues assisting us with our ongoing research in the Wellesley region are Dr Ian Lilley and Dr Sean Ulm of University of Queensland, Prof Nick Evans of University of Melbourne, Dr David Trigger of University of Western Australia, Dr Neville White of La Trobe University and Dr Sheila Pellekaan-Holst of University of New South Wales.

[8]35-mm SLRs, 70-mm Hasablads, panchromatic and colour film.

poses, a 'fish trap' was defined *as a constructed rock wall enclosing a space for the purposes of trapping fish and other marine animals through the action of tidal movement.* The enclosed space can be termed the 'pen'. A 'fish-trap site' was then defined as *a cluster of fish-traps and/or associated rock wall features in close proximity.*

A trap therefore according to this definition is identified primarily by the space enclosed in the pen, not the number of walls. However, the incorporation of natural features into trap construction, and the weathering, deterioration and burial by sand of some walls can make the identification of individual pens difficult in some cases. But why should we be identifying pens as opposed to walls as units? And what of those cases where there are common walls to several pens? Should we identify more complex units?

One issue is how to identify two segments of a wall which line up with one another, but are separated by a small gap. Is this one wall, perhaps with a gate, or two discretely separate walls? Is it one wall with a portion dismantled, destroyed or covered with sand? Or do the two pieces of wall belong to two partially destroyed traps? Should it be identified as one unit or two units? Given the possible range of reasons here for the gap (if it is a gap), it would be methodologically wiser to count such a feature as two wall units.

The problems of identifying what constitutes a pen are even more difficult. Although we assume we can easily identify a pen when we see a roughly semi-circular shape with walls characteristically extending from the high tide line at right angles to the shore, there are nevertheless many walls or sets of walls which do not readily conform to this pattern. For example consider figure 4.3. At fish trap Site No. 1, the two more southerly walls form only semi-enclosed pens, not being oriented to the shore; whereas at Site No. 2 there is an undulating wall that could be interpreted as one large pen or several smaller pens. In figure 4.4, two lengths of wall (one of which is again undulating with three or four corners) are connected into a stand of mangroves. From the aerial photo it is not possible to discern whether this is one continuous wall, but if so, is it one trap, two traps or five traps?

Another methodological problem of trap identification is the confusion that arises when a natural reef or rock formation forms a natural trap and the tendency for local Aboriginal people to identify this as a 'trap' made by their ancestral heroes or creator people. This can be termed a 'naturefact' in material culture theory, after Oswalt's (1973, 1976) taxonomy of material culture items. Naturefacts are unmodified natural objects that are used consistently by a group to obtain food or water. In the case of a naturally occurring fish trap, Oswalt would further classify it as an 'intended facility'; a 'facility' being defined as a part of the environment that might attract, contain, hold, restrain, or direct an animal or other resource (as opposed to an 'implement' which is defined as a natural object that is physically transported to use as an instrument or weapon).[9]

In addition, we can consider the case of a naturally occurring substrate enclosed largely by rock outcrops and/or sandbanks or spits of gravel or shell, which only requires a few short pieces of wall to be constructed to seal it up into an effective trap. In such a case it may not be possible to readily discern the human-made pieces of wall and one may easily mistake an artefact for a 'naturefact'. Such a problem of classification is evident at Bountiful Island where Lardil consultants have identified two traps as manufactured by their ancestor *Marnbil* (figure 4.5).

To sum up, trap features that can be recognized in the aerial photography could be: a single trap with a clearly identifiable pen, or; an identifiable trap site but whose division into individual units is unclear or ambiguous, because not all features form obvious pen shapes, or; a trap-like feature that may actually be a natural rock formation (table 4.1).

Understanding design and subsistence usage and how it effects interpretation of units

Lardil consultants[10] have given *derdernin*[11] as the name for the rock wall fish trap, whilst the Kaiadilt[12] call their traps *ngurruwarr.* All groups asserted that they caught not only fish, but turtle and dugong in traps. Other by-products were crabs obtained from the crevices within and underneath the rock walls, oysters from on the rocks themselves, and a range of

[9]The usefulness of Oswalt's taxonomy is in the comparative study of the relative complexity of different material cultures.

[10]Kelly Bunbujee, Fred Jaurth, Henry Peters, Scotty Wilson and Pompey Wilson (all now deceased).

[11]The Lardil dictionary gives *derndernyin* as the term for a 'fish trap made of stones' (Ngakulmungan 1997:102) but provides *derdernyin* and *derdernin* as alternative recordings. The term *derndernin* was recorded by the linguist Ken Hale in the late 1960s from his Lardil informants (Hale *et al.* 1981). Also see Evans (1992).

[12]Pat Gabori, Maude Gabori, Roongka, Arthur Paul, Fred Bijarib. (The last three were deceased at the time of writing).

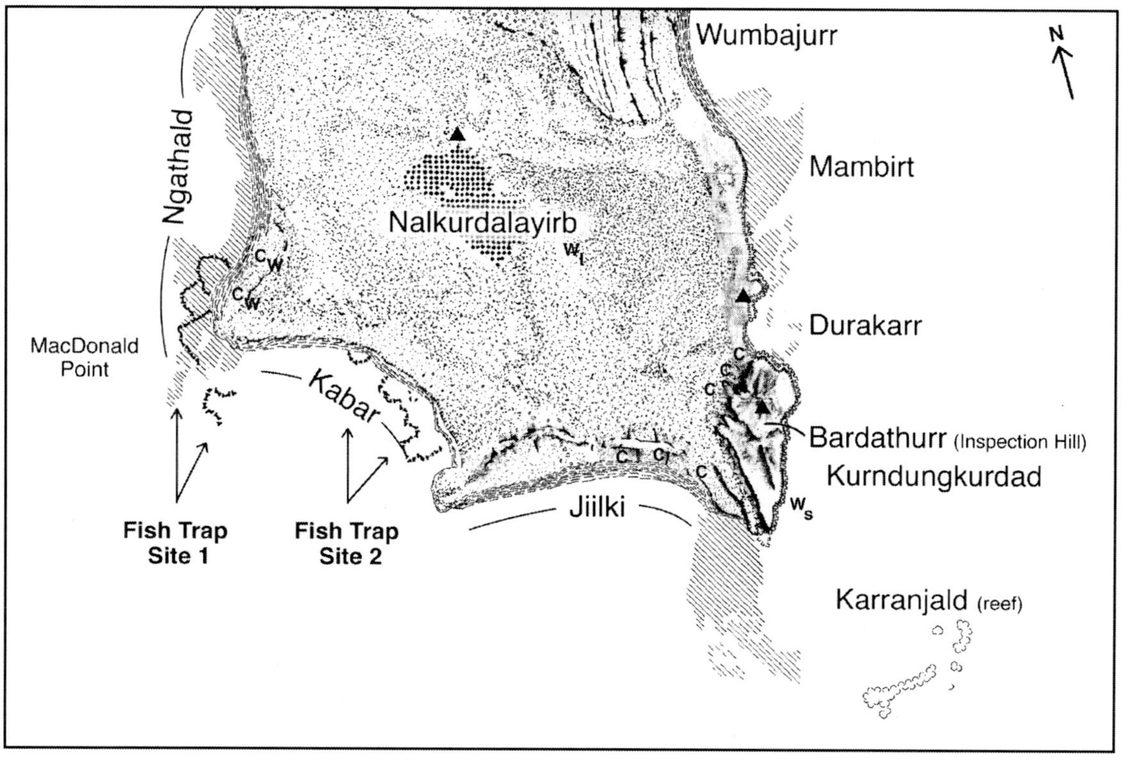

Figure 4.3: Map of southern end of Sweers Island in the vicinity of Inspection Hill and MacDonald Point showing environmental units with Aboriginal geography and the location of two fish trap sites at Ngathald and Kabar. Note the use of party walls and undulating walls in the trap designs. (Source: Map of Ringurrng by Paul Memmott 1982, made with Darwin Moodoonathi and Arthur Paul.)

Problem	Source
False positives	Geological feature that resembles rock wall
Missed positives	Traps missed during systematic search of aerial photos due to: Trap decayed and weathered, losing its linear visual coherence Trap not visible through lack of contrast of rock wall colour on substrate Trap obscured by depth of water at time of survey Trap obscured by turbid water at time of survey Trap buried by sediments Trap too small
Wrong search model	Should a 'naturefact' be included if indigenous consultants identify it as a fish trap made by Ancestral Beings? Insufficient ethnographic data to understand the design and functioning of certain trap types, resulting in erroneous interpretation
Problematic unit description	Integrity of walls lost due to weathering, burial, rock removal Integrity of pens lost due to loss of integrity of walls Observer failure to appreciate more complex traps forms and whether all of the proximate walls were in contemporaneous use

Table 4.1: Summary of problems with identifying rock wall fish traps by aerial photos at 1:25000.

Figure 4.4: Two lengths of wall connected to a stand of mangrove trees, Bayley Island. Is this one continuous wall and if so, is it one or two traps? (Photo by Connah and Jones of University of New England, 12/5/82.)

Figure 4.5: An inter-tidal rock platform on (large) Bountiful Island asserted by the Lardil to be a fish trap created by the Ancestor Marnbil. Signs of constructed rock walls are not obvious from aerial reconnaissance or visual ground inspections. There are however many rock pools which would yield a rich catch of food resources.

species of shellfish from the muddy and sandy substrates of the traps.

The use of rock wall traps was under the direction of the local patriclan country custodian. With his (or her) permission, traps could be used whenever tidal conditions were suitable. In the most advantageous conditions, numerous fish were said to be left stranded on the floor of the trap when the water had completely run out. They were easily collected by hand (by men or women) or with a pronged fishing spear (by men), and carried ashore in bark containers or small hand-nets (figure 4.6). Dugong and turtle could be stranded in the same way but this was more likely during the biggest tides when the tops of the walls were covered with a substantial height of water. Such tides were associated with the sea surges of cyclones and/or the large influx of freshwater from the Gulf streams during the wettest summers. Lardil consultants stated that there were different types of traps for catching dugong, turtle and fish, and referred to a 'double trap' at the northern end of Mornington Island,[13] one part being for fish and one for dugong. This may refer to the combination of an inner and outer wall situated on a higher and lower substrate contour respectively.

By all appearances the Kaiadilt traps seem identical to those of the other groups. However Lardil consultants noted that many Kaiadilt traps encompassed 'bigger paddocks', containing a 'lot of weed'. This implies that the Kaiadilt have in some cases built large traps on areas of muddy substrate where sea grasses grow, which attract grazing dugongs and turtles; and further suggests that the Kaiadilt were more dependent on stone-wall traps than the Lardil to catch dugong and turtle. We intend to investigate this hypothesis with our ongoing research through a comparison of fishing technology repertoires. The Lardil employed a series of hibiscus rope nets for hunting dugong, a material culture item not used by Kaiadilt. We also note that the Aboriginal Protector, Dr Walter Roth (1901b) was informed of an 'alleyway' constructed by the Kaiadilt somewhere on the south-west side of Bentinck Island, to trap dugong. One of the highest densities of fish traps in the study region occurs at the south-west corner of Bentinck Island (see figure 4.12).

A second method of harvesting fish could be used prior to the trap becoming emptied, once the tide had fallen below the top of the wall. Fish were 'herded' into schools and towards one or more spearsmen by individuals hitting the surface of the water. Another technique was to simply walk along the walls spearing fish, as the tide fell. Similarly, it would be possible for dugongs to be chased by men on rafts into the shallower water where they became grounded and/or could be easily speared.

Numerous changes in wall direction can be explained by the stratagem of reducing construction labour by taking advantage of the highest area of the irregularly contoured substrate on which to position the wall. In this manner, various pieces of naturally protruding rock substrate could be designed into a wall under construction, as well as areas of elevated mangroves being used as part of the side (or wing) walls. However there is a further possible design criterion influencing the trap layout, evident from a preliminary examination of data, that of positioning an outer apex (i.e. the point of a V shape, albeit noting that seldom is it a perfect V shape), to coincide with a channel system of small runlets of water (visible at very low tide when the last waters recede), and which flow together to make a larger channel, draining out to sea. When the tide is rising or falling these streamlets become channels where the current is strongest, and through their velocity naturally direct and carry fish travelling inshore and offshore when the tide is changing (figures 4.7 and 4.8). Thus as the water in the trap falls with the tide, it will drain to the V point of the trap leaving a large pool in the lower part of the V, thereby concentrating the fish into one area. Several choices are then presented as to how to harvest the fish and there is clear evidence that at least three methods were employed. First the fish could be speared (by men). Secondly one could wait until the trap fully drains and pick up by hand any fish that might be stranded on the substrate in the V point.[14] Thirdly the fish could be netted by either men or women, facilitated by a gate in the wall at the V point formed by removing a few rocks. The gate was left open on the rising tide, and then shut on the falling tied often with mangrove foliage (*Avicennia eucalyptophylla* or *Rhizophora mucronata*). Gaps were found in the foliage 'doors' and hunters could station themselves at these points with small purse-shaped hand nets, whilst others chased the fish in the emptying trap towards these points.

[13] In the country of Sandy and his son Maurice Sandy (both deceased) near White Cliff.

[14] However the authors know insufficient about fish behaviour to understand whether a fish caught within a trap or a receding tide has a behavioural capacity to seek to escape from the trap whilst the sea level is sufficiently high and before it drops below the level of the wall. This needs to be tested in an experimental usage of a trap. (This will initially take place at a Kaiadilt outstation at *Nyinyilki* on south Bentinck Island.)

Figure 4.6: Ronnie Jupiter spearing a crab, Bayley Point, on a falling tide, September 1983. Note that the fishing spear was a male-specific implement. The wall is cemented together with oysters. (Photo from Trigger 1985:127.)

Figure 4.7: Trap wall at low tide, showing a drainage channel and indicating a likely point to design in a gate. (Ref. RB 5668/7, AERC L8/2-158).

Figure 4.8: Trap wall designed to take advantage of a rocky substrate and a drainage channel system. (Photo by Errol Stock, July 2004.)

Lardil Elder Kelly Bunbujee has provided one of the authors (PM) with a typology of traps as shown in figure 4.9. The enclosed wall or *derdernin* is used for harvesting on a receding tide by spearing or picking up, the two-walled race or *baljan* for chasing and spearing, and the gated semi-circular trap *yilin* for chasing and netting.

A geometric method of classifying enclosed pens is as V shapes, curved shapes, or rectilinear shapes. One of our colleagues[15] has pointed out that a circle is the most effective (least lineal distance) in enclosing a large area. However, we have rarely seen a perfect semi-circular trap shape.[16] A further consideration is that rectilinear shaped traps are mechanically more susceptible to wave damage than V-shaped, pointed or curved ones (Bowen and Rowland 1999:3). Although Kelly's three categories occur in the Wellesleys, there are many more units and complexes which do not conform to these categories.

A refinement of the V corner with gate, evident in the Kaiadilt data, is what we have termed a 'pocket' whereby the corner gate leads the fish into a small fully enclosed pen, a contained area where they might remain alive for some time before the water drains away (figure 4.10). This would be a useful technique if only one or two individuals were attempting to harvest a large number of fish (a dozen or more), allowing time to transport a portion of the catch to land to a protected storage point without scavenging birds or dogs stealing any dead fish.

Yet another technique of trapping fish that can be discerned from an examination of the aerial photographs involves a curved trap wall, but rather than being oriented to the shore to catch a falling tide, it is oriented to catch fish travelling with currents flowing parallel to the shore. An example can be seen in the map in figure 4.11 at *Murarri* on Sweers Island (which also appears to have a pocket trap in its southern corner).

One of our later fieldwork tasks will be to address the various reasons why irregular and/or complex wall patterns occur. As noted, a key issue is likely to be substrate properties—undulations in sediment, small channels, rocky protuberances and outcrops and local run-off patterns—which in turn affect the design layout of a trap, with trade-off decisions being made to maximize yield but simultaneously minimize energy in construction.[17]

Measuring trap density and its significance

In his arguments for the scientific and cultural significance of the Kaiadilt, Tindale placed strong emphasis on his finding of high population density in relation to coastal exploitation. He calculated that the total area of the Kaiadilt islands[18] is about 181.3 square kilometres, with about a quarter of this area being littoral. Some 137.2 square kilometres is land, of which about a third is barren interior claypan. Tindale then calculated for those South Wellesley Islands which were regularly occupied and readily accessible in 1940, the last year when all Kaiadilt were present in their traditional lifestyle, that only about 14 square miles of reef were in constant use, giving a population density of over eight persons obtaining food on each square mile of reef, equivalent to over three persons per square kilometre. Tindale concluded that these figures were "remarkably high for a 'stone age' people", and that in "the southern parts of Australia, even in areas of high rainfall the figures for the most dense populations seemingly went no higher than about one person per two square miles..." Later Tindale (1977:249) suggested that the population density in the South Wellesleys was "one of the highest known for a living stone tool using people depending on foraging for their existence". Our revized calculations of fish trap units and densities will enable a reconsidered critique of Tindale's arguments on density.

One of Tindale's Kaiadilt consultants explained how the various South Wellesley clan groups had differing potential to build and exploit traps depending upon certain geographic properties of their estates. Tindale interpreted his informant's information as follows:

> Dolnoro S, U, and X people [the estates of Oak Tree Point, Raft Point and the western point/Albinia Is. groups] have reef areas which they can work throughout both the NW and SE trade wind seasons, their NW season fish

[15] Mike Rowland, Queensland Environmental Protection Authority, Brisbane.

[16] Another colleague has observed Torres Strait Island traps are neater in shape than Wellesley ones but they are all located on relatively flat substrate (Dr Ian Lilley).

[17] A further issue that is a possible reality given the local traditional practice of increased rituals at 'Story Places', is that a small number of wall features may be constructed as increase centres; they may not actually be used as traps.

[18] Tindale includes Bentinck, Sweers, Allen, Horseshoe, Albinia, Douglas, Bessie and Margaret (or McCarthy) Islands in his calculation.

Figure 4.9: Three categories of rock wall fish traps according to Lardil consultant Kelly Bunbujee (deceased): the enclosed wall, the race, the gated semi-circle (1983).

Figure 4.10: Part of a complex of rock wall fish traps off the south-west corner of Bentinck Island, with a pocket trap in the apex of a large pen. (Photo by Richard Robins.)

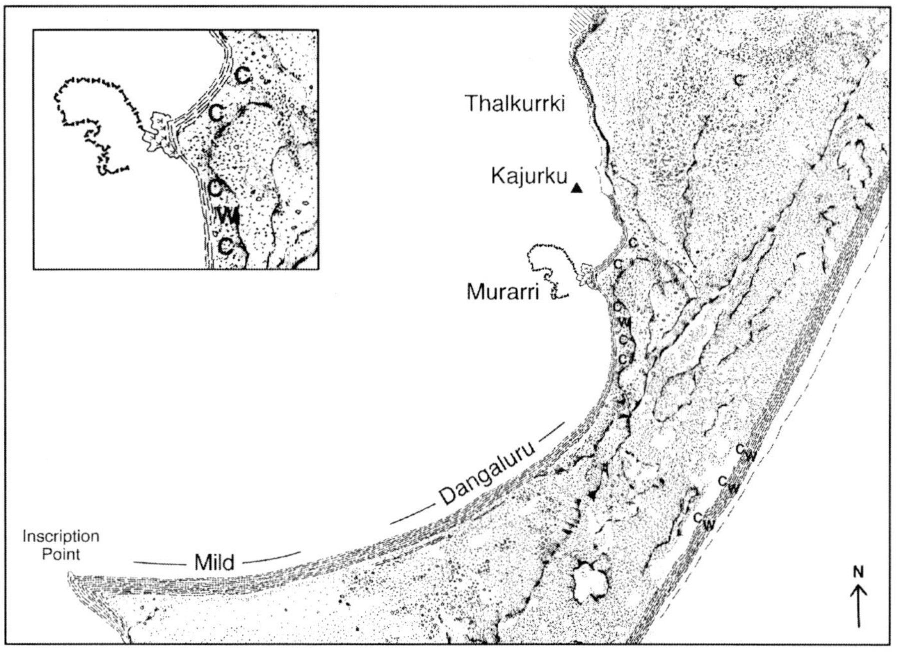

Figure 4.11: Map of Central Sweers Island in the vicinity of Inscription Point showing environmental units with Aboriginal geography and the location of a fish trap at Murrarri. Note that the trap is designed to catch fish moving with currents parallel to the shore. (Source: From map of Ringurrng by Paul Memmott 1982, made with Darwin Moodoonathi and Arthur Paul).

> traps, etc., being built on the lee side, and so protected, and the rest protected during the opposite season. Some other hordes-people can only be sure of fish supplies for about one-half of the year because fishing is often difficult on a windward shore in boisterous weather. Such folk have to depend to a larger extent on estuaries and the foods in mangrove swamps. The people of Dolnoro S have hard rock reefs and can build very substantial fish traps denied to some others who have only fragile coral to work with. (Tindale 1962b:304)

These geographic determinants can be summarized as availability of rock supply and availability of sheltered trap sites for both prevailing winter SE winds and summer NW monsoons. However Kaiadilt consultant, Arthur Paul (dec), provided Memmott with an interesting anecdote of animal behaviour which represents an alternate hypothesis. He described the walls as windbreaks, and explained that when surface conditions were choppy, schools of fish sought out the fish-trap walls against which to shelter from those prevailing boisterous winds. If on a windy day a hunter visited a piece of windward coastline, he could expect to find fish sheltering inside traps close to the walls.

Notwithstanding this latter proposition, a significant theoretical question then is whether the relative densities of fish traps were determined by geographic and environmental factors (e.g., protection from prevailing winds, regular habitat usage by marine fauna), or whether additional cultural factors come into play in explaining the distribution of traps (e.g., limits of estate boundaries). The biophysical and cultural factors determining fish trap location are complex. Our research will identify which factors are the critical ones, but this is the subject of a separate analysis. What can generally be said at this early stage is that there is a widespread availability of suitable rock on all islands, and that lack of rock availability does not appear to be a key determinant in explaining the wide variation in trap densities between the Kaiadilt and Yangkaal on the one hand and the Lardil on the other.

Measuring fish trap density might provide a measure of the degree of local productivity of the fish trap technology and enable comparisons between islands, cultural groups and local clan groups on their degree of economic reliance on this technology. For example, we have already suggested the Kaiadilt relied more intensely on fish traps than the Lardil. In support of this hypothesis, refer to figure 4.12, an aerial photo of the south-west corner of Bentinck Island with most (but not all) trap walls enhanced for clarity. Features include both small and large traps, many inner and outer traps, party walls, lineal walls (no obvious pens), a pocket trap and one on-shore to off-shore sequence of four successive pens. A similar density of traps pertains for the southern end of Fowler Island (to the south-east of Bentinck). These two examples represent the highest densities of traps in the Wellesleys from our preliminary visual inspection of aerial photos, even though the precise calculation of such remains a methodological difficulty.

In order to explain such a high density, let us reflect on the dynamics of demography and fish traps. A possible simple scenario is as follows. A small Kaiadilt clan builds a trap and it yields an adequate supply of fish supporting the group. Over generations the clan population grows. The threshold of a sustainable yield is reached and then passed, resulting in the necessity to build another trap. There thus may be a direct relation between population growth and fish trap reproduction.

However in the case of the Lardil with the advent of large net fishing and dugong net hunting, there was a capacity to obtain a larger yield by drawing several clans together for multiple large net practices, as an alternative to the use of traps, albeit only under suitable climatic and social circumstances. The Kaiadilt could not so readily achieve this option, having only a smaller population on which to draw (notwithstanding feuds) and no way to readily draw mainland people over to Bentinck (in contrast to Mornington).

The above raises the methodological question, how should we measure the density of traps? Do we calculate numbers of wall units, trap units, or fish trap sites, per lineal coastal kilometre, or the summated area of pens per coastal kilometre?

The density of traps per kilometre of coast, however we measure it, may not necessarily be a useful measure of productivity. There is insufficient evidence in the Aboriginal Australian ethnography to argue that larger traps yield a directly proportional quantity of fish and hence support a larger population (Bowen and Rowland 1999:34–35). Nevertheless if we cannot assume a direct relationship between pen area and yield, we still may be able to make comparative analyses of densities for similar off-shore environments in the same geographical locale where one would expect yields to be more or less consistent.

The temporal properties of traps

Another pertinent issue when considering fish trap density is whether, amongst those traps evident at the time of early colonial contact, there were old disused traps. If so, it would be inappropriate to include them in any measure of trap numbers in synchronic use. We cannot assume all traps were for synchronic use. In fact Tindale has hypothesized the converse as recorded in his 1963 journal at *Rukuthi*[19] or Oak Tree Point on the northern peninsula of Bentick Island.

> There are several generations of fish traps. Old ones are preserved in part because oysters have sealed them to the basement rock and where there is sufficient water have been able to prevent their being swept away. The oldest ones with roots up to 3 or 4 feet [90–120 cm] above present day effective use were formerly covered with live oysters but these are now just dead oyster rock with boulders incorporated...
>
> At [Birrmuyi[20]] I was able to measure the depth below high tide mark of the principal fish traps with a margin of error of about a foot [300 cm]. The older ones nearer the shore are now ineffective through drop in sea level by up to 2 or 3 feet [60–90 cm] or a little more. (Tindale 1963:241)

Tindale concluded that functioning traps were over 2.1 m below highest tide (1963:237-8). He assumed the inshore traps were from a period of higher sea level but had no effective technique to date those traps. According to Stock's extrapolation from the Australian east coast data (Larcombe and Carter 1998; Baker et al. 2001), the peak mean sea level in the late Holocene occurred around 5300 BP then dropped about 1.7 m to the present position between 4000 and 3400 BP. It is possible that as mean sea level rose, trap walls were extended higher and/or further inshore. Conversely, as mean sea level dropped some construction effort may have been concentrated at offshore sections.

An alternative hypothesis to the in-shore traps remaining from a period of higher sea levels is that a complex of inner and outer rock wall fish traps were used in the same harvesting event upon a falling tide (figure 4.13). We note in the Ganggalida data of our colleague David Trigger,

---

[19]His 'Rokoti' or 'Lukuti'. Unfortunately we have not obtained any drawings by Tindale of these fish traps from the S. A. Museum, if indeed he made any.

[20]Tindale's 'Berumoi'.

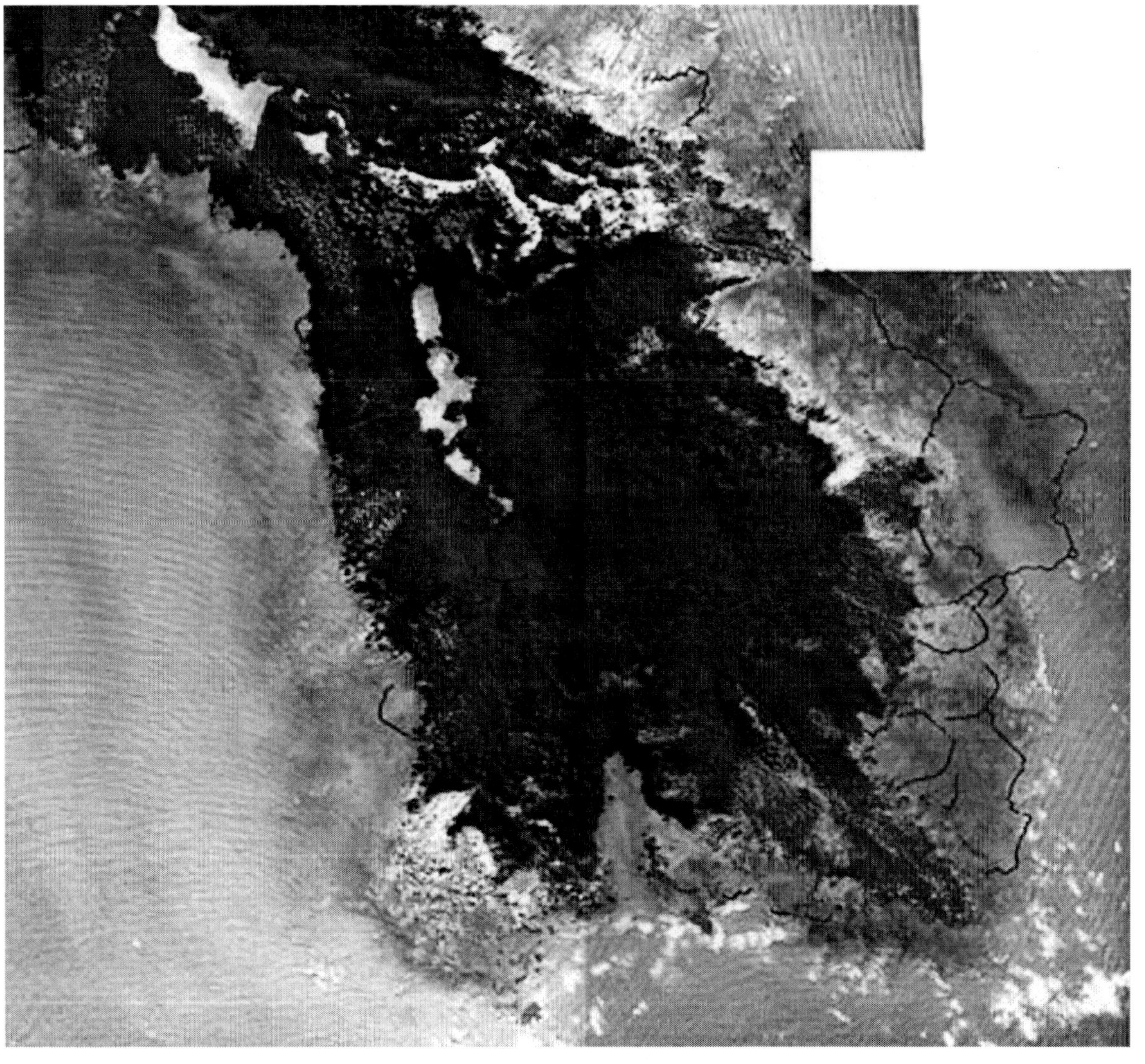

Figure 4.12: The south-west corner of Bentinck Island showing part of the large rock wall trap complex adjacent to a mangrove forest (aerial photo enhanced with line).

a reference to people building up the walls of their traps when the tidal range becomes higher in the wet season. Traps of varying heights may have also been employed under the varying seasonal tidal regimes that occur in the Gulf of Carpentaria.

An interesting alternate explanation was suggested by another colleague Ian Lilley based on John Terrell's theory that the traps may not be very efficient at all but may be indispensable to obtaining food in an unstable environment particularly after a catastrophe, for example a tidal surge associated with a cyclone that destroys other food sources. Here is a possible hypothesis as to why the Kaiadilt and Yangkaal had more traps. They lived on relatively small islands with a comparatively smaller range of edible plant and animal species and hence potential food yield in the interior land systems. On Mornington where there is a greater inland food harvest, there would be less dependence on littoral and marine food sources at such a critical time.[21]

The problem of dating fish traps

There is the further issue of whether, if a temporal sequence of wall construction based on changing sea levels was hypothesized, some sort of dating techniques could be used to verify such. It is reasonable to assume proximate camps would have been frequently occupied to carry out exploitation, surveillance and maintenance of the traps. Is there a possibility of correlating fish trap age with dates obtained from archaeological remains in nearby camps? For example Ganggalida Elders have demonstrated the use of a freshwater well approximately one kilometre inland from the Bayley Point traps. They stated this well site was commonly used as a camping place whilst using the traps, and the large mound of shellfish that remains was evidence of such use.[22] A critical related factor is fish bone and shell disposal, for unless such refuse is stored in a midden, a random scattering of bone and shell will weather relatively speedily in the local environment and leave no clear archaeological record of species type. Thus if there was a change in pattern of fish species consumption due to environmental change, it may not be visible in an excavation of a nearby campsite.[23] Dating midden/campsites can only provide an inferred approximation of the age of an adjacent fishtrap.

An alternate approach to dating is to look for trap sites where the possibility may exist of excavating the buried 'arm' walls of traps where they are found to be buried by later sediments. Tindale (1963:237-238) noted this possibility:

> Description of fish traps on point at Windjarukarru. 'Roots' of trap are on basement rock and cemented by oysters. One arm goes under the beach and thus established "when the sea level was a little higher than now".... These fish traps are ancient according to Percy [Loogatha].[24]

Excavations in accumulated sediment and beside the trap walls may provide materials suitable for dating (figure 4.14). These could include carbon-bearing materials deposited in the sediment. Calcareous faunal remains (sub-fossil relics) attached in growth positions on the rock wall could be candidates but suitable species in this region have yet to be determined. Luminescence dating of sediments would probably not be attempted unless they could be shown to be aeolian deposits.

Again Tindale (1963:249) speculated on the relation between age and oyster remains on the old traps at *Rukuthi* (his 'Rokoti'): "Where the rocks were less than 3 feet [90 cms] below high tide mark all the oysters on them were dead ones. At 4 feet [120 cm] below high tide line an old one or two were alive but most were dead and eroded...".

There is also the more remote possibility of finding an artefact within an existing fish trap wall that may have some datable property suitable for modern techniques (e.g., residue analysis). Tindale (1963:241) noted the use of walls to store artefacts on Bentinck Island:

> The wall directly off Berumoi spring extends up to only 1 ft. [30 cm] below highest tide in the protected bay. The inner end was once cemented by

[21] To explore this hypothesis we would need inter-island comparisons of topographic heights, tidal surge inundation areas, edible interior fauna and flora, and to carry out interviews with consultants on post-cyclone food collection traditions.

[22] Similarly, in the Point Parker area, a major freshwater spring exists about one kilometre south from the location of the traps, but very close to the beach. People are said to have camped on the beach in the dry season, but to have sought shelter further inland in the stormy wet season.

[23] From an anthropological perspective we also need to understand if there were any rules of refuse disposal in addressing the interpretation of the existence or the non-existence of middens in the archaeological record.

[24] Referring to Percy Lugutha, a Kaiadilt Traditional Owner for this area, who was also known to one of the current authors (P. M.).

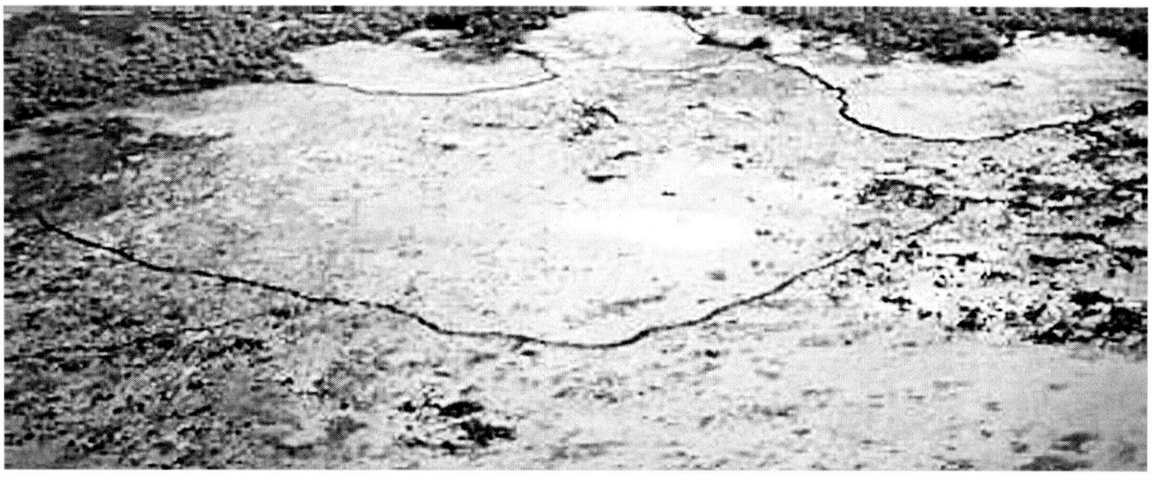

Figure 4.13: Illustration of a complex of inner and outer rock wall fish traps. Were these used in the same harvesting event upon a falling tide, or were they used under different tidal regimes, or are the inner ones remaining from a period of higher sea level? (Photo by Richard Robins.)

Figure 4.14: A rock wall of a fish trap at Bayley Point that appears very ancient. Oyster growth has forced the individual boulders apart. It is within an outer pen which had been used in relatively recent time. (Photo by Richard Robins).

oysters but these have in large part decayed away leaving most of the stones loose again. Wedged in this wall at 3 ft. [90 cm] I found an oyster [hammer] stone and at 2 ft. [60 cm] a loose flake of jasper was present inside the wall..."

Conclusions

The material cultures of the Wellesley groups have cultural commonalities as well as obvious differences. An accurate ethnographic and archaeological description of the fish traps and their properties must be made before contributing to an understanding of cultural change and isolation amongst the Wellesley groups. For example, how exactly do the traps work? What is the relationship between traps and nets? Is one a surrogate for the other or did they perform different economic and social roles? We hypothesize that the Kaiadilt developed a greater dependence on fish traps and may have specialized this technology in particular ways (fish herding). Our study will attempt to make comparisons between the four groups on the extent of trap use (versus other fishing and hunting technologies), trap numbers, sizes, layouts and shapes, usage under different seasonal tidal conditions, fish herding techniques inside traps, usage of trap gates and pockets, and other such properties. A specialized Kaiadilt trap technology would possibly explain why they had a less complex material culture repertoire than those of their neighbours.

As for our methodological problems, our interim working decision is to treat each single wall feature as a unit and defer abstract interpretation (of, for example, what is a pen?) until after more fieldwork. Despite the complexity outlined above, our final view on the way forward is to measure as much as possible, qualify our assumptions and guesstimates, correlate all of our physical units and trust some outstanding patterns will emerge from the analysis.

Our intention is to use these data on rock wall fish traps as one essential baseline for a reactivated research programme centred on the pre-history and cultural change of the Wellesley Islanders. The traps have the potential to play a vital role in the story of the people of the southern Gulf of Carpentaria. This potential will be greatly enhanced if we can answer the question—what exactly is a fish trap?

References cited

Baker, R.G.W., Haworth, R.J. and Flood, P.J. (2001). Intertidal fixed biological indicators in Australia: a multi–dimensional measure of coastal characteristics for the mid to late Holocene. *Quaternary International* 83–85:257–273

Bowen, G.M. and Rowland, M.J. (1999). *Indigenous Fish Traps and Weirs of Queensland: An Overview, Assessment and Recommendations.* Report to the Australian Heritage Commission and Environmental Protection Agency. Brisbane and Canberra.

Boyd, A J. (Maj). (1896). Narrative of Captain G. Pennefather's exploration of the Coen, Archer and Batavia Rivers, and of the islands on the western coast of the Gulf of Carpentaria in 1880. In *Proceedings and Transactions of the Royal Society of Queensland* XI: 46–60.

Evans, N. (1992). *Kayardild dictionary and thesaurus: a vocabulary of the language of the Bentinck Islanders, North-west Queensland.* Parkville, Victoria: Department of Linguistics and Language Studies, University of Melbourne.

Evans, N. (2005). East across a narrow sea: micro-colonization and synthetic prehistory in the Wellesley Islands, Northern Australia. In T. Osada (ed.), *Linguistics, Archaeology and the Human Past: Occasional Paper No. 1.* Kyoto: Research Institute for Humanity and Nature.

Hale, K., Farmer, A., Nash D., Simpson, J. (1980). *A preliminary dictionary of Lardil.* Massachusetts: M.I.T., Typescript.

Larcombe, P. and Carter, R. (1998). Holocene bay sedimentation and sea levels, Magnetic Island. In D. Johnson, and R. Henderson (eds.) *Short Geological Field Trips in the Townsville-Charters Towers Region*, pp. 72–79. Geol. Soc. of Aust.

Memmott, P. (1979). Lardil Properties of Place: An Ethnological Study in Man Environment Relations. Unpublished Ph. D. Thesis, University of Queensland, St Lucia.

Memmott, P. (1982). The South Wellesley Islands and the Kaiadilt. Unpublished MS. Aboriginal Data Archive, University of Queensland, St Lucia.

Memmott, P. (1985). Aboriginal Fish Traps in the Gulf of Carpentaria and Wellesley Islands Research Project. Report on Data Collection and Compilation. Unpublished report, Aboriginal Data Archive, University of Queensland, St Lucia.

Memmott, P., Evans, N. and Robins, R. (2004). Understanding isolation and change in island human populations through a study of indigenous cultural patterns in the Gulf of Carpentaria. [Paper presented to "Islands of the World VIII Conference: Changing Islands—Changing Worlds", Kinmen Island, Taiwan, November 1–7, 2004], submitted to *Transactions of the South Australian Museum* invited for submission, January 2005).

Memmott, P., Robins, R. and Trigger, D. (1984). Aboriginal fish traps in the Gulf of Carpentaria. Unpublished poster, Australian Archaeological Association Conference, Valla Beach, New South Wales.

Memmott, P. and Trigger, D. (1998). Marine tenure in the Wellesley Islands region, Gulf of Carpentaria. In N. Peterson and B. Rigsby (eds.), *Customary Marine Tenure in Australia*, pp. 109–124. Oceania Monograph No 48, University of Sydney.

Ngakulmungan Kangka Leman (Language Projects Steering Committee). (1997). *Lardil Dictionary.* A vocabulary of the language of the Lardil people, Mornington Island, Gulf of Carpentaria, Queensland. Mornington Island: Mornington Shire Council.

Oswalt, W.H. (1976). *An Anthropological Analysis of Food-Getting Technology.* New York: Wiley and Sons.

Oswalt, W.H. (1973). *Habitat and Technology, The Evolution of Hunting.* New York: Holt, Rinehart and Winston.

Robins, R.P. (1983). Gulf of Carpentaria fish trap survey summary. Unpublished MS, Queensland Museum, Brisbane.

Robins, R.P. Stock, E. C. and Trigger, D. S. (1998). Saltwater people, saltwater country: Geomorphological, anthropological and archaeological investigations of the coastal lands in the southern Gulf of Carpentaria. In *Memoirs of the Queensland Museum, Cultural Heritage Series.* 11:75–125.

Robins, R.P. (1998). Literature Survey of Information Relating to the Archaeology of the Southern Gulf of Carpentaria and Wellesley Islands: Wellesley Island Sea Claim. Report for Andrew Chalk and Associates, Solicitors. Brisbane: Queensland Museum.

Robins, R.P. (2000). *Map of Fish Traps of the Southern Gulf of Carpentaria, Explanation of Methods of Compilation.* Brisbane: Queensland Museum.

Roth, W.E. (1901a). The Wellesley Islands—visit by Dr. Roth. *The Education Office Gazette* 1901: 320–322.

Roth, W.E. (1901b). The Carpentaria blacks: the Wellesley Islands visit by Dr. Roth. In *The Observer* 24 August 1901.

Tindale, N.B. (1962a). Geographic knowledge of the Kaiadilt People of Bentinck Island, Queensland. In *Records of the South Australian Museum* 2/14: 259–296.

Tindale, N.B (1962b). Some population changes among the Kaiadilt people of Bentinck Island, Queensland. In *Records of the South Australian Museum* 2/14: 297–319.

Tindale, N.B. (1963). *Journal of Field Trip to Wellesley Islands.* Adelaide: South Australian Museum.

Tindale, N.B. (1977). Further report on the Kaiadilt people of Bentinck Island, Gulf of Carpentaria, Queensland. In J. Allen, J. Golson, and R. Jones (eds.) *Sunda and Sahul: Prehistoric Studies in Southeast Asia, Melanesia and Australia*, pp. 247–273. London: Adademic Press.

Tindale, N.B. (1981). Prehistory of the Aborigines: some interesting considerations. In A. Keast (ed.), *Ecological Biogeography of Australia* 3, pp. 1762–1797. The Hague: W. Junk Publishers.

Trigger, D.S. (1985). A selective photographic record of Aboriginal stone wall fish traps at Bayley Point, Gulf of Carpentaria, northwest Queensland. Unpublished MS, Queensland Museum, Brisbane.

Trigger, D.S. (1987). Inland, coast and islands: traditional Aboriginal society and material cultural in a region of the southern Gulf of Carpentaria. In *Records of the South Australian Museum* 14: 69–84.

Chapter 5

Between the Australian and Melanesian realms: the archaeology of the Murray Islands and consideration of a settlement model for Torres Strait

Melissa Carter, University of Sydney
Ian Lilley, University of Queensland Brisbane

Introduction

> The foliage screened villages below, each facing its own tiny beach, with a fleet of big fighting-canoes drawn right up to the big front avenues of palms. Each village was flanked by steep grassy headlands, or deep green of tangled jungle, with the intense green of banana-patches away behind. Behind, and farther back still, were the well-kept vegetable and fruit gardens climbing up the little hills... The island was so markedly different from the Great South Land, in its people, its rocks, its trees, its birds, its corals and fishes. Seven hundred feet above the sea, in the lava rocks of that old crater, are huge chunks of dead coral, proving how in ages past the volcano pushed Mer right up through the bottom of the sea. The Meriam Le were vastly different from the... Australian Aboriginal... (Idriess 1947: 21–22).

The Torres Strait Islands are located between Australia's Cape York Peninsula and the southern coast of New Guinea, and were formed between 8500 and 6500 years ago when rising post-glacial seas flooded the Sahul shelf separating the Australian and New Guinea landmasses (Barham and Harris 1983) (figure 5.1). Although redolent with the idyllic tropical island imagery often portrayed by romantic novelists, Idriess' (1947) eloquent description of the Murray Islands in the eastern Torres Strait surprisingly holds certain factual truths.

The above mid-twentieth century extract parallels the first historical accounts of the culture and environment of the Torres Strait Islands (e.g. Brierly 1848–1850 in Moore 1979; Jukes 1847; MacGillivray 1852 II). These describe and define characteristics in languages, social organization, subsistence practices, material culture and geology as originating from both the Australian and Papua New Guinea mainlands. The impressive corpus of information recorded as part of the Cambridge Anthropological Expedition to Torres Strait in 1898–1899 provided the first detailed accounts of the diverse maritime and horticulturally based subsistence practices across Torres Strait, and of the intricate trade and exchange networks between islanders and the populations of the adjacent mainlands, made possible by the use of large ocean-going double outrigger canoes (Haddon 1901–1935). Over one century later, and in recognition of the important geographical niche of the Torres Strait—positioned between the Aboriginal hunter-gatherer cultures of Australia and the Melanesian agriculturalists of New Guinea—archaeological research in Torres Strait began (Walker 1972).

The first three decades of archaeological research in the Torres Strait were directed at determining the antiquity of occupation and the nature of prehistoric subsistence economies, but remained exploratory in approach and lacking in the sorts of data that might have been derived from more systematic and intensive investigations. The first radiocarbon dates from the islands suggested around 700 years of human occupation, while researchers, heavily reliant on the documentary record, estimated that human antiquity in Torres Strait did not extend much beyond the ethnographically recorded past (Moore 1972, 1979; Vanderwal 1973a). By the

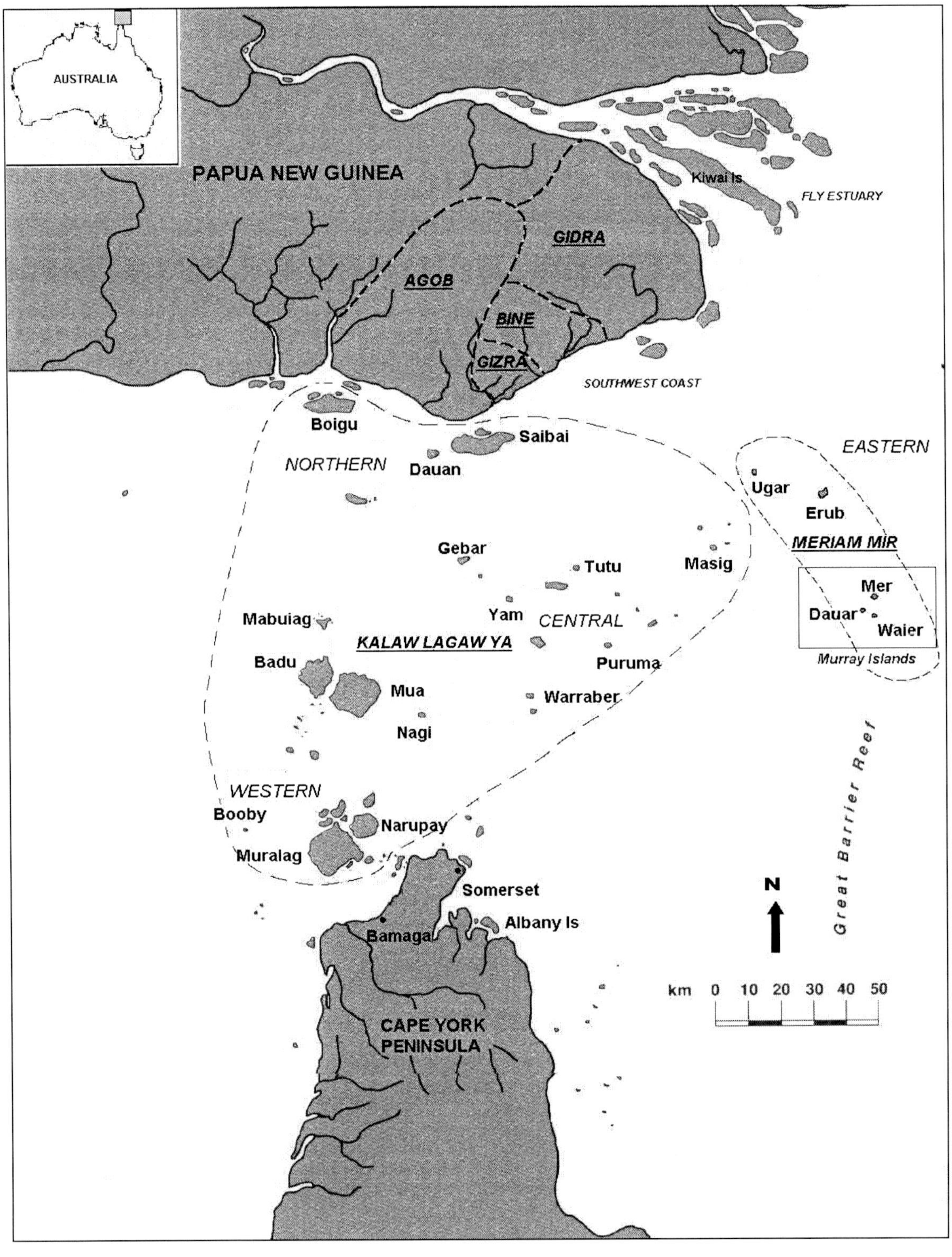

Figure 5.1: The Torres Strait Islands, showing Island groups and language divisions mentioned in text.

1990s human occupation had been tentatively broadened to 2500 years based on the problematic results of a paired shell-charcoal radiocarbon sample (Barham and Harris 1985), but archaeological evidence for pre-European trade and the emergence of the horticultural economy continued to elude investigators (Barham and Harris 1985; Ghaleb 1990; Rowland 1984, 1985).

The most recent archaeological investigations undertaken in Torres Strait have aimed to answer long-standing questions on the chronology and nature of human settlement (Barham 2000; Carter et al. 2004a; 2004b; David et al. 2004; McNiven et al. 2006), to recover and contextualise in a regional framework evidence for local subsistence development and change (Barham and Harris 1985; Barham et al. 2004; Parr and Carter 2003), and to offer explanations for cultural change observed in the region's more recent history (David et al. 2005; David and Weisler 2006; McNiven 2006). Carter's (2004) doctoral investigations on Mer and Dauar Islands were the first archaeological research conducted in the eastern Torres Strait. She confirmed a late Holocene occupation sequence of the Murray Islands, establishing first and permanent settlement of the islands from c.2800 years ago. The present paper summarises the results of selected analyses of the Murray Islands assemblages (see Carter 2004 for more details), followed by a consideration of these data in the context of the emerging settlement models for Torres Strait.

Archaeology of the Murray Islands

The Murray Islands comprise the three small volcanic islands of Mer (Murray), Dauar and Waier, and are the most eastern islands in the Strait, located 6km from the Great Barrier Reef (figure 5.2).

The Eastern Islands are the most fertile of the Torres Strait Islands, with rich soils derived from underlying lavas that support dense monsoonal vine-thicket vegetation. The traditional inhabitants of the Murray Islands are the Meriam Le. They historically inhabited all three islands, but today form a concentrated community along the northwest foreshore of Mer. Dauar and Waier are still used extensively for temporary residence, gardening, fishing and gathering activities. Traditional subsistence on the Murray Islands combined intensive horticulture of common Pacific cultigens, including banana, yam, coconut and sweet potato, with the exploitation of marine resources, including shellfish, fish and turtle (Haddon 1912: 130–136). Gardening and the procurement of marine resources are still important subsistence activities on the islands today, although store bought goods forms the bulk of the contemporary diet (Bird 1996).

The indigenous language of the Eastern Torres Strait Islanders is Meriam Mir, a non-Austronesian language that is closely related to the Bine, Gizra and Gidra languages spoken by southern Papuans, and also has some Kiwai language features (Wurm 1972: 349). Meriam Mir is distinct from Kalaw Lagaw Ya, which is the parent language spoken throughout the rest of the Torres Strait, and is structurally an Aboriginal language, though it has some non-Austronesian elements (Wurm 1972).

On Dauar excavations were conducted at the past occupation sites of Sokoli and Ormi (see figure 5.2 for locations). These sites are located on opposite coasts of the island on the margins of a low saddle between the two large hills that dominate the local topography. Situated on the northeastern coast, Sokoli is known to have been used for gardening in the recent past. The excavation revealed over 2 meters of archaeological deposit, mostly comprising marine shell remains with smaller quantities of fish and turtle bone. Contrasting to the brown colluvial matrix of the upper half of the deposit, a notable stratigraphic feature at Sokoli was a c.15 cm layer of fine grey ashy sediment 50 cm below surface level (bsl) which in the western profile of the excavation continued as a linear feature 10–20 cm in width to a depth of 160 cm bsl. The other major stratigraphic feature of the excavation was a clear transition from dark brown colluvium and dense archaeological shell in the upper half of the deposit, to a sediment matrix of coarse-grained red beach sand with considerably less shell in the lower half of the deposit.

Several material culture items were recovered during the excavation. These include a broken but intricately carved bone spatula (figure 5.3), and a single pottery sherd from the upper half the deposit. Both items are the first such artifacts to be recovered in situ in the Torres Strait (Carter 2003). There is no oral history or documentary evidence to suggest that pottery was ever traditionally manufactured in the Torres Strait Islands. Nor were earthenware vessels part of the impressive collection of Torres Strait material culture acquired by Haddon during the late nineteenth century (Moore 1984).

The site of Ormi is located on the southeastern coast of Dauar (see figure 5.2). The reef flat on this side of the island is more extensive than off Sokoli on the northeastern coast, and there

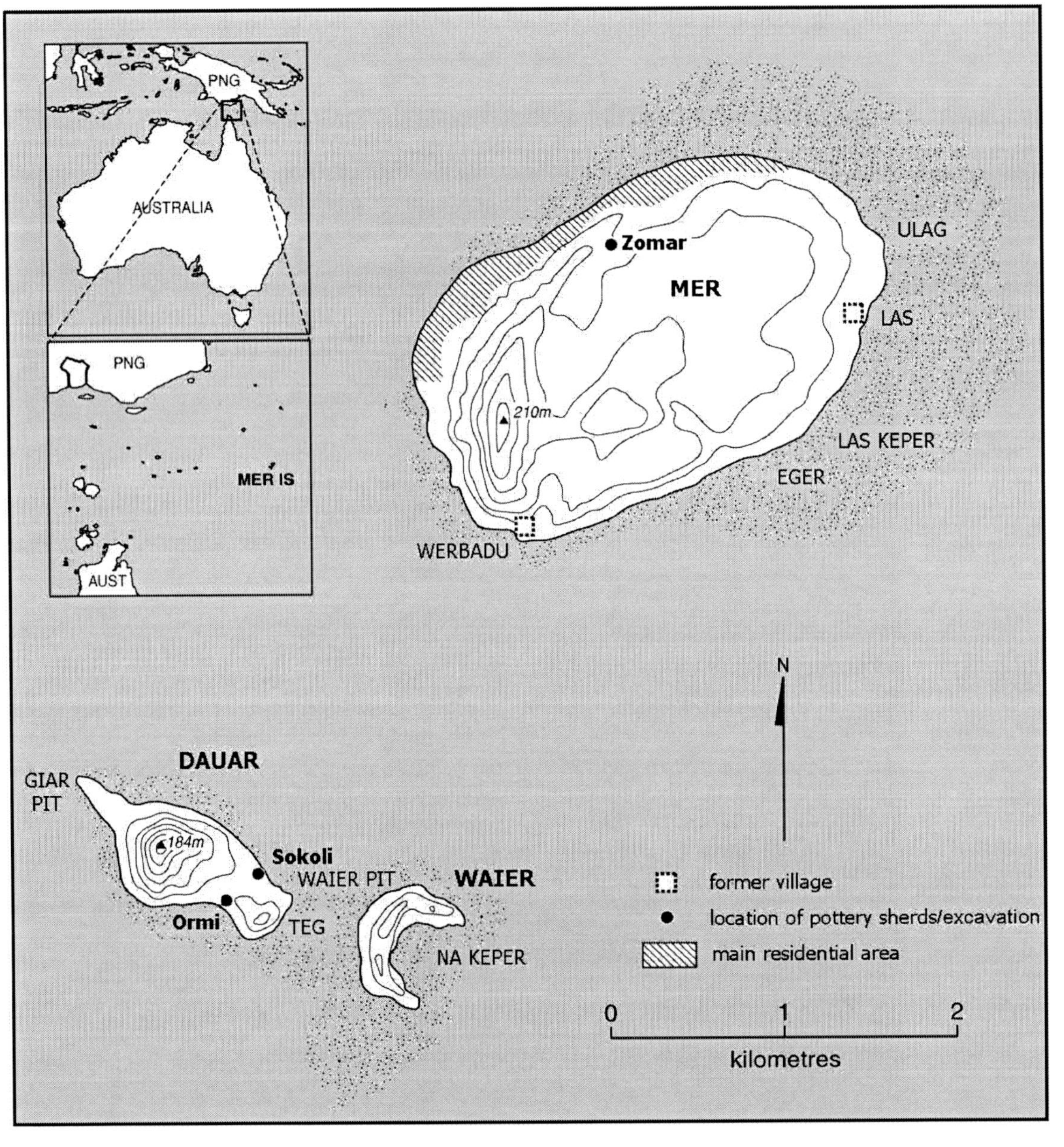

Figure 5.2: The Murray Islands, showing location of excavations and pottery finds.

Figure 5.3: Carved bone spatula from Sokoli (Dauar).

is a system of large stone fish traps that are still used and maintained today (figure 5.4). The Ormi excavation revealed the same depth of archaeological deposit recovered as Sokoli, and as at Sokoli, the upper half of the deposit contained dense archaeological shell remains within brown colluvium, while the lower half of the deposit comprised coarse red sand with considerably less shell. A total of 23 pottery sherds were excavated from Ormi. Most are very small (<1.5 cm) and were recovered from the lower stratigraphic units (between c.120 cm and 190 cm bsl), although several larger sherds were recovered from the middle to upper strata (c.80 cm bsl).

During the second season of fieldwork in 2000, another pottery sherd was bought to Carter's attention. During construction work for the installation of telecommunications towers, a local resident had found two large conjoining sherds earlier that year on the high northwestern hill at Zomar on Mer (see figure 5.2) for location). These sherds are much larger and thicker than the excavated sherds from Dauar, with the smaller sherd also displaying very faint decoration consisting of a chevron pattern of dots and a circular pattern of dashes (figure 5.5)).

The date of the excavated deposits was determined by conventional radiocarbon assay of excavated archaeological marine shells. All dates reported here were calibrated using CALIB (v4.4) (Stuiver and Reimer 1993), applying the marine calibration model dataset of Stuiver et al. (1998) for marine shells. A ΔR correction value for northeast Australia of 12 ± 7 for all marine shell samples was also applied, recommended by Ulm 2002a, 2002b. Figure 5.6 shows the calibrated age-depth curves for Sokoli and Ormi. The basal dates for each site suggest initial occupation c.2800 years ago. At both sites the rate of deposition was rapid, particularly at Ormi where over two metres of assemblage was deposited in just over 1000 years. Dating also shows that the surface deposit at Ormi dates to 1600 years ago, indicating that the most recent deposit is missing. This is most likely a result of disturbance caused by housing construction close to the excavation site several years ago. At Sokoli, an anomalous date (Wk7480) represents a sample taken from the ashy stratigraphic feature described above, and perhaps indicates some post-depositional disturbance associated with the formation of this feature. Based on the chrono-stratigraphic sequences, the pottery sherds excavated from Ormi were deposited between c.2500 and 1600 years ago, while the sherd at Sokoli was deposited considerably later at c.700 years ago.

Further archaeological insights in the Murray Islands

The excavations on Dauar have produced evidence for 2800 years of human occupation in the eastern Torres Strait. In addition, the Murray Islands excavations produced the first archaeological evidence for possible pre-European trade links between Torres Strait and New Guinea, demonstrated by a small assemblage of pottery sherds. The recovery of most of the sherds from the lower and basal stratigraphic units at Ormi suggests that these exchange networks may date from as early as c.2500 years ago. Importantly, this timeframe for pottery in the eastern Torres Strait predates the earliest known dates for the appearance of pottery on the southern Papuan coast by c.500 years (Bickler 1997). We return to this point below. The evidence for horticulture was considerably less obvious in the excavations on Dauar, with no macroscopic remains of cultigens or associated horticultural tools observed. Based on ethnographic accounts of gardening practices on Mer which describe the use of fire to prepare the ground for planting (MacGillivray 1852, II: 26), the ashy feature at Sokoli was concluded to be the only possible evidence for past horticultural practices on Dauar, although this interpretation of this feature is questionable.

Several lines of analysis were employed to obtain more detailed information regarding pre-European subsistence and trade and exchange on the Murray Islands. Exploratory phytolith and starch grain analysis of excavated sediment samples from Dauar attempted to identify the presence or absence of edible species as well as past patterns of vegetation change. This analysis was the first direct attempt to recover and identify paleobotanical remains associated with horticultural subsistence in the Torres Strait (for a full discussion of results see Carter 2004 and Parr and Carter 2003).

At Sokoli, starch grains resembling coconut (*Cocos nucifera*) palm phytoliths were identified in the lowermost deposit dating from 2800 years ago. Results also demonstrated a shift from grass species *Themeda australis* to the hardier and more fire resistant species *Imperata cyclindrica* by 1600 years ago. Banana (Musuaceae) phytoliths were extracted from the ashy sediment dating to between 1500 and 700 years ago. By 1500 years ago there was also a clear increase in carbonised phytoliths, which may be further evidence of firing and clearance at the site by this time. For Ormi the results indicate that the site was also affected by changes in vegetation patterns, possibly shortly after initial occupation,

Figure 5.4: Stone fish traps on reef flat at Ormi, Dauar Island.

Figure 5.5: Conjoining pottery sherds recovered as surface finds at Zomar, Mer Island.

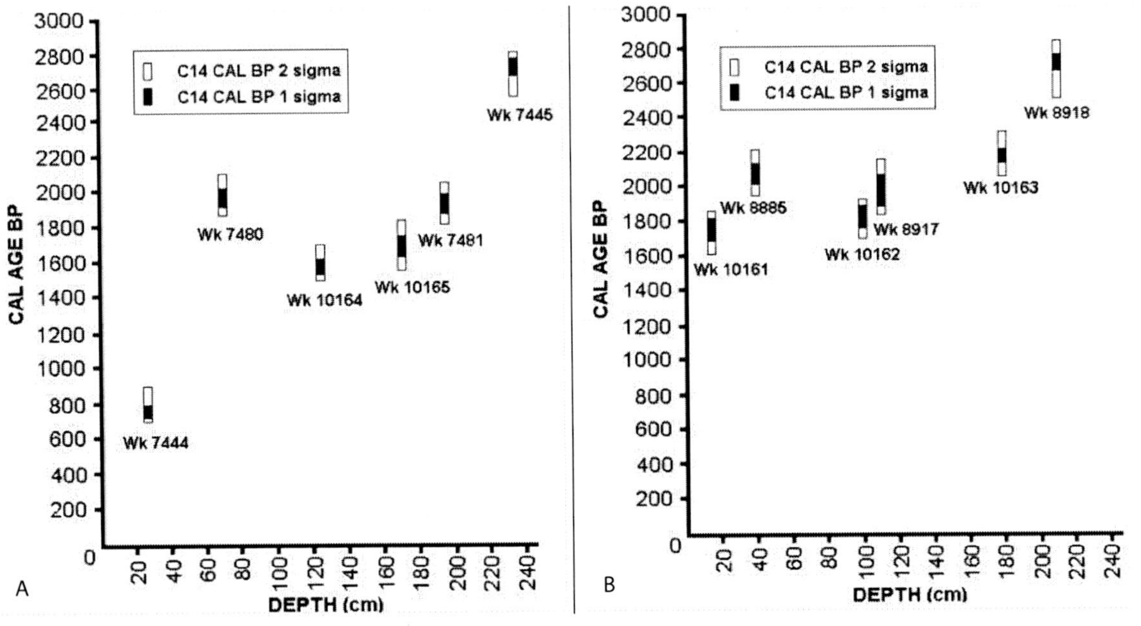

Figure 5.6: Calibrated radiocarbon age depth curves for (A) Sokoli and (B) Ormi sites, Dauar Island.

resulting in a shift from a closed habitat to a more open site after 2000 years ago. The presence of coconut and banana phytoliths as well as yam (*Dioscorea* sp.) starch grains between 2500 and 2000 years ago was also identified.

Contrary to received wisdom, recent genetic analyses of numerous tropical horticultural crops, including banana and yam, lend support to the proposition that certain species may have been domesticated in New Guinea rather than in centres of domestication in Southeast Asia (Denham et al. 2003; Lebot 1999). In addition, the suggestion has also recently been made by Denham (2004) that this and other archaeobotanical evidence supports hypotheses of sustained contact between New Guinea and Island Melanesia, as well as with South East Asia, prior to the late Holocene dispersal of 'neolithic' Oceanic Austronesian-speaking Lapita potters through Melanesia. It is argued that if isolation between these regions was not a critical factor before 3500 years before present (BP), then languages, people and material culture items could have spread throughout New Guinea and Island Melanesia at various different times (Denham 2004). Denham (2004) implies that such a situation therefore brings into question the concepts of 'neolithic packages', 'indigenous origins' and 'Austronesian expansion', so commonly used in defining the Holocene prehistory of Island Melanesia.

The implications of these hypotheses in the present context is that the presence of a range of common Pacific cultigens on the Murray Islands, as demonstrated by the results of phytolith and starch grain analyses, may provide evidence for importation of horticultural domesticates from New Guinea into eastern Torres Strait from as early as c. 2800 years ago. Although not conclusive, the results suggest that the ashy sediment at Sokoli may have been associated with gardening activities or garden preparation, or perhaps even the growth of an individual cultivated plant (see Carter et al. 2004a: 253 for further discussion). Combined with the changes in stratigraphy and archaeological deposit at both Sokoli and Ormi described earlier, the emerging chronological sequence for the eastern Torres Strait may indicate the development and intensification of a horticultural economy over a period of at least 1000 years from the time of initial occupation.

Mineralogical analyses of the archaeological pottery sherds from Mer and Dauar was undertaken to obtain further detail about the settlement of the Murray Islands and to confirm the nature of trans-Strait cultural networks. Studies included energy dispersive X-ray (EDS) micro analysis and microprobe (SEM) of thin sections by geologist Dr Mike Rubenac (James Cook University), as well as petrographic analysis of the temper inclusions by Emeritus Professor William Dickinson (University of Arizona) (Dickinson 2003). Due to restrictions in funding and an uncertainty about the stratigraphic integrity of the small lowermost pottery sherds at the time, the number of artefacts selected for analysis was limited, and consisted of the three largest sherds excavated from the upper strata of Ormi on Dauar, and the undecorated unprovenanced plain sherd from Mer.

In summary, the results revealed combinations of igneous and metamorphic inclusions in all four sherds. These included abundant fragments of plagioclase, clinopyroxene, quartz, brown hornblende, schist and basalt/andesite. Magnetite and altered olivine were also identified, though in much lower proportions. Calcareous grains were not present in any of the sherds, suggesting that the source of the pottery tempers was some distance from a reef or coralline sediment. The microprobe analysis showed that the elemental composition of the feldspars and clinopyroxenes in the pottery was similar to that of the same minerals found in the vicinity of the New Guinea Highlands (Mackenzie and Johnson 1984).

Dickinson's analysis of the temper sands identified two groups: coarser-grained sherds with significant admixtures of non-volcanic detritus from metamorphic rocks, and sherds with finer-grained temper, with recycled sedimentary debris and a lack of metamorphic detritus. Different sources for the two sherd groups have been suggested. The Kubor Anticline—nested between the volcanic clusters of the New Guinea Highlands—has been identified as the likely temper source for the coarser-grained sherds. The suggested temper source for the finer-grained sherds is the hinterland of the Kikori River system, located in the northwest of the Gulf of Papua, where sand from multiple sources is well mixed and of finer size than upstream sands (Dickinson 2003). Refer to figure 5.7 for places mentioned in the text.

The foregoing findings have been met with scepticism by Australasian archaeologists, including one of us (IL). Central to these doubts is our existing knowledge of New Guinea Highlands pottery manufacture based on previous archaeological and ethnographic research in the region. This knowledge has established that there was a limited tradition of pottery manufacture in the Highlands dating to within the last 1000 years (White 1972: 148). The antiquity of the analysed pottery from Dauar is in the order of 2000 cal years BP. The idea that there may have been pottery production in the Highlands that long ago remains difficult to accept. The sherds analysed to date represent only a small fraction of the pottery assemblage recovered from the Murray Islands. Mineralogical testing of a much larger sample is required before any definitive conclusions regarding the origin of the Murray Islands pottery sherds can be drawn. Future research would benefit from a comprehensive review of all previous mineralogical studies of pottery both from the Highlands and the southern coast of New Guinea. For a more detailed account and discussion of the Murray Islands pottery analysis see Carter (2004).

The question of Austronesian influence in Torres Strait

The results of the Murray Islands investigations have provided significant information on the pre-European archaeology of the eastern Torres Strait, with important insights into the antiquity of human occupation, the emergence of traditional subsistence economies and the nature of regional trade and exchange networks (Carter 2002; Carter et al. 2004a, 2004b). The first occupation of the islands is dated to 2800 years ago, with evidence that common Melanesian crops such as banana, yam and coconut may have been imported from New Guinea by the original colonisers. As the population on the islands increased, subsistence related activities left their signature in the form of increases in the density of marine resource remains, along with possible destabilization of up-slope soils and the infilling of coastal embayments caused by on-site horticultural clearance regimes. Based on the radiocarbon chronologies of both Dauar sites, these effects are clearly noticeable in the archaeological record around 2000 years ago. The presence of pottery on Dauar is dated to between 2500 and 700 years ago, with the results of preliminary mineralogical analysis of the selected sherds confirming their manufacture in New Guinea.

The question of the possible intensification of activity in the eastern Torres Strait from 2000 years ago was first proposed by Lilley (1999) in an unpublished conference paper. He suggested that intensification of economic activity in the wider Torres Strait-Gulf of Papua region may have 'pulled' Oceanic Austronesian-speakers westwards along Papuan coast, while the occupants of the Eastern Islands were strengthening their position in regional trade networks that were emerging within and beyond Torres Strait (Lilley 1999). Ethnohistorical records and oral history indeed testify to movements of coastal Kiwai from the Fly Estuary to the Papuan coast facing the Strait. This allowed the strategic establishment of local villages where they were able 'to dominate the older established exchange movements ... [through] their dominance of maritime and fishing technology' (Lawrence 1998: 13). It is

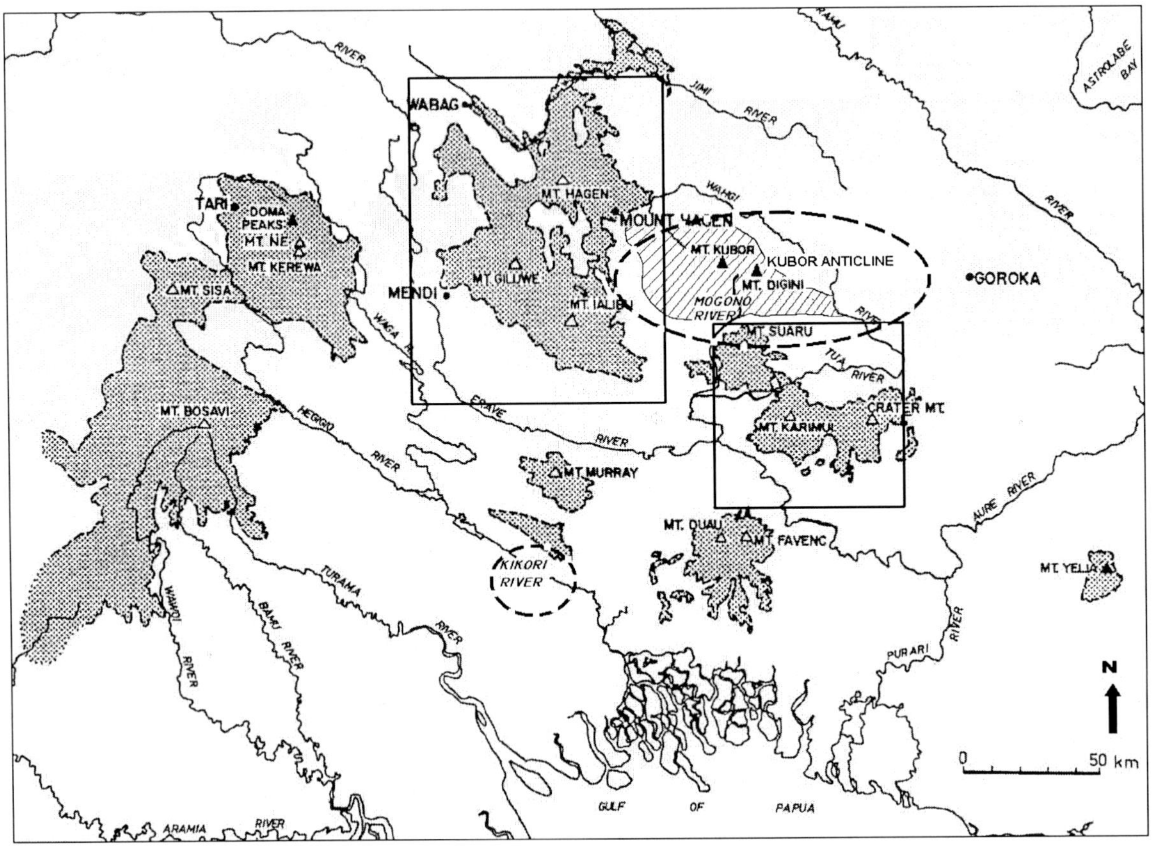

Figure 5.7: New Guinea Highlands showing major volcanic clusters (boxed) and possible source locations of Murray Islands pottery tempers (dashed circles) (after Mackenzie and Johnson 1984).

this activity in the Eastern Islands that Lilley hypothesized may have been the 'pull factor' that attracted Austronesian speakers along the Papuan south coast.

We suggest that the presence of New Guinea earthenware on the Murray Islands, and the nature of the excavated archaeological suites from Dauar (including the evidence for a dual marine and horticulturally-based subsistence system) may support the hypothesis that human settlement of the eastern Torres Strait c.2800 years ago may have occurred as a result of movements of Non-Austronesian-speakers within the trans-Strait region during the late Holocene, rather than as a direct or even indirect result of the migration of Oceanic Austronesian speakers into Island Melanesia from about 3300 years ago (Spriggs 1995; Kirch 2001). During the extended period in which the present paper remained in press, a broadly similar hypothesis constructed around a comprehensive synthesis of the by-then much better known mid-late Holocene archaeology of Torres Strait was developed (McNiven et al. 2006).

McNiven et al.'s (2006) model rests largely on the results of excavations in the western Torres Strait, including the islands of Badu, Mabuiag and the smaller Pulu Islet. Excavations in Mask Cave on Pulu recovered 16 red-slipped pottery sherds and small quantities of shell and fish, turtle and dugong bone. On the basis of the detailed radiocarbon chronologies obtained from the Mask Cave excavations, McNiven et al. (2006) propose an 'Australian-Papuan' settlement model for the Torres Strait.

Phase 1 of this model dates from 2900-3800 years ago and is associated with a 'major demographic expansion of local populations' linked to mainland Australia (McNiven et al. 2006: 66). Among several lines of evidence offered in support of this Australian-derived occupation of western Torres Strait, it is argued that the date of 3800 years ago coincides with the intensification of Aboriginal activity archaeologically documented across northern Australian (i.e. David and Lourandos 1998; O'Connor 1999) and on the Queensland coast (i.e. Rowland 1999; Ulm 2006).

In contrast, Phase 2 of this model (dating from 2100 to 2600 years ago) is concluded to be 'associated with the immigration of Papuan maritime, horticultural and pottery-making peoples to the eastern and western islands of Zenadh Kes [Torres Strait] commencing 2600 years ago' (McNiven et al. 2006: 49). Remarkably, in view of the sourcing results of the Dauar ceramics, petrographic analyses of the Mask Cave sherds suggest the pottery was made on the Western Islands, thus representing 'the first identified Indigenous pottery tradition for Australia' (McNiven et al. 2006: 66).

From their review of known sites of red-slipped pottery on the southern Papuan coast, McNiven et al. (2006: 69–72) identify two significant implications for their model of Papuan expansion into Torres Strait c.2600 years ago. First, the comparison of form and age of ceramic traditions, broad-scale cultural connections and shared cultural traditions within southern Papuan assemblages strongly support the translation and association of those traditions southwards into Torres Strait. Second, and perhaps most critically, the existence of a pre-2000 year old ceramic tradition on the Papuan south coast cannot be ruled out on the bases of the existing archaeological chronologies and sequences established for known sites (for example, Irwin 1985; Allen 1972; Vanderwal 1973b; Rhoads 1982). This latter finding in particular, along with the intriguing results of the Murray Islands pottery sourcing, demonstrates the need for further archaeological research on the Papuan south coast.

McNiven et al.'s (2006) model contrasts to an earlier occupation model for the region based on the results of excavations on Badu, in which David et al. (2004) offered a 'Papuan-Austronesian' settlement model for Torres Strait. Radiocarbon dates for the Badu excavations confirm permanent human occupation of the island from 3500-3000 years ago (David et al. 2004). It was concluded that the contemporaneity of occupation in the Western Islands and the Lapita expansion in Island Melanesia pointed to either a direct or indirect association 'with expansionist Austronesian influences from further to the east' (David et al. 2004: 73). A number of bodies of evidence are offered in support of this view, including linguistic data, oral tradition and rock art. The researchers claim that 'some words in many (if not all) of the languages of the Torres Strait area have strong resemblances to Austronesian vocabulary, in some cases to that of Papuan Central District and South-East Papuan Austronesian languages' (David et al. 2004: 74). However, this linguistic evidence, along with suspected Austronesian traits in Islander oral histories and legends as well as in rock art stylistic affinities, lacks chronological control and may reflect the influx of Pacific Islanders to Torres Strait in colonial times.

As the current and most comprehensive settlement model proposed for Torres Strait, Mc Niven et al.'s. (2006: 73) scenario attributes the permanent occupation of the islands from 2600 years ago 'not to an Austronesian migration, but to a Papuan migration'. We agree with this suggestion and reiterate that to propose that the capacity of Islanders to occupy their remote island homes on a permanent basis, or to engage in trade with their neighbours derives from Austronesian speakers fails to recognize the capabilities of the Non-Austronesian inhabitants of Torres Strait and their obvious seafaring skill. In line with archaeological results from the Admiralty Islands, Bismarck Archipelago and the Solomon Islands, Lilley (1999) suggested that there is more than enough evidence to show that some Non-Austronesian speakers were very capable seafarers who occupied remote islands and traded a range of goods over long distances for tens of millennia. Denham's (2004) recent critique of the chronologically and culturally laden terms often associated with post-Pleistocene human occupation of New Guinea reflects a similar outlook on the dynamic nature of late-Holocene occupation and cultural emergence in southern New Guinea and Torres Strait.

Conclusions

Recent archaeological research into the settlement of the eastern Torres Strait has revealed a complex archaeological scenario. This scenario sees a multitude of regional interactions and cultural transformations occurring on the doorstep of the Oceanic Austronesian world, at a time when that world was itself first emerging. Given this geo-historical context, Denham's (2004) speculation that isolation may not have played a significant role in the development of the cultural complexes that emerged in Island Melanesia becomes all the more inviting. On this basis, we argue that the Australian-Papuan settlement model for Torres Strait recently proposed by McNiven et al. (2006) contributes very usefully to emerging debate concerning the multifaceted dynamism that characterized the mid- to late-Holocene prehistory in Australia and the Indo-Pacific.

Acknowledgments

The Murray Islands research presented in this paper was undertaken by MC between 1999 and 2004 at the School of Anthropology, Archaeology and Sociology at James Cook University, Townsville. MC would like to acknowledge the Meriam Le for their support of the Murray Islands Archeological Project, and Chairman Ron Day and Lloyd Maza (Mer Ged Kem Le) for granting permission for the project. The Passi family (Sokoli) and the Kaddy family (Ormi) are especially thanked for allowing the excavations to take place on Dauar. Sunny Passi and Aris Kaddy are acknowledged for their assistance in the field. Thanks also to Mike Rubenac and William Dickinson for their time and skill in undertaking the pottery analyses. Jeff Parr (Southern Cross University) undertook the phytolith and starch grain analyses, which was funded by the Doctoral Merit Research Scheme at James Cook University. Radiocarbon dating was undertaken at the Waikato Radiocarbon Dating Laboratory and was partially funded by the Waikato Radiocarbon Dating Fund awarded to MC. In addition to funding the pottery analyses (grant G2001/6541 awarded to MC), the Australian Institute of Aboriginal and Torres Strait Islander Studies provided Conference Support funds to MC to attend the Global Perspectives on the Archaeology of Islands conference in Auckland, December 2004, where the ideas presented in this paper were first presented.

References cited

Allen, J. (1972). Nebira 4: An early Austronesian site in central Papua. *Archaeology and Physical Anthropology in Oceania* 7(2): 92-124.

Barham, A.J. (2000). Late Holocene maritime societies in the Torres Strait Islands, northern Australia – cultural arrival or cultural emergence? In S. O'Connor and P. Veth (eds.), *East of Wallace's Line: Studies of Past and Present Maritime Cultures of the Indo-Pacific Region*, pp. 223–314. Rotterdam: AA Balkema Press. *Modern Quarternary Research in south-east Asia*, series No. 16.

Barham, A.J. and D.R. Harris (1983). Prehistory and palaeoecology of Torres Strait. In P.M Masters and N.C Flemming (eds.), *Quaternary Coastlines and Marine Archaeology: Towards the Prehistory of Land Bridges and Continental Shelves,* pp. 529–557. London: Academic Press.

Barham, A.J. and D.R. Harris (1985). Relict field systems in the Torres Strait region. In I.S Farrington (ed.), *Prehistoric Intensive Agriculture in the Tropics,* pp. 247–283. Oxford: British Archaeological Reports, Internationsal Series 232 (Vol. 1).

Barham, A.J, M.J Rowland and G. Hitchcock (2004). Torres Strait bepotaim: an overview of archaeological and ethnoarchaeological investigations and research. In I.J McNiven and M. Quinnell (eds.), *Torres Strait Archaeology and Material Culture.* Memoirs of the Australian Museum, Cultural Heritage Series 3(1): 1–72.

Bickler, S.H. (1997). Early pottery exchange along the south coast of Papua New Guinea. *Archaeology in Oceania* 32:151–162.

Bird, D.W. (1996). Intertidal Foraging Strategies among the Meriam of the Torres Strait Islands, Australia: An Evolutionary Ecological Approach to the Ethnoarchaeology of Tropical Marine Subsistence. Unpublished PhD Thesis, Department of Anthropology, University of California, Davis.

Carter, M. (2002) Recent results of excavations on the Murray Islands, eastern Torres Strait and implications for early links with New Guinea: bridge and barrier revisited. In S. Ulm, C. Westcott, J. Reid, A. Ross, I. Lilley, J. Pragnell and L. Kirkwood (eds.), *Barriers, Borders and Boundaries. Proceedings of the 2001 Australian Archaeological Association Annual Conference*, pp. 1–10. Brisbane: Anthropology Museum, University of Queensland. Tempus No. 7.

Carter, M. (2003). The Murray Islands Archaeological Project: results of recent archaeological analyses. *Australian Aboriginal Studies* 2002(2): 75-77.

Carter, M. (2004). North of the Cape and South of the Fly: The Archaeology of Settlement and Subsistence on the Murray Islands, Eastern Torres Strait. Unpublished PhD thesis, School of Anthropology, Archaeology and Sociology, James Cook University, Townsville, Queensland.

Carter, M., P. Veth, A. Barham, D. Bird, S. O'Connor, and R. Bird (2004a). Archaeology of the Murray Islands, Torres Strait: implications for a regional prehistory. In R. Davis (ed.), *Woven Histories, Dancing Lives: Torres Strait Islander Identity, Culture and History*, pp. 234–258. Canberra: Australian Aboriginal Studies Press.

Carter, M., A. Barham, P. Veth, D. Bird, S. O'Connor, and R. Bird (2004b). The Meriam Islands Archaeological Project: excavations on Mer and Dauar, eastern Torres Strait. In I. McNiven and M. Quinnell (eds.), *Torres Strait Archaeology and Material Culture*, pp. 163–182. Memoirs of the Queensland Museum, Cultural Heritage Series 3(1).

David, B. and H. Lourandos (1999). Landscape as mind: land use, cultural space and change in north Queensland prehistory. *Quaternary International* 59: 107–123.

David, B., I. McNiven, R. Mitchell, M. Orr, S. Haberle, L. Brady and J. Crouch (2004). Badu 15 and the Papuan-Austronesian Settlement of Torres Strait. *Archaeology in Oceania* 39:65–78.

David, B., J. Crouch and U. Zoppi (2005). Historisizing the spiritual: *Bu* shell arrangments on Badu Island, Torres Strait. *Cambridge Archaeological Journal* 15(1): 71–91.

David, B., and M. Weisler (2006). Kurturniaiwak (Badu) and the archaeology of villages in Torres Strait. *Australian Archaeology* 63: 21–34.

Denham, T. (2004). The roots of agriculture and arboriculture in New Guinea: looking beyond Austronesian expansion, Neolithic packages and indigenous origins. *World Archaeology* 36(4):610–620.

Denham, T, S.G. Harberle, C. Lentfer, R. Fullagar, J. Field, M. Therin, N. Porch ane B. Winsborough (2003). Origins of agriculture at Kuk swamp in the Highlands of New Guinea. *Science* 301: 189–193.

Dickinson, W.R.H. (2003). Nature of Sand Tempers in Sherds from the Murray Islands East of Torres Strait. Unpublished report produced for Melissa Carter.

Ghaleb, B. (1990). An Ethnographic Study of Mabuiag Island, Torres Strait, Northern Queensland. Unpublished PhD Thesis, Institute of Archaeology, University of College London, London.

Haddon, A.C (ed.) (1901-1935). *Reports on the Cambridge Anthropological Expedition to Torres Straits*, 6 volumes. Cambridge: Cambridge University Press.

Haddon, A.C. 1912 *Reports on the Cambridge Anthropological Expedition to Torres Straits*. Volume 4, Art and Crafts. Cambridge: Cambridge University Press.

Idriess, I. (1947). *Drums of Mer*. London: Angus and Robertson Ltd.

Irwin, G. (1985). *The Emergence of Mailu as a Central Place in Coastal Papuan Prehistory*. Terra Australis 10. Canberra: Department of Prehistory, Research School of Pacific and Asian Studies, The Australian National University.

Jukes, J.B. (1847). *Narrative of the surveying voyage of H.M.S Fly during years 1842-1846*. London: T&W Boone.

Kirch P.V. (2001). *Lapita and its transformations in Near Oceania: archaeological investigations in the Mussau Islands, Papua New Guinea, 1985-88*. Volume 1, Introduction, Stratigraphy, Chronology. Archaeological Research Facility Contribution No. 59. Berkeley: University of California.

Lawrence, D. (1998). Customary exchange in Torres Strait. *Australian Aboriginal Studies* 2:13–25.

Lebot, V. (1999). Biomolecular evidence for plant domestication in Sahul. *Genetic Resources and Crop Evolution* 46(6):619–628.

Lilley, I. (1999). Late Holocene Transformations in Torres Strait (or the long arm of Lapita...). Paper presented at the Australian Archaeological Association conference, Mandurah, Western Australia.

MacGillivray, J. (1852) *Narrative of the Voyage of HMS Rattlesnake*. Volume 2. London: T&W Boone.

Mackenzie, D.E. and R.W. Johnson (1984). Pleistocene volcanoes of the western Papua New Guinea Highlands: Morphology, geology, petrography and modal and chemical analyses. *Bureau of Mineral Resources, Australia Report* 246.

McNiven, I.J. (2006). Dauan 4 and the emergence of ethnographically-known social arrangements across Torres Strait during the last 600-800 years. *Australian Archaeology* 62: 1–12.

McNiven, I, W.R Dickinson, B. David, M. Weisler, F. Von Gnielinski, M. Carter and U. Zoppi (2006). Mask Cave: Red-slipped pottery and the Australian-Papuan settlement of Zenadh Kes (Torres Strait). *Archaeology in Oceania* 41: 49–81.

Moore, D.R. (1972). Cape York Aborigines and Islanders of the western Torres Strait. In D. Walker (ed.), *Bridge and Barrier: The Natural and Cultural History of Torres Strait*, pp. 327–344. Canberra: Research School of Pacific Studies, Department of Biogeography and Geomorphology, Australian National University.

Moore, D.R. (1979). *Islanders and Aborigines at Cape York. An ethnographic reconstruction based on the 1848–1850 'Rattlesnake' Journals of O.W. Brierly and information he obtained from Barbara Thompson.* Canberra: Australian Institute of Aboriginal Studies.

Moore, D.R. (1984). *The Torres Strait Collections of A.C. Haddon: A Descriptive Catalogue.* London: British Museum Publications.

O'Connor, S. (1999). A diversity of coastal economies: shell mounds in the Kimberley region in the Holocene. In J. Hall and I.J. McNiven (eds.), *Australian Coastal Archaeology*, pp.37–50. Research Papers in Archaeology and Natural History No. 31. Canberra: ANH Publications.

Parr, J.F. and M. Carter (2003). Phytolith and starch analysis of sediment samples from two archaeological sites on Dauar Island, Torres Strait, northeastern Australia. *Vegetation History and Archaeobotany* 12(2): 131–141.

Rowland, M.J. (1984). Archaeological survey and excavations on Moa and Naghi Islands, Torres Strait, November 1–25. 1981. Unpublished manuscript, Department of Community Services, Brisbane.

Rowland, M.J. (1985). Archaeological investigations on Moa and Naghi islands, Western Torres Strait. *Australian Archaeology* 21: 119–132.

Rowland, M.J. (1999). The Keppel Islands—'a 3000 year' event revisited. In J. Hall and I.J. McNiven (eds.), *Australian Coastal Archaeology*, pp.141–155. Research Papers in Archaeology and Natural History No. 31. Canberra: ANH Publications.

Spriggs, M. (1995). The Lapita culture and Austronesian prehistory in Oceania. In P. Bellwood, J.J. Fox and D. Tryon (eds.), *The Austronesians: Historical and Comparative Perspectives*, pp. 112–133. Canberra: Department of Anthropology.

Stuiver, M, and P.J. Reimer (1993). Extended 14C data base and revised CALIB 3.0 14C age calibration program. *Radiocarbon* 35(1):215–230.

Stuiver, M., P.J. Reimer, E. Bard, J.W. Beck, G.S. Burr, K.A. Hughen, B. Kromer, G. McCormac, J. van der Plicht and M. Spurk (1998). INTCAl98 Radiocarbon age calibration, 24,000–0 cal BP. *Radiocarbon* 40(3):1041-1083.

Ulm, S. (2002a). Marine and estuarine reservoir effects in central Queensland, Australia: determination of δR values. *Geoarchaeology* 17(4):319–348.

Ulm, S. (2002b). Calibrating marine radiocarbon dates: a guide to Australian δR values. *Australian Association of Consulting Archaeologists Inc. Newsletter* 89:10–14.

Ulm, S. (2006). *Coastal Themes: An Archaeology of Southern Curtis Coast, Queensland.* Terra Australis 24. ANU E Press.

Vanderwal, R.L. (1973a). The Torres Strait: protohistory and beyond. In P. Lauer (ed.), *Occasional Papers in Anthropology 2,* pp. 157–194. St Lucia: Anthropology Museum, University of Queensland.

Vanderwal, R.L. (1973b). Prehistoric Studies in Central Coastal Papua. Unpublished PhD thesis, Australian National University, Canberra.

Walker, D. (ed.) (1972). *Bridge and Barrier: The Natural and Cultural History of the Torres Strait.* Canberra: Research School of Pacific Studies, Department of Biogeography and Geomorphology, The Australian National University.

White, J.P. (1972). Ol Tumbuna. *Archaeological excavations in the eastern Central Highlands, Papua New Guinea.* Canberra: Department of Prehistory, Research School of Pacific Studies, Australian National University. *Terra Australis* No. 2.

Wurm, S.A. (1977). Torres Strait—a linguistic barrier? In D. Walker (ed.), *Bridge and Barrier: The Natural and Cultural History of Torres Strait*, pp. 345–366. Canberra: Research School of Pacific Studies, Department of Biogeography and Geomorphology, Australian National University.

Chapter 6

Colonization, environment and insularity: prehistoric island use in the Great Barrier Reef Province, Queensland, Australia

Michael J. Rowland, Cultural Heritage Coordination Unit, Natural Resources and Water, Queensland

Introduction

Archaeological research has been undertaken on a limited number of islands throughout the Great Barrier Reef Province, Queensland, from Torres Strait to Facing Island (figure 6.1). Significant studies are limited to the Torres Strait, Whitsunday, Keppel and Northumberland Islands (see Rowland 1996: 193, table 11.2). I commenced archaeological investigations on islands in the Great Barrier Reef Province in 1978, predominately on the Keppel Islands (Rowland 1980, 1981, 1982a, 1982b, 1983a, 1985a, 1992, 1996, 1999a, 2002), more briefly in the Torres Strait (Rowland 1985b; Barham et al. 2004), on the Percy Islands (Rowland 1984), Dunk Island (Rowland 1989) and the Whitsunday Islands (Rowland 1986). At this time there was little evidence to suggest when, why and how the islands were used. Archaeological data from offshore islands elsewhere in Australia suggested they might have been occupied largely on a seasonal basis for specialized food gathering and perhaps only within the last few thousand (ca. 2000) years. However, this did not prove to be the case on islands of the Great Barrier Reef Province.

Coastal Queensland was inhabited by at least 40,000 years ago. The majority of islands within the Great Barrier Reef Province, however, did not form until about 6000 years ago at which time they can be considered uninhabited islands potentially colonisable by human groups (Rowland 1996: 193). On this basis I previously used an island colonization model as a heuristic device to direct discussion on how islands of the Great Barrier Reef Province were used (Rowland 1996). This is a useful approach (see also Anderson 2004a) that lead to consideration of a series of basic questions such as when were the islands first occupied, why were they used, how did people get to them and what population structures and economies developed on them? The same approach is taken in this paper, incorporating information that has become available in the last decade. Based on this review and other evidence I then critique recent views that deconstruct the role of insularity and environmental change in island research. Jared Diamond at an earlier islands conference in Auckland in 1989 remarked that 'New Zealand is as close as we will get to the opportunity to study life on another planet' (Diamond 1990: 2). The otherwise eloquent 1998 Pulitzer Prize winning author (Diamond 1997), might on this occasion have been seriously misinterpreted. Diamond of course was not referring to New Zealanders as aliens, but was alluding to two key concepts in island studies—island environments and isolation—that are also a focus of this paper.

Islands of the Great Barrier Reef Province

The Queensland coastline extends over ca. 10,000 km through 19° of latitude. From 9°S to 24°S it is dominated by the juxtaposition of the Great Barrier Reef. Between the reef and the mainland is an area of ca. 230,000 km^2 and numerous islands. North of the Tropic of Capricorn there are over 600 continental and reef islands ranging in size from 50,000 hectares to less than 1 hectare totalling in excess of 2,500 km^2. Most islands are smaller than 400 hectares and tend to be clustered into small groups of large islands surrounded by a scatter of smaller islands (Rowland 1996: 191, table 11.1). My focus is on the

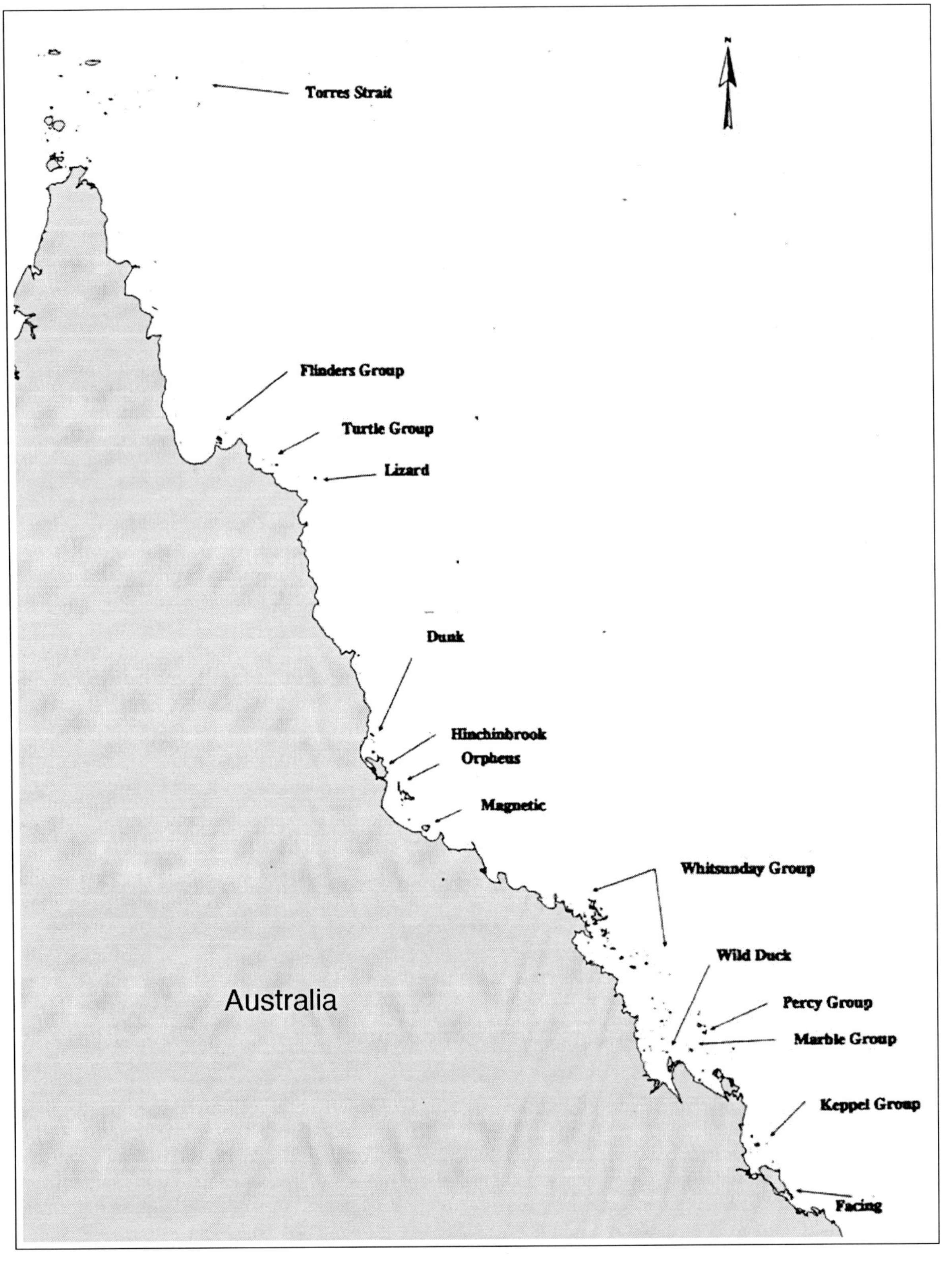

Figure 6.1: Great Barrier Reef Province and islands.

continental islands north of the Tropic of Capricorn not the sand barrier islands of Southern Queensland.

The Great Barrier Reef Province is characterized by high levels of micro-environmental diversity. Tropical and sub-tropical climatic processes, the relative absence of rocky coastlines, the presence of tidal plain coasts in restricted areas and the dominance of reefs and islands are major features of the Province. Environmental changes during the Pleistocene would at times have been dramatic and impacted substantially on human populations occupying the coast. Unfortunately, such changes would also have destroyed a substantial part of the archaeological record so that impacts cannot be easily determined. Holocene environmental changes are generally thought to have been less pronounced but fluctuations occurred in sea-levels, climatic regimes, tidal patterns and coastal resources (Rowland 1996: 191-192). I have proposed for some time that the *scale* and *amplitude* of Holocene environmental changes and their impacts on human populations have been underestimated in the Australian context (e.g., Rowland 1983b, 1989, 1996, 1999a, 1999b) and return to that issue later in this paper.

When were islands of the Great Barrier Reef Province first occupied?

Sea levels on the Queensland coast were last at, or close to present levels about 125,000 years ago, then oscillated, reaching a low of −150 m about 18–20,000 years ago with present levels reached about 6000 years ago. Sea-level change over this period was generally gradual but also punctuated by periods of more rapid retreat and advance. The Queensland coast was occupied by at least 40,000 years ago but sea-level change has destroyed most of this evidence. A date from Hook Island in the Whitsunday Island group of over 9000 years and from Border Island of ca. 7000 years ago (Barker 2004: 63, table 6.1, 107, table 8.1), and recent dates suggesting the Torres Strait Islands were permanently occupied between 8000 and 6000 years ago (David et al. 2004) support the view that the coast was being used prior to the phase of major island development. The majority of Australian offshore islands, including those of the Great Barrier Reef Province, however, were not used until after 6000 years ago.

The earliest evidence for occupation of any of the offshore islands is that from the open site of Mazie Bay, North Keppel Island (Rowland 1981, 1983a, 1992, 1999a). At this site charcoal associated with an unexcavated cultural layer at a depth of 3.5 m is dated at 4274 ±94 BP (NZ 456). This date is 1500–2000 years later than the orthodox view of sea level stabilization. Dates for coastal sites on the adjacent mainland are predominately more recent (Ulm and Reid 2000). Given this pattern of dates it is perhaps not surprising that Beaton (1985) hypothesized a time-lag in the development of coastal economies. However, there are alternative views. It is unlikely that coastal resources would have taken long to establish. It is also unlikely that Pleistocene and early Holocene coastal resources would have been overlooked, a view initially accepted by Beaton (1978: 146) but subsequently rejected (Beaton 1985). It can also be no surprise that more sites are preserved from recent times and that some types of sites and remains (e.g., shells) might not survive particular chronological thresholds.

The orthodox view is that Australian sea-level changes over the last 6000 years were relatively minor (Chappell 1982: 71-73). In fact, the scale and magnitude of change in both climate and sea-level may have been greatly underestimated and the impacts of cyclones, storm surges and other erosive effects would also have had substantial impacts on the coast (Rowland 1989, 1996, 1999a). Relative sea level and coastline stability might only be inferred for the last 2000 years or less. Normal decay factors might account for less Pleistocene sites than Holocene sites and while an increase in site numbers throughout the Holocene may have resulted from a gradual growth in population there is no suggestion that major population transformations occurred. A need for populations to disperse more widely in response to fluctuating environmental conditions might also account for an increase in site numbers after 3500 years ago (Rowland 1996: 195). Evidence from the Keppel's adds to dilemmas relating to sea levels and time lags. North Keppel was not used for around 1500–2000 years after the time of accepted sea level stabilization. None of the ten sites excavated on South Keppel Island date to older than 800 BP. The fact that no early site has been located on South Keppel is surprising given the close proximity of all islands in the group. Perhaps a date of 6000 BP for generalized sea level stabilization does not apply. There is some evidence for a high level at 4500 years and a transgression at 3500 BP. The high level at 4500 BP might equate with the earliest date from Mazie Bay of ca. 4200 and the transgression at 3500 BP might equate with the hiatus in dune structure and occupation deposits at this time (Rowland 1980, 1981, 1983a, 1999a).

Newly available habitats and the attraction of economic resource packaging on islands would have provided inducement to exploit the islands whether formed at 6000 years ago or later. Open site evidence may however have rarely survived. While a time lag in the development of coastal economies cannot be easily dismissed it is based on missing evidence and certainly no generalized pattern is evident at this stage for the entire coastline. Contrary to Beaton (1985) then, site numbers, antiquity and distribution of sites on islands of the Great Barrier Reef Province and adjacent mainland coast must, at least in part, be a reflection of environmental selection pressures over the last 6000 years. Better preservation of more recent sites and minor population increases through time might lead to the unremarkable observation that more sites would be preserved through time. It is now more widely accepted that apparent 'intensification' might in part be a result of research focus and preservation (e.g., Bourke 2004; Ward 2004). Many sites that once existed must now be lost and there is a need to focus on factors that determine the characteristics of early coastal sites and where they might be located (Rowland 1996).

Why were islands of the Great Barrier Reef Province occupied?

People exploiting the Pleistocene coastline may simply have retreated to the new islands as sea levels rose. Hook Island in the Whitsunday group appears to have been continuously occupied from 9000 years ago (i.e., prior to it becoming an island) (Barker 2004) and it has been proposed that the Torres Strait Islands were occupied in three phases commencing around 8000 BP (David et al. 2004). Most islands however, were not occupied until well after 6000 years ago. Initially few islands appear to have been permanently occupied and may simply have been incorporated into the coastal economies of adjacent mainland groups. However, within the last 2000 years or earlier Magnetic, Hinchinbrook, Goold, the Keppel, the Whitsunday and Torres Strait Islands at least were permanently occupied (Rowland 1996: 193).

Coastally adapted populations may have undergone little change to adapt to an island environment. Seasonal use might represent an adaptive phase prior to full occupation on some islands. Reasons for dispersal to and permanent occupation of islands include warfare, disease, population growth and even a supposed natural tendency to explore. Mid- to late- Holocene population restructuring and increasing use of marginal environments are favoured general explanations for the use of islands (see Lourandos 1997) but this is far from proven.

Beaton (1985: 18) proposed that occupation of the islands of Princess Charlotte Bay at 2500 BP ago had to 'await' the arrival of the outrigger canoe of Papua-Melanesian origin. There are cause and effect issues with this explanation, besides which bark canoes were in use in the Reef Province by at least 4200 BP years ago that had the capacity to reach the islands of Princess Charlotte Bay (Rowland 1984, 1987, 1995).

For the New South Wales coast, Sullivan (1982: 16) proposed that islands were visited only in summer and were the focus of fishing and bird gathering. She argues this pattern was established only within the last 2000 years and was related to a widespread change in technology facilitated by an exchange of knowledge between local groups of an expanding population. Sullivan's model also suffers cause and effect problems and is not compatible with evidence from islands off the Queensland coast, many of which were occupied on a permanent or semi-permanent basis for some thousand of years. The range of shellfish, fish, sea mammals and other resources exploited on the Keppel and Whitsunday Islands indicate broad-based, permanent rather than seasonal economies (Barker 2004; Rowland 1996).

Sullivan (1982: 10) also proposed that people travelled further to islands in the Reef Province than on the New South Wales coast because it is more sheltered. A more complex explanation is however required. For example, the Percy Islands are a long way off the coast, they cannot be seen from the mainland, and are not well protected from oceanic conditions nor tides and winds emanating from the mainland. They can be reached by island hopping but a minimum distance of 27 km is still required. The Percy Islands have adequate to abundant coastal resources, minimal terrestrial resources, a few fine-grained stone resources and turtles nest on the islands. Based on the small size of the islands and limited resources it is difficult to imagine the islands being particularly attractive given the lengthy and difficult sea crossing. Therefore, to suggest people travelled further within the Reef Province because it was more sheltered falls short of a complete explanation in motivational and ecological terms (Rowland 1984).

Jones (1976, 1977), following MacArthur and Wilson used *size* and *distance* to define a dichotomy between Tasmanian islands occupied permanently and those occupied seasonally. He found that the minimum island size necessary to support an autonomous population in Tasmania was 77–90 km^2 and that residential islands tended to be closest to the mainland. In the Reef Province much smaller islands supported residential populations. Limiting distances for Tasmanian watercraft were in the vicinity of 5–8 km, but evidence from the Percy Islands suggest limiting distances at least four times greater within the Reef Province and other types of watercraft may have substantially extended these distances (Rowland 1984, 1987, 1995). The Percy Islands are the most distant islands off the east coast of Australia and since they were visited no effective limiting distance for watercraft can be identified in the area. Islands as small as 3 hectares and 13 km from the mainland were used, as were continental islands 50 km from the mainland (Rowland 1996: 202, table 11.5).

Population pressure (Beaton 1983), changes in social relationships and their intensification (Barker 2004; Lourandos 1997), changes in technology (Beaton 1985; Sullivan 1982), environmental and ecological changes (Border 1999; Rowland 1996, 1999a and b) and culture contact (Rowland 1987, 1995) are among a range of factors that explain the use of islands in the Great Barrier Reef Province but a convincing holistic explanation has yet to be achieved.

How did people get to the offshore islands?

'No boats, no island colonisation' (Anderson 2004b: 264). Archaeological evidence for watercraft is unfortunately rare and inferences must therefore be drawn from indirect evidence. A variety of watercraft was used in the Great Barrier Reef Province. Single-piece bark canoes were sighted along most of the coastline, three-piece sewn bark canoes were known around the Whitsunday Islands, double outriggers extended as far south as Cape Melville and single outriggers were in use at least as far south as the Whitsunday Islands and possibly beyond.

Little is known regarding the antiquity or capabilities of the various canoe types. However, North Keppel Island was being visited by at least 4200 years ago (13 km distant) and evidence from the Percy Islands demonstrates that 50 km trips from the mainland or at least a 27 km island-hopping journey to these islands was possible (Rowland 1984, 1987, 1995).

Bowdler (1995) was surprised that people colonized Australia 40,000 years ago across a 65 km barrier but only began to re-cross minimal water barriers in the last 4000–3000 years and with regularity only in the last 1000 years. The coast was important in the colonization of Australia (see Bowdler 1977, 1990), and remained so. It may therefore be the case that the apparent lack of watercraft use may to some degree simply reflect missing evidence. Nevertheless, it is likely that there were changes in watercraft types and use throughout prehistory and new watercraft technologies may have been introduced along the Queensland coastline within the last few thousand years (Rowland 1987, 1995).

Population development and survival on the offshore islands

Population extinction may occur on islands where colonising populations are small, where island area is small and/or where distance from a source of new migration is considerable. Islands in the Pacific and elsewhere in Australia were occupied and then abandoned indicating a failure to establish viable populations. No similar islands have to date been located on the Queensland coast and are unlikely to be (Rowland 1996). Following colonization of new areas, population biologists propose two basic models of population growth. The first involves initial rapid population growth that reduces the time over which the chance of random extinction is highest and prior to controls imposed by environmental resistance. The second model involves a slow initial growth rate followed by a faster rate at greater population size until a ceiling is imposed by density dependent factors. The first model has at least one significant implication for the intensification debate (Lourandos 1997). It suggests that in response to new resources or habitats, populations may respond in the short-term by increasing growth rates. Thus instead of seeing increased use of offshore islands in the Holocene as part of an Australian-wide trend toward population growth and/or intensification it might be more appropriate to see the islands as special cases of new habitats capable of supporting relatively rapid population growth. 'Intensification' might therefore be seen as a regional response to new habitats/resources and may have occurred periodically as new areas became available. In this way 'intensification'/population increase/changes in social relationships, might all be seen as a response to new economic opportunities which varied regionally and through time (Rowland 1996: 201).

Jones (1976: 259-260) proposed that isolated populations whose numbers exceed 500 could survive indefinitely, while those with fewer than

300 would become extinct. Other studies, albeit mainly computer simulations, suggest much smaller populations are capable of surviving isolation. Many forms of population control are available to human groups to maintain stability and even major demographic disturbances may not severely disrupt demographic structures provided they are not too abrupt or last too long. There is little reason to believe that small populations could not survive independent of contact with other groups for substantial periods of time. The range of types of watercraft available in the Reef Province suggests that distance was not a factor that would have lead to the permanent isolation of any island population. Nevertheless, evidence from the Keppel Islands suggests a degree of isolation in a relatively small population over a period of at least 700 years (Rowland 1980, 1982b, 2002).

Genetic changes in small populations can be deleterious but others may have no obvious negative effects and in some cases deleterious genes may be eliminated. Furthermore, inbreeding in hunter-gatherer populations may not be uncommon since breeding populations are much smaller than actual populations, while inbreeding may also be avoided by a diversity of social rules (Rowland 1996: 200). Some physical differences in the Keppel Islanders such as the high incidence of auditory exostoses in skulls was initially thought to be evidence of inbreeding in a small population (Rowland 1980: 5). However, recent worldwide research suggests diving in cold water to be the cause of this pathology. Nevertheless, the Keppel Island skeletal remains show a number of physical differences that require explanation. The differences may be attributed to isolation, historical factors, adaptations to an island environment, and/or external contacts but currently the issue remains unresolved (Rowland 2002).

For Tasmanian Islands, Jones (1976: 254) estimated that the minimum area required to support an autonomous population was 48–56 km^2 and that the smallest unit of population that could function as an independent unit was 30–60 people. Within the Reef Province permanent populations were supported on much smaller islands. Population densities in the Reef Province have been estimated at 1 person/25–125 hectares (Rowland 1996: 201, table 11.4). These are high densities by Australian standards but can be explained in terms of the relationship between length of coastline and island area, not island area *per se*. Seashores and islands arguably offer a territorial advantage of between 3 and 4 times compared with mainland areas. Early explorers on the Queensland coast confirm that the offshore islands were more densely populated than many mainland areas (Rowland 1986: 82). Sedentism is an outcome of high population densities and this is a feature of islands of the Reef Province. Huts and groups of huts were noted on Dunk, Palm, Goold and Middle Percy Islands. On South Keppel Island remains of a hut structure have been excavated (Rowland 1980) and Roth photographed substantial huts on South Keppel in 1897 (figure 6.2).

The massive fish traps of Torres Strait and Hinchinbrook Island also indicate a degree of permanency on these islands (Rowland 1996: 201). Not all islands of the Reef Province however supported permanent, sedentary populations. Some may have supported 'sedentary-cum-mobile' economies, in which substantial periods may have been spent on the islands and seasonal visits then made to other islands and the mainland. This pattern in particular is suggested for the Marble Islands (Border 1999).

The development of island economies in the Reef Province

The Reef Province is characterized by fairly constant broad ecological variables but also considerable habitat diversity. It is an area of high biomass, with 'upwelling' throughout the area, ensuring high productivity near reefs and islands. The islands generally have a reduced terrestrial flora and fauna but high density of easily accessible coastal resources. Sea birds are a significant feature of many continental islands and most coral cays, while three species of turtle, the dugong and over 1000 species of tropical fish inhabit the waters of the region (Rowland 1996).

Limiting factors

As noted above, Jones (1976) concluded that for Tasmania island *size* and *distance* from the mainland strongly correlated with island use, thus approximating patterns predicted by island biogeographical models. In the Reef Province by comparison neither island size nor distance were limiting factors to the same extent. Islands as small as 3 hectares and 13 km from the mainland were used, as were continental islands 50 km from the mainland. Size or distance may have determined the extent to which islands were seasonally or permanently exploited but insufficient data is available at present for this to be demonstrated (Rowland 1996: 202, table 11.5).

Figure 6.2: South Keppel in 1897.

Economic systems on the islands

Neither ethnohistoric, nor archaeological investigations can at present reveal a complete picture of economic adaptations made to islands of the Reef Province. Ethnohistoric sources tend to emphasis unique and male dominated activities, for example, while archaeological studies reveal the importance of shellfish, fish and mammals but remain largely mute on the question of plant use. Research on the Keppel Islands and Whitsunday Islands provide some insight into 'typical' island economies of the area.

The Keppel Islands are continental islands, at 23°S, just north of the Tropic of Capricorn. There are two major islands in the group, North Keppel (588ha) and South Keppel (1454ha), together with seven smaller islands comprising a total land area in excess of 20 km^2. Coastal and reef areas associated with the islands were the major focus of resource use, increasing the exploitable area by a factor of 3 to 4 times. However, the islands terrestrial resources cannot be overlooked. The islands comprise a range of vegetative structural communities and these communities contain at least 197 species of plants representing 61 families, of which at least 57 species have been documented as Aboriginal foods. Included among these were a number of potential staples (Rowland 1996: 203. table 11.6). Terrestrial mammals and reptiles are present but limited in diversity and numbers. Bats, sand goannas, the blue-tongued lizard, common tree snakes and carpet pythons are all present today. Wallabies and koalas may have been present in the past and at least 41 species of birds have been recorded on the islands of which at least 29 are known to nest there. Marine mammals include dugong. Turtles nest on the island and include the green turtle, loggerhead, hawksbill and leatherback. Molluscs are abundant in a full range of ecological zones and are well represented in middens. The range of fish available to the Keppel Islanders is unknown, but is generally large in reef waters (Rowland 1996: 203). The Whitsunday Islands comprise 73 islands ranging from 0.4 to 10,935 hectares. A number of excavations have revealed extensive use of coastal resources from 9000 years ago. Shellfish predominate by number but fish and turtle in terms of meat weight estimations. The quantity of faunal terrestrial remains is small but the range extensive. Plant foods are also present including after ca. 520 BP the poisonous *Cycas media* and *Bruguiera gymnorhiza* (Barker 2004: 140ff).

Specialized island economies of the Great Barrier Reef Province

Islands in Australia, with their reduced diversity, have generally been seen only as seasonal resource bases for fishing, birding or turtling or more uniquely as social retreats (McNiven 2000). In the Reef Province however diversity was such that permanent or semi-permanent populations were supported on the islands. This would seem to be the case for at least the Keppel Islands, Percy Islands, Whitsunday Islands, Hinchinbrook/Dunk Island groups and of course the Torres Strait Islands. In these cases there is evidence of more frequent movement within the island groups themselves rather than between the islands and the mainland (see Barker and Schon 1994: 56). Archaeological evidence from the islands points to a broad-based exploitation of island resources focusing on marine foods. Artefacts reflect a similar focus on marine raw materials. Shell artefacts (including fishhooks) made from local sources appear to have been commonly used throughout islands of the Province. Stone toolkits are generally restricted in nature and made of local stone. Other features of these island-based economies include evidence for huts and fishtraps on many of the islands and the presence of a range of watercraft types, and the importance of harpoons (figure 6.3), fishhooks, and nets (Rowland 1996).

Based on the above evidence I was able to define a widespread and unique adaptation to islands of the Great Barrier Reef Province (Rowland 1996) that has subsequently been seen as applicable to other northern Australian islands (Barker 2004; O'Connor 1992). Nevertheless, the timing of the development of these island based economies remains problematic. Mazie Bay on North Keppel Island for example was occupied from at least 4200 years ago to 3500 BP (Rowland 1981: 64-65, figure 1; 1999a). On South Keppel Island a number of sites all date within the last 800 years BP (Rowland 1982a: 45, table 1, 1999a) so that the development of true island based economies might only be seen to date from the last 1000 years or less. Barker (2004: 147) for the Whitsunday Islands argues for a generalized coastal economy from 9000 years ago, replaced by a more specialized marine economy from ca. 3000 BP. Importantly, Ulm (2000: 66) has identified the emergence of 'focused marine resource exploitation' from 3500 BP on the mainland coast. More recently, David et al. (2004) has recognized this date as marking a significant increase in human activity on the Torres Strait Islands. I have argued for some time that sig-

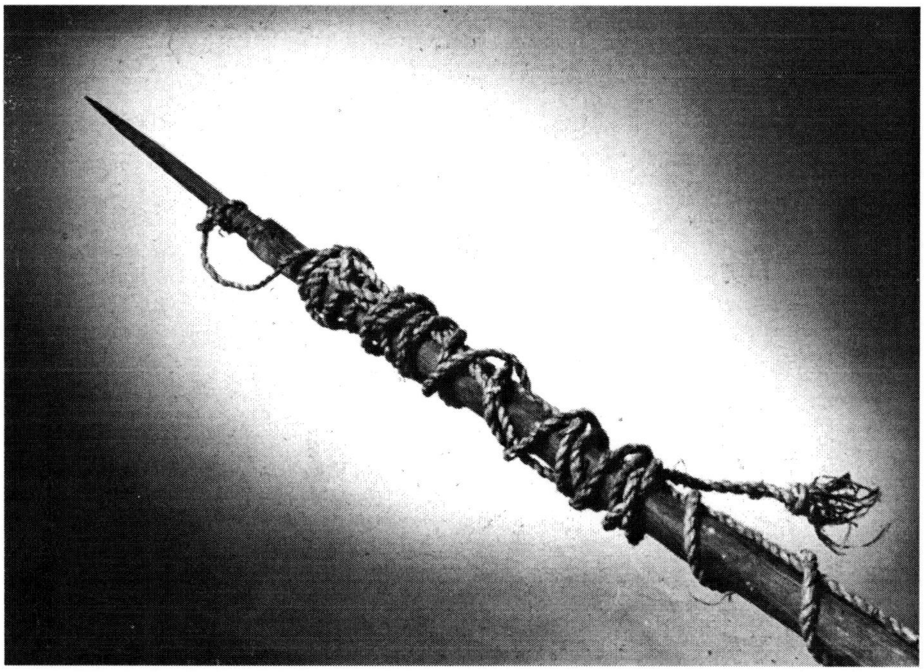

Figure 6.3: Keppel harpoon.

nificant change occurred in the archaeological record at 3500 BP (Rowland 1980, 1983b, 1989, 1996, 1999a and b) and will address this issue again below.

Myths about isolation and other 'straw men'

Terrell and his colleagues (Terrell et al. 1997: 156; Terrell et al. 2001; Rainbird 1999a and b, 2002, 2004) have attacked what they call the 'myth of the primitive isolate'. They have suggested that '[we] should adopt as a working hypothesis the universality of contact and influence as a fundamental feature of human existence' (Terrell et al. 1997: 174) and despite critical comment (e.g., Bellwood 1997; Kirch 1997; Broodbank 1999; Keegan 1999) have gone so far as to suggest that people in the Pacific 'kept more or less in touch with one another ever since the first arrival of people at least 45,000 years ago' (Terrell et al. 2001: 107). Terrell's views in particular appear to have become more entrenched and at times contradictory:

> Nowadays it is generally acknowledged that social, economic, political and sexual intercourse among people near and far has been so much a part of human existence in most places on earth that it would be difficult—and hence pointless?—to continue framing our story as a species as a tale about the origins and descent of separate and distinct human 'populations', 'peoples', 'ethnic groups' or 'races'. (Terrell 2004a: 588)

Elsewhere, he makes the sweeping statement:

> the platitude that being surrounded by water makes island populations inherently more isolated is a tired cliché that ought to be retired from active service not only in anthropology and archaeology, but also in ecology and evolutionary biology. (Terrell 2004b: 219)

It appears in part at least that we have his mother to thank for this perspective who advised him 'it really isn't a good thing to be cut off from other people' (Terrell 2004b: 218)![1]

Ironically, in the collection of papers for which Terrell provides an overview, isolation is a key in

[1] MacArthur and Wilson (1967) and many writers before them have noted that much of the biotic world is insular in the form of habitat patches. I have long found Littlewood's (1972: 22) definition of an isolate useful: 'An isolate is a region characterized by reduced information exchange with other regions. This information can be linguistic, cultural, in the form of matter or energy, or genetic.' Usually the reduction in the rate of information exchange is not complete and isolation can be viewed as nodes of central tendency in the genetic or diffusional landscape. Of course where there is a striking discontinuity in information the 'island' model is approached.

his own 'sleeping giant' hypothesis concerning New Guinea's place in Pacific prehistory:

> Evidently few people lived on the coasts of this huge island/continent during most of human time in this part of the world. Earth and sea conspired to isolate New Guinea, like a sleeping giant, from contact with their neighbors for the first 25,000-35,000 years of settlement. (Terrell 2004c: 605)

Rainbird's (1999b) response to criticism in title and brevity is superficial. While his and Terrell's views are clearly in conflict with the vastness and isolation of the Oceanic world, the human differentiation observed at the time of European contact and evidence that interaction at this time may in fact have been declining. In his recent book on 'fluidity' and 'fusion' in the context of Micronesia, Rainbird ends with a typical post-modern flourish: "I have been attempting to construct a long-term history of a place that I have in the material presented in this volume tried to show does not exist. So, what can I say I have achieved?" (Rainbird 2004: 253).

His deconstruction, based on so many 'straw men' do not offer the anticipated range of fresh insights. Rainbird (2004: 1) refers to the now "discredited" 'islands as laboratories' concept and cites his 1999 paper as confirmation of this. While there is support for his view (e.g., Lape 2004), the general concept is far from discredited and while recognising some of its previous limitations this and related island concepts have been used in recent studies across a range of fields to produce powerful new insights into island use (e.g., Armstrong 2004; Bass 1998; Blackburn et al. 2004; Burney 1997; Dewar 1997; Fitzhugh and Hunt 1997; Fort 2003; Grove 1995; Johnson and Wroe 2003; Kirch 2004; Nunn 2004; Patton 1996; Rolett and Diamond 2004; Sax et al. 2002; Stapp et al. 1999; Steadman et al. 1999; Vitousek 2002; Wardle 2002; Walter 2004, to mention just a few).

Terrell and Rainbird have clearly overstated their case (see Anderson 2004b: 254–257). Whether it is in relation to the Pacific, Australia or elsewhere, I take the view that isolation, geography and history remain important in understanding the past. Even the more moderate critics of isolation that attempt, for example, to reposition New Zealand prehistory within the wider Polynesian context accept that things took off on different trajectories in New Zealand and that insularity and distance were critical factors (Walter 2004). History followed different trajectories for different peoples because of geography and differences among people's environments not because of biological differences among people themselves (Diamond 1997: 25, 2005). While this approach can be caricatured as 'environmental determinism', it is another 'straw man' or cliché of limited on going useful debate (Rowland 1996, 1999a, 1999b).

Geography, history and in particular isolation were important in shaping the human populations of Sunda and Sahul, since as Storm (2001: 377) notes, two clearly distinguishable cranial patterns, an East Asian (Sunda) and Australian (Sahul) one can be defined. Beaton (2000: 182–184; see also Diamond 1993, 1997), has bluntly noted:

> that Australia, as we know, was not on any important ancient trade route. Then, as now, it was not exactly on the way between anywhere and anywhere else. For the traders of the world it was a *cul de sac*, if not entirely a dead end.

Beaton then argues that Tasmania can be seen as an extreme version of Australia. Tasmania was thus in an awkward position in the wider world of cultural systems long before it was isolated by post-glacial sea level rise. Tasmania, like the tips of the three southern continents, all shared, to some degree, the '*cul de sac*' effect and the roots of isolation were therefore likely to have been well established before sundering of the Bassian Plain. Nevertheless, from about 12,000 years ago, the Aboriginal people who occupied Tasmania, Kangaroo Island and some other islands of Bass Strait found themselves cut off by rising sea levels. Because they lacked seaworthy vessels they were then completely isolated. Flannery (1994: 263; see Jones 1977 for the original version) argues at this point that the relatively small size of their island homes drove their evolution in a particular, inexorable and fateful direction, which was towards extinction. However, there is other compelling evidence to suggest that the Tasmanian universe was in fact expanding at the time of European contact (e.g., Bowdler 1995). Gamble (2000: 125) hints that the isolation of Tasmania may have been a deliberate social construction, but over a period of 12,000 years this seems unlikely. The Tasmanians missed out on innovations that did become widespread across southeast Australia (Mulvaney and Kamminga 2001). Elsewhere, the Tiwi of Bathurst Island were not completely isolated but semi-isolated (Morris: 2001). Likewise, the Keppel Islanders were semi-isolated physically, culturally and linguistically from people on the adjacent mainland. Isolation, adaptation to an island ecosystem and the possibility of external contacts, may account for this (Rowland 2002). The Torres Strait Islands have long been seen as a 'bridge or barrier' (e.g., Walker 1972) but recent work (Carter 2001, 2002, Carter et al. 2004a and b, Barham 2000) and especially that of David et al. (2004) confirm my view that history, geography and at various times, isolation played significant roles.

MacArthur and Wilson's (1967) theory of island biogeography has been critically reviewed from the time of its publication. Recent commentators continue to be critical and seek modifications to the theory (Brown and Lomolino 2000; Burney 1997; Lomolino 2000; Whittaker 1992, 2000). Nevertheless, there remains a corpus of island characteristics (including 'isolation') that provide a key to understanding island use. History, geography and isolation therefore will remain key concepts in attempting to understand Australian prehistory. Isolation, whether at the continental or individual island scale, be it physically or culturally imposed is of particular importance. There is no need to discard isolation as an explanatory concept in Australian prehistory in favour of the 'universality of contact' proposed by Terrell and his colleagues for the Pacific. In fact, in Australia isolation will, in a less polemical frame than Flannery has espoused, remain important in explaining both the deep and recent past and the future. At the same time, however, at different times and scales, both internal and external contacts were also significant (Rowland 2002). It is ironic but not surprising that modern technology, air over sea travel, is imposing a new and different pattern of isolation on Pacific Island populations (Anderson 2004b: 263).

Like most people in recent times I have never thought of islands purely as 'a piece of land surrounded by water' (cf. van Dommelen 1999: 250; also Terrell et al. 1997), but real islands do exist and have distinct characteristics. Equally, I have never (cf. Terrell et al. 1997; Terrell et al. 2001) thought of people who occupy islands as 'primitive'.[2] However, to deny the significant characteristics of 'real' islands would be to discard so many things of considerable significance both to the development of science and to our understanding of the relationship between humans and their environments. The importance that the Galapagos and also 40-odd other islands made to the development of Darwin's theory of evolution cannot be denied (Armstrong 2004: 246-247). Further, while Terrell et al. 2001 attempt to deny physical differences between Pacific islanders we may note that while racism, can be competitive, exclusive and derogatory, it can equally be a positive celebration of our diversity (Oppenheimer 2004: 3). To deny human diversity would be to undermine the very basis of anthropology/archaeology. It would also clearly be irresponsible to ignore the fact that inhabitants of small Pacific islands are already close to the limits of existence because of their small size, isolation and the effects of modern sea level rise. Thus we still have much to learn from the so-called 'Mystery Islands' of the Pacific (Anderson 2001; Diamond 2005). That relative isolation played an important role in human history until recent times (whether on islands or continents) can be seen most clearly in the impact of introduced diseases on populations around the world including Australia (Campbell 2002). The major period of impact was only in the last five centuries and reached its height in the 16th, 17th and 18th centuries (McMichael 2001; see also Diamond 1997; Karlèn 1995).

While there has been overemphasis on some aspects of island biogeography, Terrell and Rainbird, in particular, have set up a number of 'straw men' that have little relevance to recent island studies. Before reducing true islands to a metaphor, we should recall (while noting some outdated language) the very basic features of islands so well expressed by Fosberg (1963: 5; see also Carlquist 1965: 3-6 and Simberloff 1974: 162):

> Relative isolation, limitation in size (space resource); limitation in, or even absence of certain other resources; limitation in organic diversity; reduced interspecies competition; protection from outside competition and consequent preservation of archaic, bizarre, or possibly ill-adapted forms; tendency toward climatic equability; extreme vulnerability, or tendency toward great instability when isolation is broken down; and tendency toward rapid increase in entropy when change has set in. The limited size makes relatively small changes capable of rather profound general effects; in other words, the buffering effects of great size and density are lacking.

[2] Of course the interesting question remains in island contexts as to whether or not people tend to retain cultural items that are outdated elsewhere or conversely whether changes are more readily accepted because of small population size (e.g., Vayda and Rappaport 1963).

Ironically, so many of these characteristics relate to Terrell's (1997) 'postponed agenda' of human biogeography that still has much to offer.

Not casting around for an environmental explanation—more 'straw men'

Barker (1999: 120) has stated that "archaeologists, casting around for an environmental explanation to fit the late human adaptation to coasts and late Holocene change, have seized upon sea level change as an obvious correlate to mid/late Holocene coastal occupation".

Barker nominates me (specifically Rowland 1983b, 1999b) as one of the most consistent advocates of environmental processes acting as agents of change in Australia's coastal archeological record. In contrast, he sees little significant correlation between environmental change and change in the archaeological record. He argues that marine ecosystems, especially in the tropics, are extremely resilient, diverse and adaptable. The archaeological evidence from the Whitsunday Islands he argues clearly demonstrates that a broad range of coastal resources were available throughout the Holocene and were largely unaffected by the documented environmental changes recorded for the last 10,000 years (Barker 1999: 121, 1996). The evidence for this however is unconvincing (Beaton 1995: 804; Cane 2001: 157–158; Rowland 1996: 195) and given the extent of changes at the Pleistocene/Holocene transition unlikely (see papers in Allen and O'Connell 1995).

Barker (1999: 13–14) argues that marine resources are inherently flexible, resilient and quick to re-establish in the face of environmental changes in both the short and long term. Likewise, he argues prehistoric hunter-gatherers probably dealt with both short- and long-term dynamic changes to their landscape. In common with Lourandos (1997) Barker considers late Holocene archaeological changes are best explained in terms of a complex set of sociocultural factors for which environmental change is but one guiding influence. He relates the changes to a demographic restructuring of populations in the region after ca. 3000 BP that can be viewed in terms of increasing 'complexity' of sociocultural relationships, and possibly as the outcome of population increase. He then suggests that maritime/coastal inter-regional interaction may have placed the Whitsunday area on the periphery of wider dynamic systems of exchange and interaction, with links to north-eastern Cape York Peninsula, Torres Strait and Melanesia (Barker 1996). He emphasizes that the Aboriginal people of the Whitsunday Islands were not passive receivers of ideas and material from the north but that their participation should be viewed in the context of a greater degree of regional interaction, in which the flow of ideas and materials was probably mutual (Barker 1996: 38). While this is an impressive scenario it is difficult to link it to the archaeological record. In fact a reanalysis of stone artefact discard from Nara Inlet 1 by Lamb and Barker (2001: 42) indicates that greatest stone artefact discard occurred between 8990 BP and 7190 BP and thus significantly prior to the late Holocene (contra Barker 1996). Discard densities continued to decline steadily until after 520BP when the pattern is reversed and there is a subsequent increase in discard. Stone material from the South Molle Island quarry was traded between the islands and does not show a distribution linking it with broader patterns of exchange (Barker and Schon 1994). Significantly, a detailed palaeoenvironmental record from Whitehaven Swamp, Whitsunday Island shows no evidence for increased burning coinciding with the time suggested by Barker for intensification (Genever et al. 2003). The top 3 m of sediment accumulated in Nara Inlet in the last 3000 years but at a rate that slowed toward present (Heap et al. 2001) but no reference is made to this by Genever et al. (2003) or Barker (2004).

There can no longer be any doubt that environmental changes, very often abrupt, occurred during the Holocene and had significant effects on past populations (Alley et al. 2003; deMenocal 2001; Haug et al. 2003; Hodell 2001 et al.; Lockwood 2001; Rial 2004; Rowland 1999b). Over 50 globally distributed paleoclimate records have revealed periods of significant rapid climate change characterized by polar cooling, tropical aridity and major atmospheric circulation changes (Mayewski et al. 2004). The tropics played a significant role in these changes (Hendy et al. 2002; Hughen et al. 2004; Jennerjahn et al. 2004). Sea levels also varied throughout the Holocene. Baker and Haworth (1997, 2000a, 2000b) have concluded that sea level was 1.7 m above present at ca. 3500 to ca. 3300 BP and fell rapidly about that time. They have subsequently identified this as a global rather than regional phenomenon (Baker and Haworth 2001a; Baker et al. 2001b: 269). This evidence correlates well with the stratigraphic sequence from Mazie Bay on South Keppel Island where I have long proposed a hiatus in dune development and occupation deposits at ca. 3500BP (figure 6.4) (Rowland 1980, 1983b, 1996, 1999a).

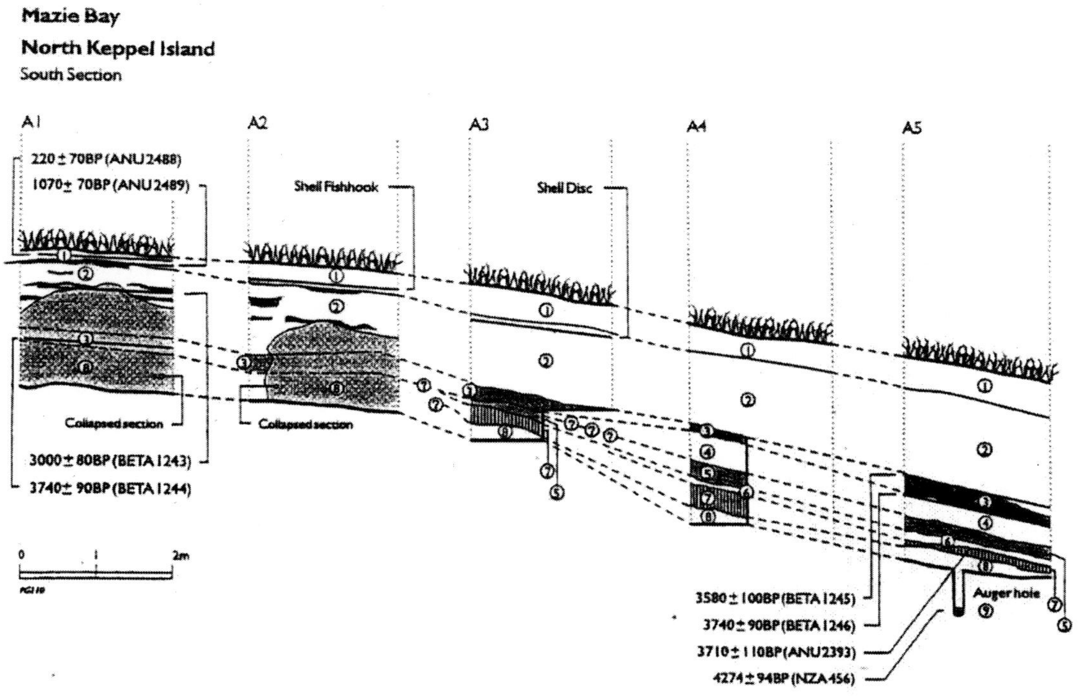

Figure 6.4: Mazie Bay soil section, showing hiatus after level 3, ca. 3500 BP.

Significant changes also occurred on the Whitsunday Islands at around this time (Barker 1996, 2004), in Tasmania (Johnson, and Wroe 2003) in the Torres Strait Islands (David et al. 2004: 74) and through many other areas of Australia and the Pacific that can be linked to environmental change.

Environmental change was a trigger for past cultural changes. However, I have never cast around for environmental explanations nor implied a direct relationship between environmental change and change in the archaeological as suggested by Barker (1996). I have in fact considered economic adaptation (Rowland (1982b), population change (Rowland 1989), culture contact (Rowland 1987, 1995) and insularity (Rowland 2002) as factors that might explain the pattern of island use within the Great Barrier Reef Province.

Conclusion

John Terrell and Paul Rainbird verge on providing a metaphorical story of the past devoid of history, geography and isolation and therefore undermining the most appealing of all aspects of the human condition, cultural diversity. It still remains tempting to study cultural diversity on and between islands and archipelagos. Bryce Barker verges on taking us from the old 'unchanging man in an unchanging environment' to 'unchanging man in a changing environment'. This is unsatisfying and unnecessary. While we continue to struggle to draw convincing correlations between cultural change and environmental change we should not avoid the issues because of some outdated 'straw man' view of past scholarship.

Aspects of a colonization model will continue to be useful in understanding how islands were used (see Anderson 2004a). Insularity should never be an *a priori* assumption of island use but it is apparent that isolation was important in human history generally. A global review of colonization demonstrates clearly that smaller and more distant islands were the last to be occupied (Gamble 1993). We all seek to avoid single factor explanations for events in the past and at a minimum must consider the synergies between history, geography and environment. However, isolation was a feature of human history and as Diamond (2005: 525) has so powerfully demonstrated archaeologists are in a unique position to address the issue.

Acknowledgements

This review is dedicated to the original inhabitants and ancestors of Tierra del Fuego, Madagascar, Easter, Pitcairn, Henderson and many other islands. They might be bemused by calls to think of islands only in a metaphorical sense and would not have had to make excuses for being fascinated by pieces of land surrounded by water. Their attempts to overcome isolation and to adapt to unique environments will continue to contribute to human history long after the passing fashions of academia.

I wish to thank the Department of Natural Resources and Water for supporting my attendance at the islands conference in Auckland. Special thanks as always are due to the Woppaburra of the Keppel Islands.

References cited

Allen, J. and J.F. O'Connell (1995). Transitions. Pleistocene to Holocene in Australia and Papua New Guinea. *Antiquity* 69 Special Number 265.

Alley, R.B., J. Marotszke, W.D. Nordhaus, J.T. Overpeck, D. M. Peteet, R.A. Pielke, Jr, R.T. Pierrehumbert, P.B. Rhines, T.F. Stocker, L.D. Talley and J.M. Wallace. (2003). Abrupt climate change. *Science* 299: 2005–2010.

Anderson, A.J. (2001). No meat on the beautiful shore: The prehistoric abandonment of subtropical Polynesian Islands. *International Journal of Osteoarchaeology* 11: 14–23.

Anderson, A.J. (2004a). It's about time: the Indo-Pacific Colonisation Project. In T. Murray (ed.), *Archaeology From Australia*, pp. 3–17. Melbourne: Australian Scholarly Publishing Pty Ltd.

Anderson, A.J. (2004b). Islands of Ambivalence. In S.M. Fitzpatrick (ed.), *Voyages of Discovery. The Archaeology of Islands*, pp. 251–273. Westport, Connecticut: Praeger.

Armstrong, P. (2004). *Darwin's Other Islands.* London: Continuum.

Baker, R.G.V. and R.J. Haworth. (1997). Further evidence from relic shellcrust sequences for a late Holocene higher sea-level for eastern Australia. *Marine Geology* 141: 1–9.

Baker, R.G.V. and R.J. Haworth. (2000a). Smooth or oscillating late Holocene sea level curve? Evidence from cross-regional statistical regressions of fixed biological indicators. *Marine Geology* 163: 353–365.

Baker, R.G.V. and R.J. Haworth. (2000b). Smooth or oscillating late Holocene sea level curve? Evidence from the palaeozoology of fixed biological indicators in southeast Australia and beyond. *Marine Geology* 163: 367–386.

Baker, R.G.V., R.J. Haworth and P.G. Flood. (2001a). Warmer or cooler marine late Holocene palaeo-environments? Interpreting southeast Australian and other Southern Hemisphere sea level curves using changes in fixed biological indicators and oxygen 18 content. *Palaeogeography, Palaeoclimatology, Palaeoecology* 168(3–4): 249–272.

Baker, R.G.V., R.J. Haworth and P.G. Flood. (2001a). Inter-tidal fixed indicators of former Holocene sea levels in Australia: a summary of sites and a review of methods and models. *Quaternary International* 83–85: 257–273.

Barham, A.J. (2000). Late Holocene maritime societies in the Torres Strait Islands, northern Australia—cultural arrival or cultural emergence? In S. O'Connor and P. Veth (eds.), *East of Wallace's Line: Studies of Past and Present Maritime Cultures of the Indo-Pacific Region*, pp. 223–314. Modern Quaternary Research in southeast Asia 16. Rotterdam: A.A. Balkema.

Barham, A.J., M.J. Rowland and G. Hitchcock (2004). Torres Strait *bepotaim*: an overview of archaeological and enthoarchaeological investigations and research. *Memoirs of the Queensland Museum, Cultural Heritage Series* 3(1): 1–72.

Barker, B.C. (1996). Maritime hunter–gatherers on the tropical coast: a social model for change. In S. Ulm, I. Lilley and A. Ross (eds.), *Australian Archaeology* 95, Proceedings of the 1995 Australian Archaeological Association Annual Conference. Tempus 6: 31–43.

Barker, B.C. (1999). Coastal occupation in the Holocene: environment, resource use and resource continuity. In J. Hall, and I. McNiven (eds.), *Australian Coastal Archaeology*, pp. 119–127. Research Papers in Archaeology and Natural History. ANH Publications, Department of Archaeology and Natural History, Research School Pacific and Asian Studies, Australian National University, Canberra.

Barker, B.C. (2004). *The Sea People. Late Holocene Maritime Specialization in the Whitsunday Islands, Central Queensland. Terra Australia 20* Pandanus Books, Research School of Pacific and Asian Studies, Australian National University Press, Canberra.

Barker, B.C. and Schon, R. (1994). A preliminary assessment of the spatial distribution of stone artefacts from the South Molle Island Aboriginal quarry, Whitsunday Islands, central Queensland coast. *Memoirs of the Queensland Museum* 37(1): 5–12.

Bass, B. (1998). Early Neolithic offshore accounts: Remote islands, maritime exploitations, and the Trans–Adriatic cultural network. *Journal of Mediterranean Archaeology* 11(2): 165–190.

Beaton, J.M. (1978). Archaeology and the Great Barrier Reef. The Northern Great Barrier Reef. Part B of a discussion. *Philosophical Transactions of the Royal Society of London* B284: 141–147.

Beaton, J.M. (1983). Does intensification account for changes in the Australian Holocene archaeological record? *Archaeology in Oceania* 18(2): 94–97.

Beaton, J.M. (1985). Evidence for a coastal occupation time-lag at Princess Charlotte Bay (north Queensland) and implications for coastal colonisation and population growth theories for Aboriginal Australia. *Archaeology in Oceania* 20(1): 1–20.

Beaton, J.M. (2000). Tasmania—An extreme version of Australia. In A. Anderson and T. Murray (eds.), *Australian Archaeologist: Collected Papers in Honour of Jim Allen*, pp. 182–186. Canberra: Coombs Academic Publishing, Australian National University.

Bellwood, P. (1997). Comments. *Current Anthropology* 42(1): 107–108.

Blackburn, T.M., P. Cassey, R.P. Duncan, K.L. Evans and K.J. Gasten (2004). Avian extinction and mammalian introductions on Oceanic Islands. *Science* 305: 1955–1958.

Border, A.P. (1999). Aboriginal settlement of offshore islands in the Southern Great Barrier Reef Province, central Queensland. In H.J. Hall and I.J. McNiven (eds.), *Australian Coastal Archaeology*, pp. 129–139. Research Papers iin Archaeology and Natural History 31, Research School of Pacific and Asian Studies. ANH Publications, Department of Archaeology and Natural History, Australian National University, Canberra.

Bourke, P.M. (2004). Three Aboriginal shell mounds at Hope Inlet: evidence for coastal, not maritime Late Holocene economies on the Beagle Gulf mainland, northern Australia. *Australian Archaeology* 59: 10–22.

Bowdler, S. (1977). The coastal colonisation of Australia. In J. Allen, J. Golson and R. Jones (eds.), *Sunda and Sahul. Prehistoric Studies in South East Asia, Melanesia and Australia*, pp. 205–246. London: Academic Press.

Bowdler, S. (1990). Peopling Australasia: The "Coastal Colonization" hypothesis reexamined. In P. Mellars (ed.), *The Emergence of Modern Humans: An Archaeological Perspective*, pp. 327–343. Edinburgh: University of Edinburgh Press.

Bowdler, S. (1995). Offshore islands and maritime explorations in Australian prehistory. *Antiquity* 69: 945–58.

Broodbank, C. (1999). The insularity of island archaeologists: Comments on Rainbird's 'Islands out of time'. *Journal of Mediterranean Archaeology* 12(2): 235–239.

Brown, J.H. and M.V. Lomolino. (2000). Concluding remarks: historical perspective and the future of island biogeography theory. *Island Biogeography Special Issue. Global Ecology and Biogeography* 9: 87–92.

Burney, D.A. (1997). Tropical islands as palaeoecological laboratories: gauging the consequences of human arrival. *Human Ecology* 25(3): 437–457.

Campbell, J. (2002). *Invisible Invaders. Smallpox and Other Diseases in Aboriginal Australia 1780–1880.* Melbourne: Melbourne University Press.

Cane, S. (2001). The Great Flood: eustatic change and cultural change in Australia during the late Pleistocene and Holocene. In A. Anderson, I. Lilley and S. O'Connor (eds.), *Histories of Old Ages: Essays in Honour of Rhys Jones*, pp141–165. Research School of Pacific and Asian Studies, Australian National University, Canberra: Pandanus Books.

Carter, M. (2001). New evidence for the earliest human occupation in Torres Strait, northeastern Australia. *Australian Archaeology* 52: 50–52.

Carter, M. (2002). Recent results of excavations on the Murray Islands, Eastern Torres Strait and implications for early links with New Guinea: bridge and barrier revisited. In S. Ulm, C. Westcott, J. Reid, A. Ross, I. Lilley, J. Prangnell and L. Kirkwood (eds.), *Barriers, Borders, Boundaries.* Proceedings of the 2001 Australian Archaeological Association Annual Conference. Tempus Volume 7, pp1–10. Anthropology Museum, The University of Queensland, Brisbane.

Carter, M., P. Veth, A. Barham, D. Bird, S. O'Connor and R.B. Bird (2004a). Archaeology of the Murray Islands, eastern Torres Strait: implications for a regional Prehistory. In R. Davis (ed.), *Woven Histories Dancing Lives. Torres Strait Islander Identity, Culture and History,* pp. 234–258. Canberra: Aboriginal Studies Press.

Carter, M., A.J. Barham, P. Veth, D.W. Bird, S. O'Connor and R.B. Bird (2004b). The Murray Islands Archaeological Project: excavations on Mer and Dauar, eastern Torres Strait. *Memoirs of the Queensland Museum, Cultural Heritage Series* 3(1): 163–182.

Carlquist, S. (1965). *Island Life: A Natural History of the Islands of the World.* New York: Natural History Press.

Chappell, J. (1982). Sea-levels and sediments–some features of the context of coastal archaeological sites in the tropics. *Archaeology in Oceania* 17(2): 69–78.

David, B., I. McNiven, R. Mitchell, M. Orr, S. Haberle, L. Bardy and J. Crouch (2004). Badu 15 and the Papuan-Austronesian settlement of Torres Strait. *Archaeology in Oceania* 39: 65–78.

Drake, D.R., C.P.H. Mulder, D.R. Towns and C.H. Daugherty. (2002). The biology of insularity: an introduction. *Journal of Biogeography* 29: 563–569.

deMenocal, P.B. (2001). Cultural responses to climate change during the late Holocene. *Science* 292: 667–673.

Dewar, R.E. (1997). Does it matter that Madagascar is an island? *Human Ecology* 25(3): 481–489.

Diamond, J.M. (1990). New Zealand as an Archipelago: an international perspective. In D.R. Towns, C.H. Daugherty, and I.A.E. Atkinson (eds.), *Ecological Restoration of New Zealand Islands.* Conservation Sciences Publications No. 2, Department of Conservation, Wellington.

Diamond, J.M. (1993). Ten thousand years of solitude. *Discover* March 49–57.

Diamond, J.M. (1997). *Guns, Germs, Steel. A Short History of the Last 13,000 Years.* London: Jonathon Cape.

Diamond, J.M. (2005). *Collapse: How Societies Choose to Fail or Succeed.* New York: Penguin Books.

Faulkner, P. and A. Clarke (2004). Late Holocene occupation and coastal economy in Blue Mud Bay, northeast Arnhem Land: Preliminary archaeological findings. *Australian Archaeology* 59: 23–30.

Fitzhugh, B. and T.L. Hunt (1997). Introduction: Islands as laboratories: Archaeological research in comparative perspective. *Human Ecology* 25(3): 379–383.

Flannery, T. (1994). *The Future Eaters: An Ecological History of the Australasian Lands and People.* Sydney: Reed New Holland.

Fort, J. (2003). Population expansion in the western Pacific (Austronesia): a wave of advance model. *Antiquity* 77: 520–530.

Fosberg, F.R. (1963). The island ecosystem. In F.R. Fosberg (ed.), *Man's Place in the Island Ecosystem,* pp. 1–6. Tenth Pacific Science Congress, Honolulu 1961. Honolulu: Bishop Museum Press.

Gamble, C. (1993). *Timewalkers: The Prehistory of Global Colonization.* London: Penguin Books.

Gamble, C. (2000). The Maxwell Nine and the peopling of southwest Tasmania. In A. Anderson and T. Murray (eds.), *Australian Archaeologist. Collected Papers in Honour of Jim Allen,* pp. 122–130. Canberra: Coombs Academic Publishing, Australian National University.

Genever, M., J. Grindrod and B. Barker (2003). Holocene palynology of Whitehaven Swamp, Whitsunday Island, Queensland,

and implications for the regional archaeological record. *Palaeogeography, Palaeoclimatology, Palaeoecology* 201(1–2): 141–156.

Grove, R.H. (1995). *Green Imperialism. Colonial expansion, tropical island Edens and the origins of environmentalism, 1600–1860.* Cambridge: Cambridge University Press.

Haug, G.H., D. Günther, L.C. Peterson, D.M. Sigman, K.A. Hughen and B. Aeschlimann (2003). Climate and the collapse of Maya civilization. *Science* 299: 1731–1735.

Heap, A.D., G.R. Dickens, and L.K. Stewart (2001). Late Holocene sediment in Nara Inlet, central Great Barrier Reef platform, Australia: sediment accumulation on the middle shelf of a tropical mixed clastic/carbonate system. *Marine Geology* 176: 39–54.

Hendy, E.J., M.K. Gagan, C.A. Alibert, M.T. McCulloch, J.M. Lough, and P.J. Isdale (2002). Abrupt decrease in tropical Pacific sea surface salinity at end of Little Ice Age. *Science* 295: 1511–1514.

Hodell, D.A., M. Brenner, J.H. Curtis and T. Guilderson (2001). Solar forcing of drought frequency in the Maya lowlands. *Science* 292: 1367–1370.

Hughen, K.A., T.I. Eglinton, L. Xu and M. Makou (2004). Abrupt tropical vegetation response to rapid climate changes. *Science* 304: 1955–1959.

Jennerjahn, T.C., V. Ittekkot, H.W. Arz, H. Behling, J. Pätzold and G. Wefer (2004). Asynchronous terrestrial and marine signals of climate change during Heinrich events. *Science* 306: 2236–2239.

Johnson, C.N. and S. Wroe (2003). Causes of extinction of vertebrates during the Holocene of mainland Australia: arrival of the dingo, or human impact? *The Holocene* 13(6): 941–948.

Jones, R. (1976). Tasmania. Aquatic machines and offshore islands. In G. de G. Sieveking, I.H. Longworth and K.E. Wilson (eds.), *Problems in Economic and Social Archaeology*, pp. 235–263. Cambridge: Duckworth.

Jones, R. (1977). Man as an element of continental fauna. The case of the sundering of the Bassian bridge. In J. Allen, J. Golson and R. Jones (eds.), *Sunda and Sahul. Prehistoric Studies in Southeast Asia, Melanesia and Australia*, pp. 371–386. London: Academic Press.

Karlèn, A. (1995). *Man and Microbes. Disease and Plagues in History and Modern Times.* New York: Simon and Schuster.

Keegan, W.F. (1999). Comment on Paul Rainbird, 'Islands out of time: towards a critique of island archaeology'. *Journal of Mediterranean Archaeology* 12(2): 255– 258.

Kirch, P.V. (1997). Comment. *Current Anthropology.* 38(2): 181–182.

Kirch, P.V. (2004). Oceanic Islands. Microcosms of "Global Change". In C.L. Redman, S.R. James, P.R. Fish and J.D. Rogers (eds.), *The Archaeology of Global Change. The Impact of Humans on their Environment*, pp. 13–27. Washington: Smithsonian Books.

Lamb, L and B. Barker (2001). Evidence for early Holocene change in the Whitsunday Islands: a new radiocarbon determination from Nara Inlet 1. *Australian Archaeology* 53: 42–43.

Lape, P.V. (2004). The isolation metaphor in island archaeology. In S.M. Fitzpatrick (ed.), *Voyages of Discovery. The Archaeology of Islands*, pp. 223–232. Westport, Connecticut: Praeger.

Littlewood, R.A. (1972). *Physical Anthropology of the Eastern Highlands of New Guinea.* Vol II. Anthropological Studies in the Eastern Highlands of New Guinea. Seattle: University of Washington Press.

Lockwood, J.G. (2001). Abrupt and sudden climatic transitions and fluctuations: A review. *International Journal of Climatology* 21: 1153–1179.

Lomolino, M.V. (2000). A call for a new paradigm of island biogeography. *Island Biogeography Special Issue. Global Ecology and Biogeography* 9: 1–6.

Lourandos, H. (1997). *Continent of Hunter–Gatherers: New Perspectives in Australian Prehistory.* Cambridge: Cambridge University Press.

MacArthur, R.H. and E.O. Wilson (1967). *The Theory of Island Biogeography.* Princeton: Princeton University Press.

McMichael, A.J. (2001). Human culture, ecological change, and infectious disease: are we experiencing history's fourth great transition. *Ecosystem Health* 7(2): 107–115.

McNiven, I.J. (2000). Treats or retreats: Aboriginal island use along the Gippsland coast, southeastern Australia. *The Artefact* 23: 22–34.

Mayewski, P.A., E.E. Rohling, J.C. Stager, W. Karlèn, K.A. Maasch, L.D. Meeker, E.A. Meyerson, F. Gasse, S. van Kreveld, K. Holmgren, J. Lee–Thorp, G. Rosqvist, F. Rack, M. Staubwasser, R.R. Schneider and E.J. Steig (2004). Holocene climate variability. *Quaternary Research* 62: 243–255.

Morris, J. (2001). *The Tiwi: From Isolation to Cultural Change. A History of Encounters Between an Island People and Outside Forces.* Darwin: Northern Territory Press.

Mulvaney, J. and J. Kamminga (1999). *Prehistory of Australia.* Sydney: Allen and Unwin.

Nunn, P.D. (2004). Through a mist on the ocean: human understanding of island environments. *Tijdschrift voor economische en sociale geografie* 95(3): 311–325.

O'Connor, S. (1992). The timing and nature of prehistoric island use in northern Australia. *Archaeology in Oceania* 27(2): 49–60.

Oppenheimer, S. (2004). *Out of Eden. The Peopling of the World.* London: Constable and Robinson Ltd.

Patton, M. (1996). *Islands in Time. Island Sociogeography and Mediterranean Prehistory.* London: Routledge.

Rainbird, P. (1999a). Islands out of time: towards a critique of island archaeology. *Journal of Mediterranean Archaeology* 12(2): 216–234.

Rainbird, P. (1999b). Nesophiles miss the boat? A response. *Journal of Mediterranean Archaeology* 12(2): 259–260.

Rainbird, P. (2002). A message for our future? The Rapa Nui (Easter Island) ecodisaster and Pacific island environments. *World Archaeology* 33(3): 436–451.

Rainbird, P. (2004). *The Archaeology of Micronesia.* Cambridge: Cambridge University Press.

Rial, J.A. (2004). Abrupt climate change: chaos and order at orbital and millennial scales. *Global and Planetary Change* 41: 95–109.

Rolett, B. and J. Diamond (2004). Environmental predictors of pre-European deforestation on Pacific islands. *Nature* 431: 443–446.

Rowland, M.J. (1980). The Keppel Islands—preliminary investigations. *Australian Archaeology* 11: 1–17.

Rowland, M.J. (1981). Radiocarbon dates for a shell fishhook and disc from Mazie Bay, North Keppel Island. *Australian Archaeology* 12: 63–69.

Rowland, M.J. (1982a). Further radiocarbon dates from the Keppel Islands. *Australian Archaeology* 15: 43–48.

Rowland, M.J. (1982b). Keppel Island Marine Specialists. An adaptation to the Southern Barrier Reef Province. In S. Bowdler (ed.), *Coastal Archaeology in Eastern Australia.* Proceedings of the 1980 Valla Conference on Australian Prehistory, pp. 114–120. Department of Prehistory, Research School of Pacific Studies, Australian National University, Canberra.

Rowland, M.J. (1983a). A note on corrections to radiocarbon dates from the Keppel Islands. *Australian Archaeology* 17: 134–135.

Rowland, M.J. (1983b). Aborigines and environment in Holocene Australia: changing paradigms. *Australian Aboriginal Studies* 2: 62–77.

Rowland, M.J. (1984). A long way in a bark canoe. Aboriginal occupation of the Percy Islands. *Australian Archaeology* 18: 17–31.

Rowland, M.J. (1985a). Further radiocarbon dates from Mazie Bay. North Keppel Island. *Australian Archaeology* 21: 113–118.

Rowland, M.J. (1985b). Archaeological investigations on Moa and Naghi Islands, western Torres Strait. *Australian Archaeology* 21: 119–132.

Rowland, M.J. (1986). The Whitsunday Islands. Initial historical and archaeological observations and implications for future work. *Queensland Archaeological Research* 3: 72–87.

Rowland, M.J. (1987). The distribution of Aboriginal watercraft on the east coast of Queensland. Implications for culture contact. *Australian Aboriginal Studies* 2: 38–45.

Rowland, M.J. (1989). Population increase, intensification or a result of preservation? Explaining site distribution patterns on the coast of Queensland. *Australian Aboriginal Studies* 2: 32–42.

Rowland, M.J. (1992). Conservation Plan for Cultural Heritage Sites on the Keppel Island Group, Central Queensland. Unpublished report to the Livingstone Shire Council and National Parks and Wildlife Branch, Division of Conservation, Department of Environment and Heritage, Brisbane.

Rowland, M.J. (1995). Indigenous water-craft use in Australia: The 'Big Picture' and small experiments on the Queensland coast. *Bulletin of the Australian Institute of Maritime Archaeology* 19(1): 5–18.

Rowland, M.J. (1996). Prehistoric archaeology of the Great Barrier Reef Province Retrospect and prospect. In P. Veth, and P. Hiscock (eds.), *Archaeology of Northern Australia. Regional Perspectives. Tempus Volume 4*, pp. 191–211. Anthropology Museum, University of Queensland, St Lucia.

Rowland, M.J. (1999a). The Keppel Islands–'a 3000 year' event revisited. In J. Hall and I. McNiven (eds.), *Australian Coastal Archaeology.* Research Papers in Archaeology and Natural History 31, pp. 141–155. ANH Publications, Department of Archaeology and Natural History, Research School of Pacific and Asian Studies, Australian National University, Canberra.

Rowland, M.J. (1999b). Holocene environmental variability: have its impacts been underestimated in Australian pre-History? *The Artefact* 22: 11–48.

Rowland, M.J. (2002). 'Crows', swimming logs and auditory exostoses: Isolation on the Keppel Islands and broader implications. *Tempus* 7: 61–73.

Sax, D.E., S.D. Gaines and J.H. Brown (2002). Species invasions exceed extinctions on islands worldwide: a comparative study of plants and birds. *The American Naturalist* 160(6): 766–783.

Simberloff, D.S. (1974). Equilibrium theory of island biogeography and ecology. *Annual Review of Ecology and Systematics* 5: 161–182.

Stapp, P., G.A. Polis and F. Sañchez Piñero (1999). Stable isotopes reveal strong marine and El Niño effects on island food webs. *Nature* 401: 467–469.

Steadman, D.W., J.P.White and J. Allen (1999). Prehistoric birds from New Ireland, Papua New Guinea: extinctions on a large Melanesian island. *Proceedings of the Nature Academy of Science, USA* 96: 2563–2568.

Storm, P. (2001). The evolution of humans in Australasia from an environmental perspective. *Palaeogeography, Palaeoclimatology, Palaeoecology* 171: 363–383.

Sullivan, M.E. (1982). Aboriginal Shell Middens in the Coastal Landscape of New South Wales. Unpublished PhD thesis, Department of Prehistory and Anthropology, Australian National University, Canberra.

Terrell, J. (1997). The postponed agenda. Archaeology and human biogeography in the twenty-first century. *Human Ecology* 25(3): 429–436.

Terrell, J.E. (2004a). Introduction: 'Austronesia' and the great Austronesian migration. *World Archaeology* 36(4): 586–590.

Terrell, J.E. (2004b). Models of reticulate evolution: The "Ancient Lagoons" hypothesis. In S.M. Fitzpatrick (ed.), *Voyages of Discovery: The Archaeology of Islands*, pp. 203–222. Westport, Connecticut: Praeger.

Terrell, J.E. (2004b). The 'sleeping giant' hypothesis and New Guinea's place in the prehistory of Greater Near Oceania. *World Archaeology* 36(4): 601–609.

Terrell, J.E., T.L Hunt, C. Gosden (1997). The dimensions of social life in the Pacific. Human diversity and the myth of the primitive isolate. *Current Anthropology* 38(2): 155–195.

Terrell, P.E., K.M. Kelly and P. Rainbird (2001). Foregone Conclusions? In search of "Papuans" and "Austronesians". *Current Anthropology* 42(1): 97–124.

Ulm, S. (2000). Evidence for early focussed marine resource exploitation from an open coastal site in central Queensland. *Australian Archaeology* 51: 66–67.

Ulm, S. and J. Reid (2000). Index of dates from archaeological sites in Queensland. *Queensland Archaeological Research* 12: 1–129.

van Dommelen, P. (1999). Islands in history. *Journal of Mediterranean Archaeology* 12(2): 246–251.

Vayda, A.P. and R.A. Rappaport (1963). Island cultures. In F.R. Fosberg (ed.), *Man's Place in the Island Ecosystem*, pp. 133–144. Honolulu: Bishop Museum.

Vitousek, P.M. (2002). Oceanic islands as model systems for ecological studies. *Journal of Biogeography* 29: 573–582.

Walker, D. (ed.) (1972). *Bridge and Barrier: the Natural and Cultural History of Torres Strait*, Research School of Pacific Studies, Department of Biogeography and Geomorphology, Publication BG3. Canberra: Australian National University.

Walter, H.S. (2004). The mismeasure of islands: implications for biogeographical theory and the conservation of nature. *Journal of Biogeography* 31: 177–197.

Walter, R. (2004). New Zealand archaeology and its Polynesian connections. In L. Furey, and S. Holdaway (eds.), *Change Through Time. 50 Years of New Zealand Archaeology. New Zealand Archaeological Association Monograph* 26: 125–146.

Ward, I. (2004). Comparative records of occupation in the Keep River region of the eastern Kimberley, northwestern Australia. *Australian Archaeology* 59: 1–9.

Wardle, D.A. (2002). Islands as model systems for understanding how species affect ecosystem properties. *Journal of Biogeography* 29: 583–591.

Whittaker, R.J. (1992). Stochasticism and determinism in island ecology. *Journal of Biogeography* 19: 587–591.

Whittaker, R.J. (2000). Scale, succession and complexity in island biogeography: are we asking the right questions? *Global Ecology and Biogeography* 9: 75–85.

Chapter 7

Unravelling 'mystery' and process from the prehistoric colonization and abandonment of the Mediterranean Islands

Helen Dawson, University of Kent

Introduction

Researchers in the Pacific will be familiar with the notion of 'Mystery Islands', an evocative term used to refer to Pacific islands that displayed clear signs of human occupation but were found empty by Europeans upon their arrival (Bellwood 1978). Although the abandonment of these Pacific islands has now largely been understood (e.g., Irwin 1992; Weisler 1996; Anderson 2001), in the Mediterranean a certain degree of 'mystery' still surrounds the processes of prehistoric human settlement of the islands. While the Mystery islands were generally not recolonized, it is clear that islands in the Mediterranean experienced a punctuated history of abandonment and resettlement (Dawson 2005), thus their comparison raises a number of key questions with regard to the long-term sustainability of island communities. A rise in Mediterranean island-based projects in the past 10–15 years, accompanied by exciting theoretical developments, has resulted in an increasingly convincing picture of island settlement to emerge in the region, in which patterns (or their absence) can be discerned and processes adequately discussed.

More than twenty years have passed since Cherry (1981) first systematized the colonization data on Mediterranean islands available at the time (see figure 7.1, for location of main islands and island-groups mentioned in the text). In spite of substantial progress in the theories and methods of island archaeology, Cherry has recently stated that in reality we are still far from having a clear picture of what constitutes 'colonization', as opposed to 'discovery, exploration, occupation, establishment', and other activities (Cherry 2004: 239). He has also noted that Mediterranean island colonization follows biogeographical predictions, but that human agency must be responsible for a certain degree of 'noise' in the patterns (Cherry 1981; 2004). On a theoretical level, researchers have acknowledged the complexities inherent in colonization. Its ubiquitous use as a term of analysis has been confusing, and studies now emphasize the usefulness of recognising that colonization is made up of different activities (cf. Graves and Addison 1995). In practice, however, the identification of these activities in the archaeological record remains problematic. Bowdler, for example, has claimed that occupation and exploitation, 'where total isolation is not assumed', are hard to distinguish (1995: 945), and researchers in different parts of the world have expressed similar concerns (e.g., Ramis et al. 2002). This paper digs deeper into the issue of what constitutes colonization, first by reviewing the current state of affairs in island archaeology and in the Mediterranean specifically and then by proposing a new framework for analysing human presence in the Mediterranean islands. In this framework, human activities on the islands are seen as forming a broader colonization category, as past and current colonization theories are expanded to include considerations of abandonment and recolonization. This more inclusive treatment of colonization, which will be familiar to researchers working with Pacific islands, has rarely been attempted for the Mediterranean islands (with the exception of Malta and the Cyclades [Stoddart 1999; Broodbank 2000]), and never systematically on a pan-Mediterranean scale.

A few key clarifications are due at this point. 'Initial colonization' is intended here as the setting up of a human presence in a previously unoccupied territory, and 'recolonization' as setting up in either an empty but previously occupied territory (by the same community or its descendants, or by an altogether different group) or in one occupied at the time by another com-

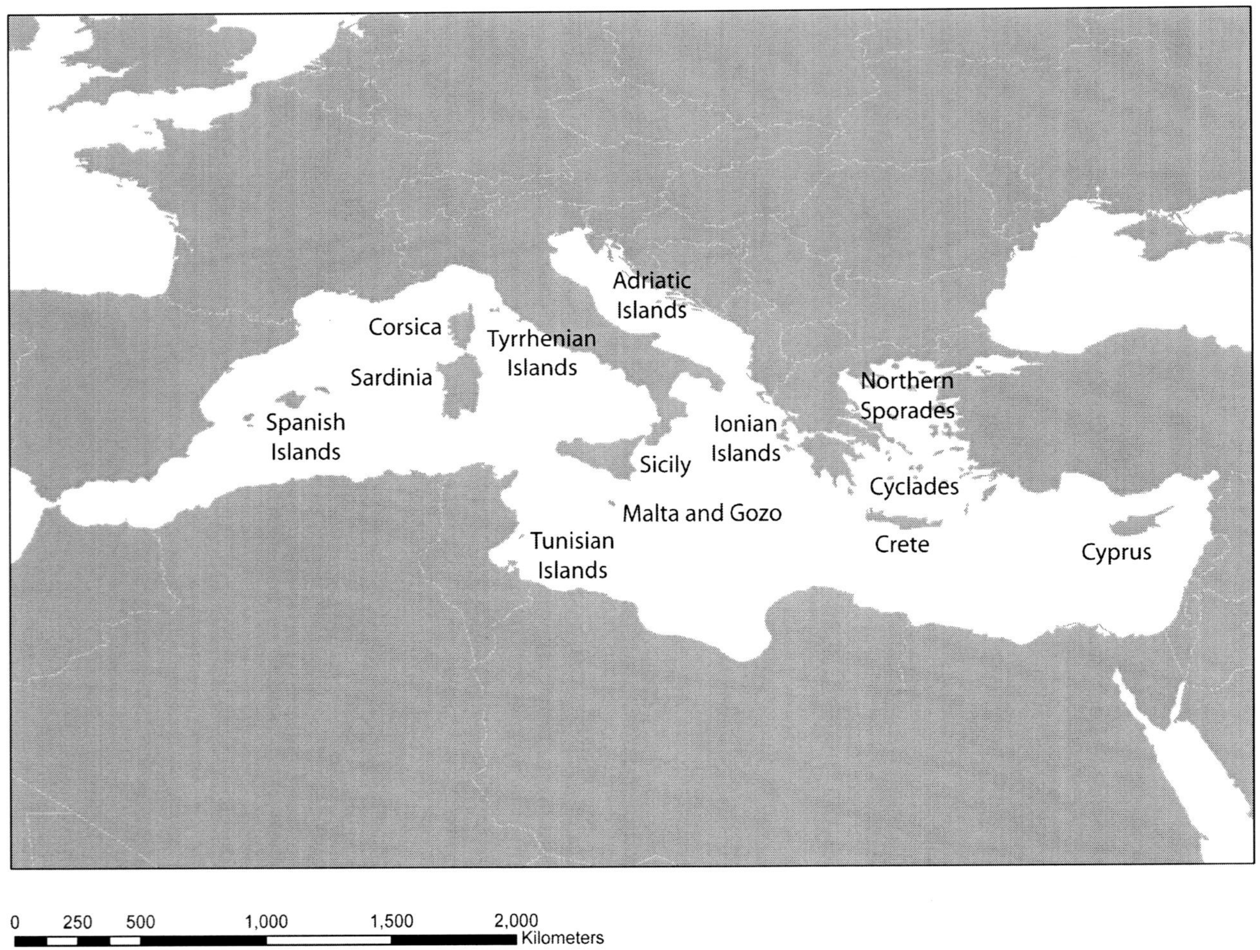

Figure 7.1: Location of main Mediterranean islands and islands-groups mentioned in the text.

munity. What that setting up involved, what motivated it, what triggered that human presence, how long it lasted, what obstacles it encountered, and what its outcomes were are all questions that need to be addressed if colonization is to be fully understood. This approach uses some key features drawn from Rockman's (2003) 'landscape learning' analysis for the timing and modalities of 'the colonization of unfamiliar landscapes' (see Rockman and Steele 2003). Rockman (2003) has made the distinction between colonization of empty spaces and of occupied spaces in terms of differences in the acquisition of knowledge. In the first case, the colonizers must learn about the new environment, whereas in the second instance, incoming colonizers have to deal with knowledge, population, and social 'barriers' (Rockman 2003: 17). The acquisition of new knowledge may be more or less visible in the archaeological record. Colonization or this 'setting up' is therefore intended as a fluid succession and in some cases overlap of activities, including the classic and well-known exploration, utilization, occupation, settlement, and even abandonment, which can follow each other in various orders (exploration, for example, can continue after settlement and abandonment and within the settlement trajectory of an island there may be multiple colonization and abandonment events). Graves and Addison (1995: 387) have noted that multiple colonizations within an archipelago or island region may increase the amount of time required for the establishment of a single community (especially if numbers are low and population growth slow) as a result of a more dispersed and in some cases mobile/fluid occupation pattern. All of these activities can and should where possible be qualified temporally and spatially, e.g., by specifying whether settlement was 'seasonal', utilization 'year-round', occupation 'place-focused' or 'resource-focused', and so on.

Mediterranean island archaeology

Mediterranean island archaeology has experienced a series of significant shifts since its first tentative steps in the early 1970s. Initially, researchers were concerned with applying the concept of island laboratory (Vayda and Rappaport 1963) to the Mediterranean, with islands such as Malta and Crete providing ideal laboratories in view of their conspicuous material culture (temples and palaces) (e.g., Evans 1973, 1977). There followed a more systematic organization of the subject, as Cherry (1981, 1990) used island biogeography (MacArthur and Wilson 1967) to understand differences in the patterns and processes of initial colonization in the eastern and western Mediterranean islands. Currently, and in parallel with island archaeology in general (e.g., Fitzpatrick 2004), Mediterranean island archaeology is enjoying a phase of stimulating growth (e.g., Broodbank 2000; Waldren and Ensenyat 2002).

For the past fifteen years since Cherry's 1990 update, researchers have had the opportunity to explore the potential and limitations of biogeographical analysis in the Mediterranean (e.g., Bass 1998; Broodbank 2000; Strasser 2003), and as a result the emphasis has continued to shift. New useful paradigms have emerged that stress usefully the interconnectedness of Mediterranean islands and that geographical parameters are culturally mediated by island communities (for example, leading to communities isolating themselves or interacting with one another) (e.g., Robb 2001): the study of configuration and the development of the 'islandscape' concept now place the attention on interaction networks (see Bass [1998] and Broodbank [2000]), and, as a result, Mediterranean island archaeology has moved from being mainly concerned with initial colonization (and thus, for reasons that we shall see, mainly the Neolithic) to focusing on subsequent island development. As both earlier colonization dates and later developments are being revealed and acknowledged, periods other than the Neolithic have come to the fore in Mediterranean island studies, so that Mediterranean island archaeology is no longer just an archaeology of initial colonization, but an archaeology that follows the long-term development of islands, and therefore a much more complete island archaeology. Advances in ideas about Mediterranean islands in the past ten years have come disguized in different kinds of publications: some have an explicitly theoretical agenda (e.g., Patton 1996) and some a practical remit (e.g., Gaffney et al. 1997), while others, these being the most useful of all, offer a combination of both (e.g., Bass 1998; Broodbank 2000). In terms of understanding the chronology of earliest settlement, views are broadly confirmed in some cases (e.g., Crete and Malta) (Broodbank and Strasser 1991; Trump 2002) while elsewhere they are changing incrementally (e.g., earlier sites in the Aegean islands) (Broodbank 1999). Major advances have concerned particularly Cyprus and the Balearics, with opposite effects on their chronologies (resulting in earlier and later colonization dates respectively) (Peltenburg 2003; Ramis *et al.* 2002); and new regional syntheses have been published (e.g., the central Adriatic islands and the Cyclades) (Bass 1998; Broodbank 1999, 2000). There remain problems, however, as the popularity of the subject (more projects, conferences, publications) has in some cases pushed it forward without allowing researchers to fully engage with key issues and problems. Ironically, what constitutes colonization and how best to study it are still a matter of great debate.

Mediterranean island colonization

Cherry has created a plot of cumulative percentages of the islands in the eastern and western Mediterranean (figure 7.2), showing evidence of occupation by a given millennium BC, and explained the differences between the two regions through their dissimilar configurations, e.g., in terms of the role of stepping-stone islands and of large islands acting as 'mainlands' (1981: 63). In 1990, in a new article, Cherry synthesized some significant developments that had taken place since 1981, but he did not update his graphs in the light of these new discoveries, so as it is, several studies of the Mediterranean (e.g., Vigne 1989; Patton 1996; Grove and Rackham 2001) still refer to the graphs contained in the original 1981 article. The main developments synthesized in the 1990 paper included a few instances of Palaeolithic occupation of true islands; an 8th mill. cal BC human presence on all the larger islands or island groups (except Crete); and human presence on Cyprus 'one to two millennia earlier' than previously believed (at the site of Akrotiri-*Aetokremnos*) (Cherry 1990: 145). Cherry also noted an increase in the number of smaller islands colonized in the Neolithic, or between the seventh and fourth millennia BP (ca. the 6th and 3rd mill. cal BC) (1990: 164).

In 1996, Vigne produced a colonization graph, which also portrays colonization as a linear trajectory. No table of data accompanied this plot but, in a footnote, the two Cherry articles are mentioned as providing the sources. In the same year, Patton presented a pan-Mediterranean graph of island colonization showing percentages of islands colonized per millennium cal. BC (largely based on Cherry's 1990 data) according to three visibility categories (rather than to the western and eastern Mediterranean distinction) (figure 7.3). This graph indicates that the timing of colonization did not follow biogeographical predictions based on the islands' visibility (since less visible islands, in category B, were colonized before ones visible from the mainland, in category A). Patton hypothesized that this might imply that the rates of discovery and colonization did not match (1996: 54–55).

Cherry's 1981 dataset was predominantly eastern Mediterranean (ca. 80 islands investigated in the eastern Mediterranean and 35 in the western Mediterranean). The database in the present review includes 145 islands (83 in the eastern Mediterranean and 62 in the western Mediterranean, tables 7.1 and 7.2). The additional data (mainly for the west, from the central Adriatic and north African Islands) have been used to amend Cherry's original graphs. Overall, the revised cumulative plot (figure 7.4) confirms the general trends noted by Cherry, i.e., a gradual and continuous 'infilling' of available land, which, generally speaking, was faster in the western than in the eastern Mediterranean, at least until the late 4th–early 3rd millennium cal. BC. The most notable difference from Cherry's original graph (apart from the fact that the temporal 'origins' of colonization have been pushed back in both east and west) is the reduction in the colonization time lag first noticed between the two areas (between the 7th/6th and the early 3rd millennia cal. BC), which is mainly the result of a set of earlier dates that have become available from the eastern Mediterranean (especially the Aegean) (Broodbank 1999). Two main issues emerge from the analysis of Cherry's cumulative plot: one conceptual, the other practical. Cherry conceived of island colonization in terms of permanent settlement: certain types of evidence, especially surface lithic finds, were discounted, since they proved human presence but not permanent occupation (Cherry 1981: 48). Considering that it was only with the Neolithic and the Bronze Age that archaeological correlates could be safely linked to settlement, Cherry concluded that island colonization was on the whole a product of such phases (*ibid.*). In the Pacific, Graves and Addison (1995: 386) have noted that Spriggs and Anderson (1993) also 'tacitly equated' the colonization of Hawai'i 'to the settlement of sites of sufficient size and duration so as to leave an unambiguous chronometric signature'. The first time that Mediterranean prehistory gets close to this type of 'colonization' is with Neolithic settlement. But to say that the 'colonization' or the Neolithic settlement of Mediterranean islands took place during the Neolithic adds nothing to our understanding of colonization: it is merely going in circles. On the other hand, Graves and Addison point out that 'colonization' or 'the placement of human populations on discovered islands... may or may not lead to the establishment of human population' (1995: 387). Thus the issue to be addressed

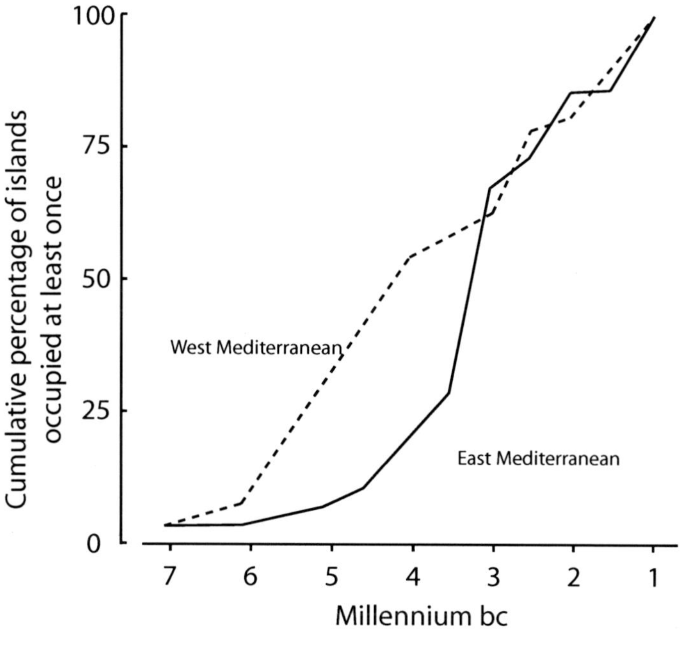

Figure 7.2: Cumulative percentage colonization plot (redrawn after Cherry 1981).

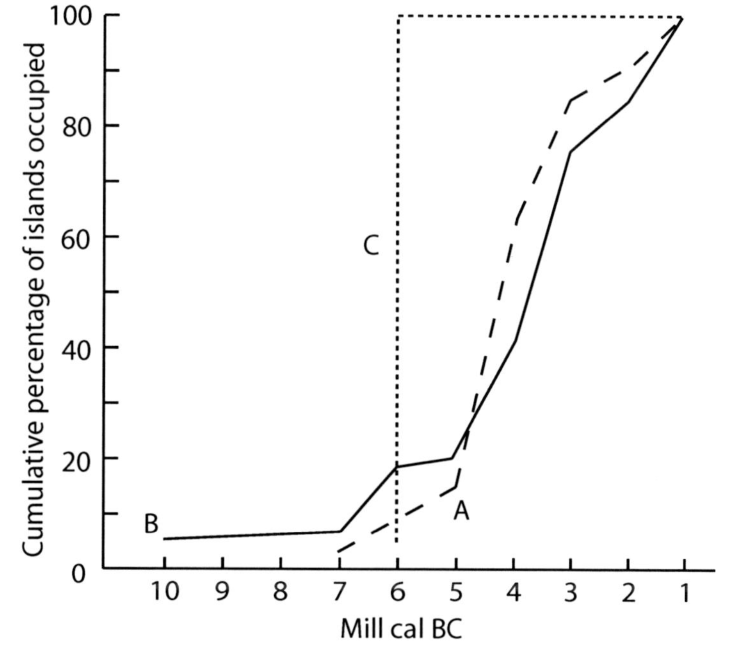

Figure 7.3: Cumulative colonization plot (redrawn after Patton 1996). A: Islands visible from the mainland; B. Islands visible from nearest other island; C. Islands not visible from nearest island/mainland.

Name	Mill. 1st col. BC	Dist. NM	Size	Name	Mill. 1st col. (BC)	Dist. NM	Size
1. Corsica	8	87	8722	32. Favignana	4	17	19.4
2. Brac	8	5.5	395	33. Marettimo	4	30	12
3. Sardinia	8	205	24089	34. Ischia	4	11	46
4. San Domino	6	25	1.5	35. Lastovo	4	25	49
5. Palagruza	6	130	0.3	36. Palmarola	4	1.4	32
6. Korcula	6	34.5	276	37. Ponza	4	33	12
7. Lipari	6	30.2	37.6	38. Zannone	4	27	4
8. Salina	6	42.9	26.8	39. Pianosa (T.)	4	35	0.11
9. Hvar	6	4.1	299	40. Ustica	3	53	8
10. Susac	6	80	4.6	41. Pantelleria	3	102	83
11. Vis	6	23.6	90.3	42. Stromboli	3	56.2	12.6
12. Filicudi	6	46.6	9.5	43. Sv Klement	3	5.6	3
13. Malta	6	80	246	44. Scedro	3	6.7	7.5
14. Gozo	6	65	67	45. Svetac	3	47.6	4.3
15. Capri	5	5	10	46. Alicudi	3	87.5	5.2
16. Elba	5	10	220	47. Majorca	3	167	3740
17. Lampedusa	5	210	20.2	48. Menorca	3	200	702
18. Giglio	5	22	15	49. Ibiza	2	86	541
19. Giannutri	5	14	3	50. Formentera	2	95	82
20. Solta	5	7.7	588	51. San Nicola	1	25	0.5
21. Pianosa	5	50	10	52. Kopiste	1	87	1
22. Iles Planes	5	5	n/a	53. Mjlet	1	18	98.6
23. Habibas	5	11	0.4	54. Comino	1	70	2.6
24. Rachgoun	5	2	n/a	55. Bisevo	1	27.8	5.8
25. Ile du Roi (Ch.)	5	11	n/a	56. Cabrera	1	175	13
26. Ile d'Isabelle (Ch.)	5	11	n/a	57. Conejera	1	178	18
27. Ile du Congres (Ch.)	5	11	n/a	58. Linosa	1	19	5.4
28. Zembra	5	10	3.9	59. Montecristo	1	10	63
29. Kuriate	5	16	12	60. Djerba	1	2.5	568
30. Panarea	5	42	3.4	61. Chergui (I. K.)	1	25	69
31. Levanzo	4	15	7	62. Rharbi (I.K.)	1	25	n/a

Note: Ch. = Chaffarinas, I.K. = Iles Kerkenna, T. = Tremiti

Table 7.1: Western Mediterranean Island data (millennium of first colonization cal BC; distance to nearest mainland (NM) km; size sq km).

Name	Mill. 1st col. BC	Dist. NM	Size	Name	Mill. 1st col. (BC)	Dist. NM	Size
1. Cyprus	10	60	9251	43. Melos	4	105	151
2. Gioura	8	70	20	44. Samothraki	4	37	178
3. Kythnos	8	39	100	45. Meganisi	4	9	20
4. Crete	7	102	8259	46. Siphnos	4	85	74
5. Alonissos	6	43	65	47. Syros	4	75	85
6. Kyra Panagia	6	59	25	48. Zakynthos	4	18	402
7. Skyros	6	33	210	49. Delos	3	112	3
8. Thasos	6	7	380	50. Dokos	3	2	20
9. Lefkas	6	0.5	303	51. Erymonisia	3	n/a	n/a
10. Corfu	5	5	593	52. Idra	3	6	50
11. Amorgos	5	105	124	53. Imbros (Gok.)	3	16	279
12. Andros	5	55	380	54. Ios	3	147	109
13. Astypalaia	5	79	97	55. Keros	3	145	15
14. Chios	5	11	842	56. Kimolos	3	106	36
15. Giali	5	18	9	57. Kouphonisia	3	160	6
16. Kalymnos	5	18	93	58. Makronisos	3	3	18
17. Karpathos	5	93	301	59. Pholegandros	3	131	32
18. Kasos	5	140	69	60. Poros	3	0.5	23
19. Kos	5	5	290	61. Sikinos	3	140	35
20. Kythera	5	15	280	62. Spetses	3	2	22
21. Leros	5	32	53	63. Tenedos (Boz.)	3	19	42
22. Lesbos	5	12	1633	64. Donoussa	3	140	14
23. Mykonos	5	112	86	65. Heraklia	3	155	18
24. Naxos	5	132	430	66. Schinoussa	3	157	9
25. Paros/Antiparos	5	115	196	67. Tinos	3	82	195
26. Psara	5	67	40	68. Nysiros	3	17	37
27. Rhodes	5	19	1400	69. Reneia	3	105	14
28. Salamis	5	0.5	96	70. Seriphos	3	62	75
29. Samos	5	5	477	71. Anafi	2	152	40
30. Saria	5	85	21	72. Lipsoi	2	37	17
31. Symi	5	8	38	73. Patmos	2	48	34
32. Thera	5	180	76	74. Skopelos	2	22	97
33. Tilos	5	20	63	75. Therassia	2	178	9
34. Ithaca	5	30	96	76. Marsa Island	2	1	7
35. Kephalonia	5	38	781	77. Antikythera	1	63	20
36. Despotiko	5	112	8	78. Arkos	1	10	5
37. Aegina	4	21	83	79. Atokos	1	8	5
38. Alimnia	4	40	7	80. Castellorizo	1	5	10
39. Chalki	4	47	28	81. Ikaria	1	47	256
40. Gavdos	4	192	30	82. Kalamos	1	2	25
41. Kea	4	22	131	83. Skiathos	1	4	50
42. Lemnos	4	62	478				

Note: Gok. = Gokceada, Boz. = Bozcaada

Table 7.2: Eastern Mediterranean Island data (millennium of first colonization cal BC; distance to nearest mainland (NM) km; size sq km).

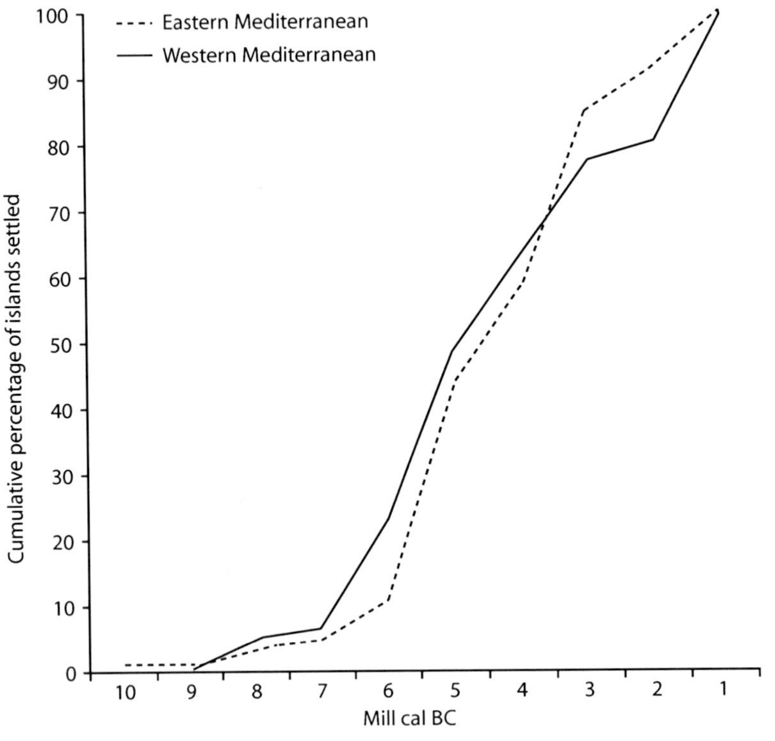

Figure 7.4: Reworked cumulative colonization plot (Dawson 2005). Unless otherwise specified, the data in all subsequent tables and figures are from Dawson (2005).

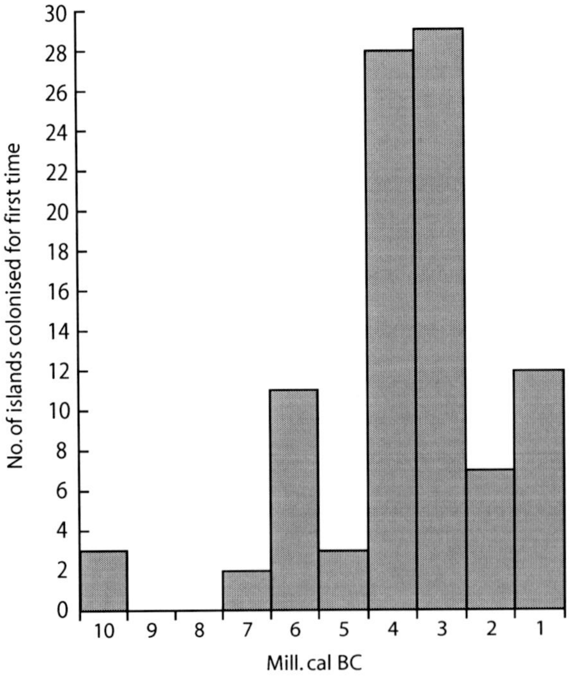

Figure 7.5: Colonization histogram (redrawn from Patton 1996).

is how human activity on islands varies through time and space. Neolithic colonization is therefore but one example of how it can vary in the Mediterranean.

From a practical point of view, cumulative curves can be deceptive, as they portray a false sense of long-term continuity: this is because they count how many islands are colonized at least once during each millennium, and then add this number to those colonized during the preceding millennium. Patton's 1996 review marked an initial departure from Cherry's cumulative representation by presenting a histogram of colonization data (figure 7.5), which counted how many islands were colonized during each period and could be used to identify distinct waves of colonization (Patton 1996: 59, 62). Patton, however, used the graph only to make some very general points about pan-Mediterranean patterns of colonization (which he linked to the Neolithic Revolution, the Secondary Products Revolution, and early state formation) (*ibid.*), and did not fully explore the implications of his observations on a regional scale.

Data analysis—part 1

Data from the western and eastern Mediterranean islands are presented in figure 7.6. This revised non-cumulative graph indicates that, on the basis of our current knowledge, the colonization pattern for the eastern Mediterranean underwent two distinct peaks (one during the 5th and the other during the 3rd mill. cal. BC); and that, in the western Mediterranean, there was a peak in the 5th millennium and a colonization wave (albeit smaller) in the 1st. Interestingly, a higher percentage of islands was colonized in the 8th than in the 2nd mill. cal. BC western Mediterranean. A series of graphs explores the roles played by distance and size, addressing whether biogeographical variables were prominent in the colonization of the islands. Figures 7.7 and 7.8 show how many islands were colonized in each millennium in the western and eastern Mediterranean respectively, based on separate distance groups.

Figure 7.7 shows that, in the west, distance was not a prominent factor overall, with far-away islands colonized both in early (8th–6th mill. cal. BC) and late (3rd–1st mill. cal. BC) periods. In the east (figure 7.8), distance appears more prominent, with only Crete, amongst the islands with distance to mainland higher than 100 km, colonized before the 6th millennium. Most western islands colonized in the 5th millennium are close to the mainland (<20 km) (figure 7.7), whereas in the east, this number is balanced out by a similar number of islands colonized that lie over 20 km away (figure 7.8). The highest number of distant islands (>100 km) was colonized in the eastern Mediterranean in the 3rd millennium (Early Bronze Age), usually via stepping-stone islands (figure 7.8 and 7.9). Turning to size, it emerges that, in the west (figure 7.10) small islands (1–10 sq km) were colonized at different times, mainly in the 5th millennium (when we saw that most islands targeted are also <20 km away), whereas no large ones (>50 sq km) were colonized in the 4th millennium and only a couple in the 5th. In the east, on the other hand, most larger islands (>50 sq km) were targeted in the 5th millennium (figure 7.11).

The combined effect of size and distance emerges more clearly from figures 7.12 to 7.16 (however, numbers are in some cases small). Figure 7.12 shows that, in the western Mediterranean, most large (>20 sq km) and distant islands (>50 km) (in black) were colonized either before or after the Neolithic (none in the 5th–4th mill. cal. BC), whereas most small nearby islands (in white) were colonized during the Neolithic. As one would expect, the colonization of small nearby islands (white in figure 12) followed roughly the same pattern as that of large nearby islands (black in figure 7.13) (i.e., Neolithic colonization, with the 8th millennium BC exception of Corsica). Figure 7.13 also shows that small far-away islands (in white) were mainly colonized from the Bronze Age onwards (with the exception of Susac, in the central Adriatic, and Filicudi, in the southern Tyrrhenian, both of which were colonized earlier and were easily reached via stepping-stone islands colonized at roughly the same time).

For the eastern Mediterranean (figure 7.14–7.15), most large islands (>20 sq km, black in both graphs) were colonized between the Neolithic and Bronze Age (5th–3rd mill.) regardless of distance. Most small and far-away islands (white in figure 7.15) were colonized from the Early Bronze Age onwards, whereas the colonization of small close-by islands (white in figure 7.14) took place gradually from the 5th millennium cal. BC onwards. Overall, figure 7.16 shows that the colonization of small far-away islands (those less favoured by biogeography) in the eastern Mediterranean seems to take place from the Bronze Age onwards, whereas in the western Mediterranean it is more evenly spread out. For the Eastern Mediterranean, the lack of colonization of larger islands in later periods (3rd mill.) is likely to reflect the fact that most of these had already been occupied by then, thus

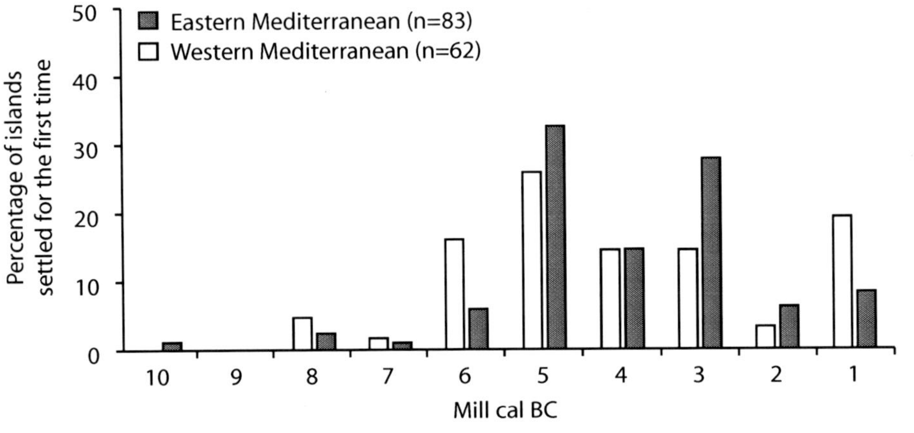

Figure 7.6: Reworked non-cumulative colonization plot.

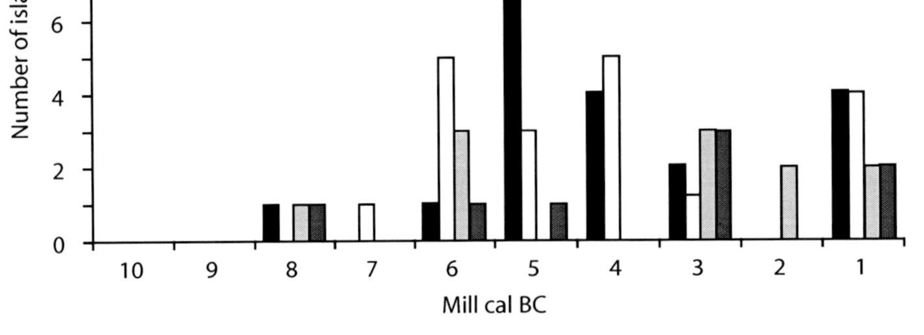

Figure 7.7: Western Mediterranean islands divided by distance to nearest mainland and colonization date.

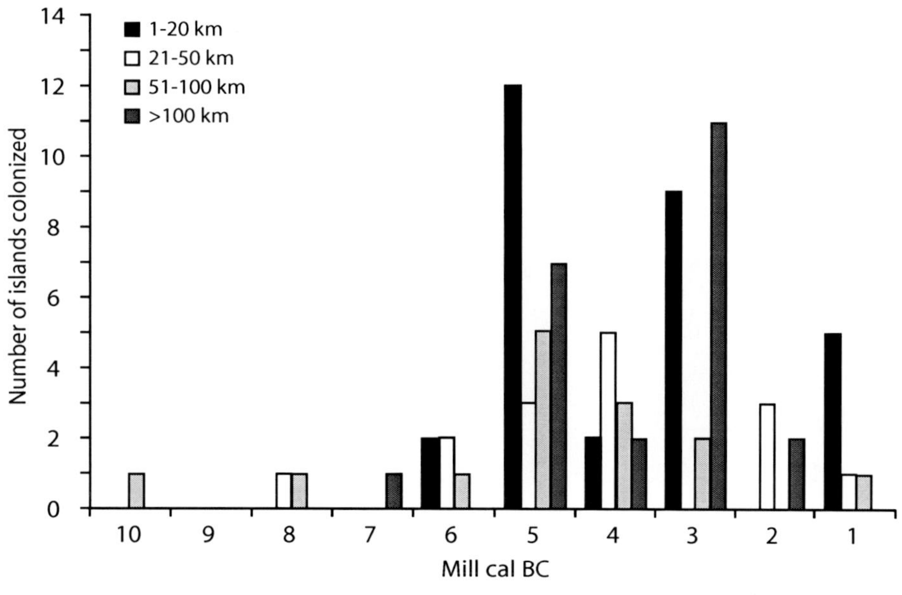

Figure 7.8: Eastern Mediterranean islands divided by distance to nearest mainland and colonization date.

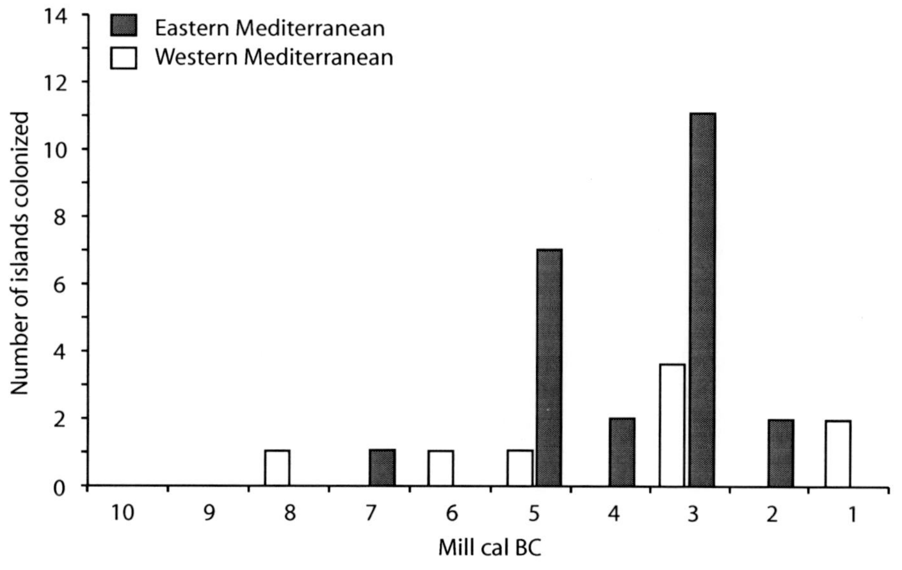

Figure 7.9: Number of islands greater than 100 km from nearest mainland colonized per millennium.

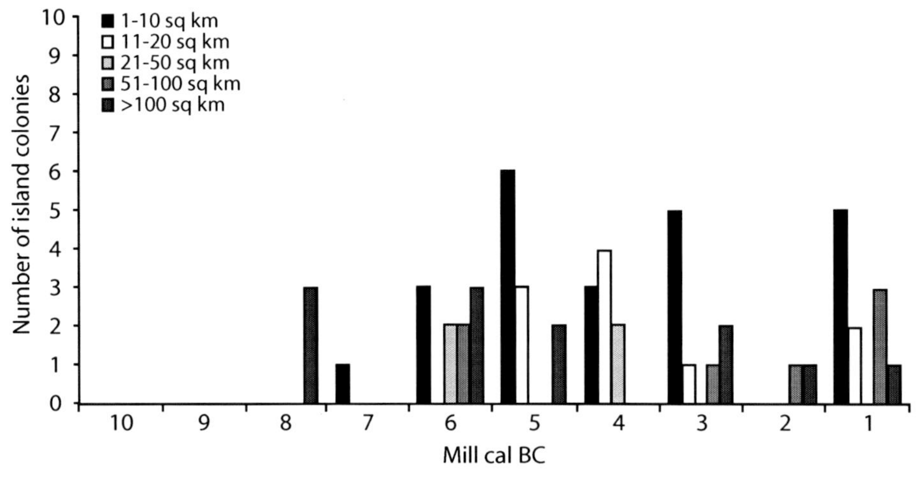

Figure 7.10: Western Mediterranean islands divided by size and colonization date.

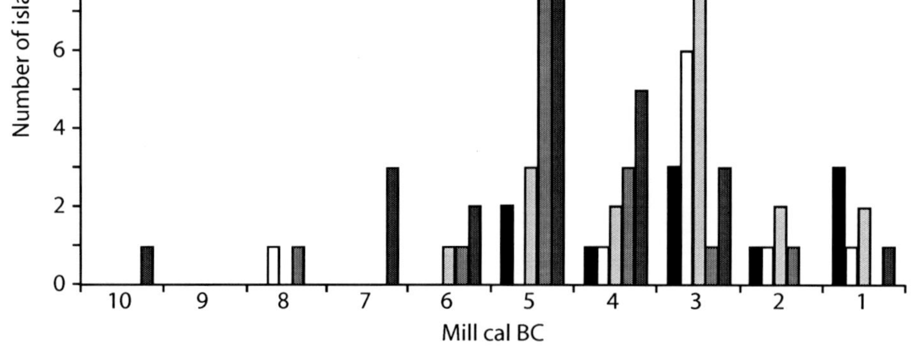

Figure 7.11: Eastern Mediterranean islands divided by size and colonization date.

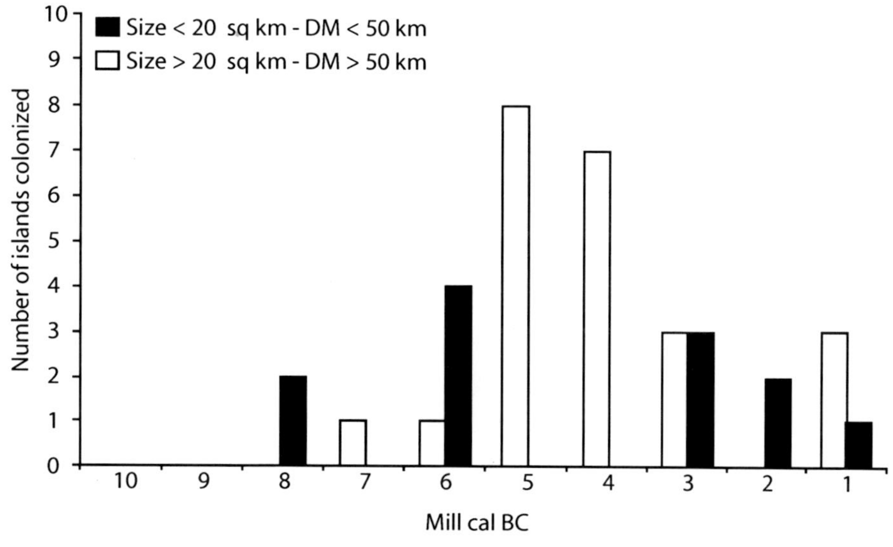

Figure 7.12: Western Mediterranean islands colonized per millennium sorted by size and distance criteria (small and close-by vs. large and far-away).

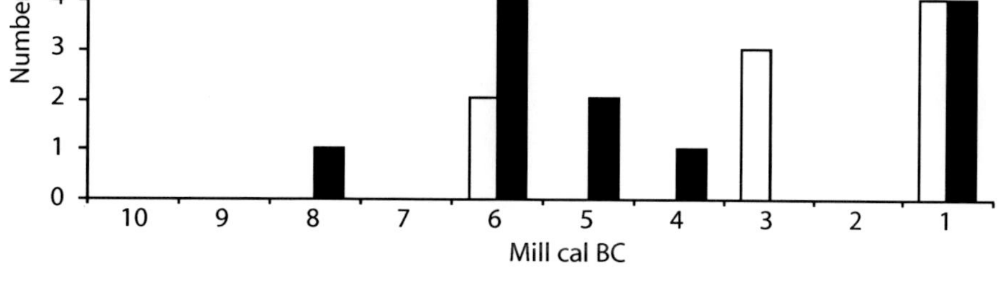

Figure 7.13: Western Mediterranean islands colonized per millennium sorted by size and distance criteria (small and far-away vs. large and close-by).

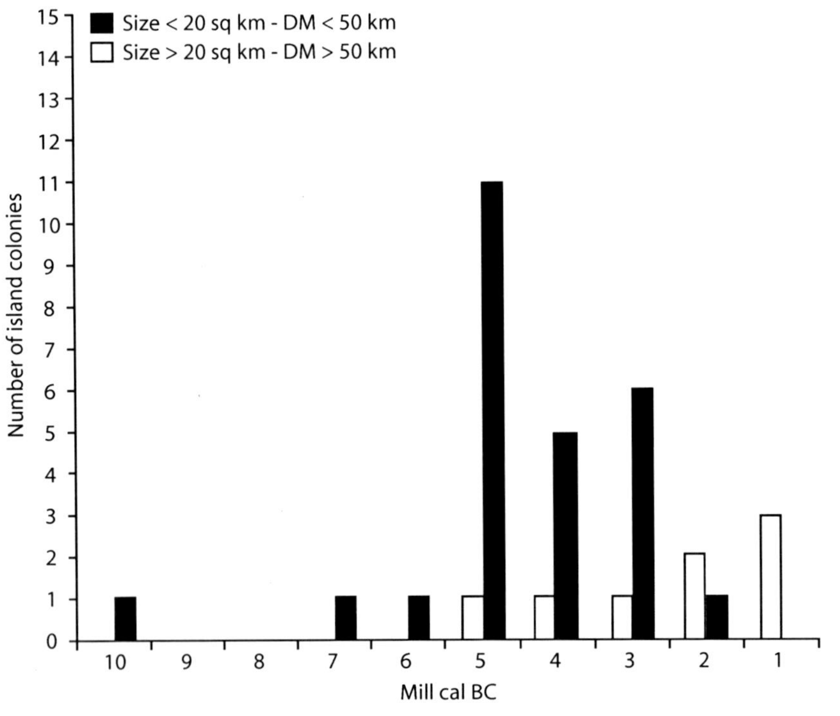

Figure 7.14: Eastern Mediterranean islands colonized per millennium sorted by size and distance criteria (small and close-by vs. large and far-away).

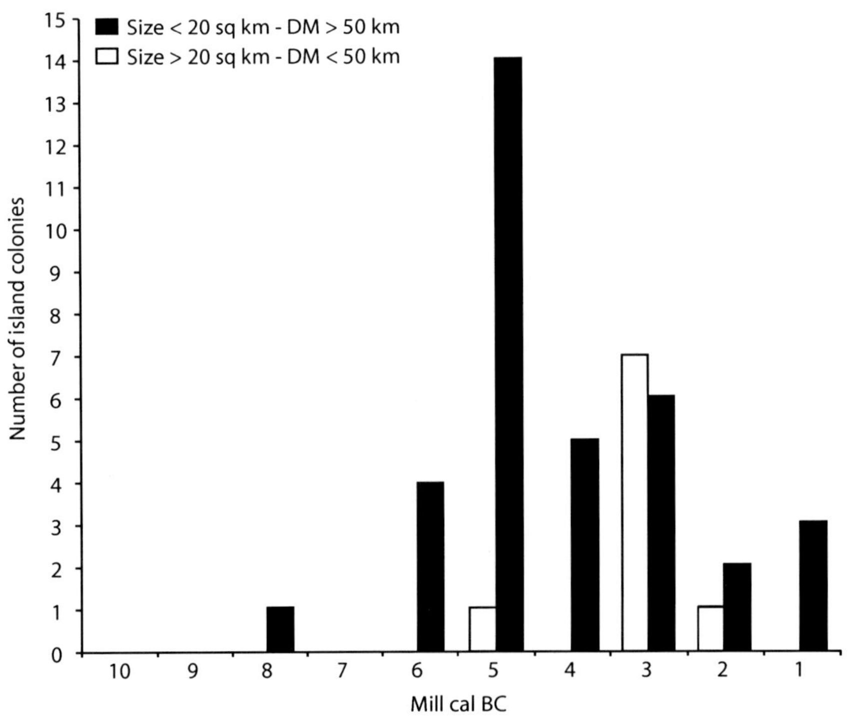

Figure 7.15: Eastern Mediterranean islands colonized per millennium sorted by size and distance criteria (small and far-away vs. large and close-by)

the pattern appears to date expansion into the smaller islands.

Human-island interaction can take place at different levels, which are reflected, somewhat problematically, in the archaeological record. Island colonization in the Neolithic does emerge as a pan-Mediterranean phenomenon from the graphs (see figure 7.17). This is because, as mentioned, Neolithic cultural remains tend to relate to built and permanent structures that are archaeologically more visible than earlier more ephemeral traces. However, there is evidence that humans were present on islands before the Neolithic and that wild species were introduced to the islands to broaden the spectrum of resources, suggesting an effective manipulation of resources. Fallow deer (*Cervus dama*), for example, was introduced to Cyprus as early as the 8th millennium cal. BC (Davis 1984; Guilaine et al. 1995, 1996; Vigne 1996: 67–69).

Apart from the well-known Upper Palaeolithic (ca. 11,000 cal. BC) material found at Franchthi Cave on the Greek mainland, indicating exploitation of obsidian from the island of Melos (Perlès 1987), there is increasing evidence for early settlement on other islands (Melos was apparently not permanently settled at this early stage). From the second half of the 9th millennium cal. BC, there were permanent settlements in Cyprus at the pre-Khirokitian sites of Kissonerga-*Mylouthkia*, Parekklisha-*Shillourokambos*, and Kalavasos-*Tenta* (but not at Akrotiri-*Aetokremnos*). In the 8th millennium, Gioura and Kythnos (two small islands in the Northern Sporadhes and Cyclades respectively) (Davis et al. 2001; Sampson 1998; Trantalidou 2004), and Korcula (Central Dalmatian islands) (Čečuk and Radic 1995; Bass 1998), were also occupied for the first time, and these may not be isolated instances, as further evidence awaiting systematic investigation has been found on several islets around Gioura (Davis et al. 2001: 79). The site of Maroulas on Kythnos spanned the Late Mesolithic-Early Neolithic transition (by then Kythnos had achieved insular status), and its excavation revealed a series of human burials, a house floor, and some circular constructions, with the remains of land and marine snails, tunny, and several other fish species (Honea 1975; Sampson 1998; Trantalidou 2004). The Cyclops Cave site on the island of Gioura (Northern Sporadhes), which was also insular at the time, also produced Mesolithic occupation under EN, MN, and LN levels (Sampson 1996), indicating increasing occupation of the cave from 8400 to 3500 cal. BC (Davis et al. 2001: 79; Trantalidou 2004).

The idea that pre-Neolithic people could not and did not colonize Mediterranean islands successfully should thus be set aside. The Neolithic itself should not be viewed as a monolithic block, as it clear that regional colonization rates varied over such a long period (see figure 7.18). For example, Sicily's satellite islands (figure 7.19) display a slight reduction in islands colonized in the 5th millennium, when one would perhaps expect a surge based on the pan-Mediterranean trends reviewed so far, and other island groups were colonized altogether much later than the inception of farming on nearby mainlands (e.g., the Spanish islands, figure 7.20) (Ramis *et al.* 2002).

The Neolithic phase tends to be favoured in discussions of island colonization in view of its pan-Mediterranean dimension. At the same time, the fragmented record of human presence on the islands (which will emerge much more strongly when the abandonment data are factored into the following analysis) offers a strong counter-argument to anyone in favour of a long-term trajectory in island colonization or of the ease of living on islands (there is no evidence of human presence on Kythnos from the 8th to the 3rd millennium cal. BC, after which the island was recolonized. Kythnos falls within the category of Cherry's [1981: 52] 'larger littoral islands' at a mere 39 km from the mainland). While the long-term impression of island colonization is that of a continuous filling of space, short-term processes differed greatly on the local scale. In the initial stages of colonization, geographic variables played a role in the discovery of the islands and in their initial use/settlement. But detailed studies of Mediterranean island basins do not always conform to geographic predictive models, suggesting that culturally driven choices were involved.

Abandonment and recolonization

Few islands in the Mediterranean were continuously occupied throughout prehistory: at least one colonization experiment appears to have 'failed' on Cyprus at the site of Akrotiri-*Aetokremnos* (Simmons 1999).[1] However, whether or not the foragers at Aetokremnos (or Kythnos) had occupation or resource exploitation in mind (or if the two were at all different to them) is open to discussion. Even if we take resource exploitation as being the more likely reason for their presence, the fact

[1] At the time of going to press, two new sites contemporary with Akrotiri-*Aetokremnos* have been identified (Ammerman et al. 2006; see Broodbank 2006 for a recent review).

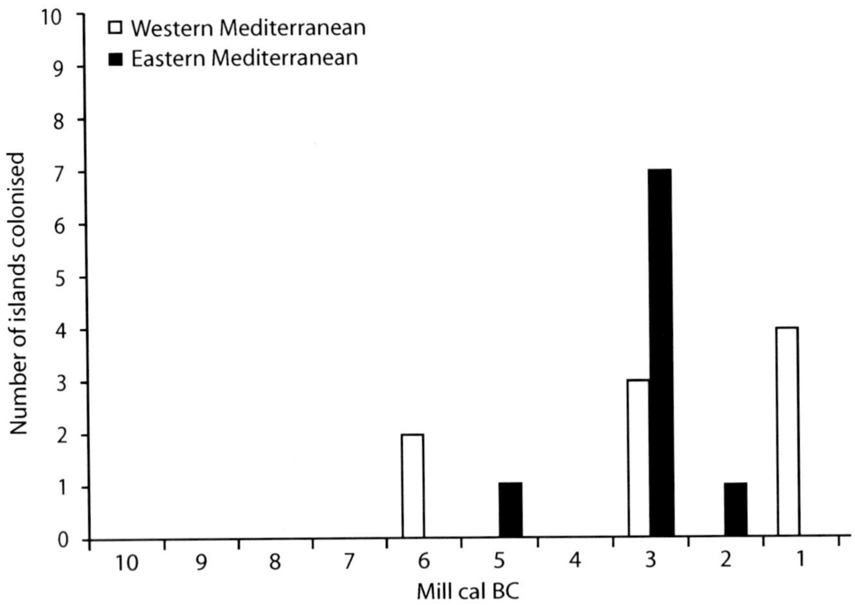

Figure 7.16: Number of small/far-away islands (<20 sq km and >50 km) from nearest mainland colonized per millennium.

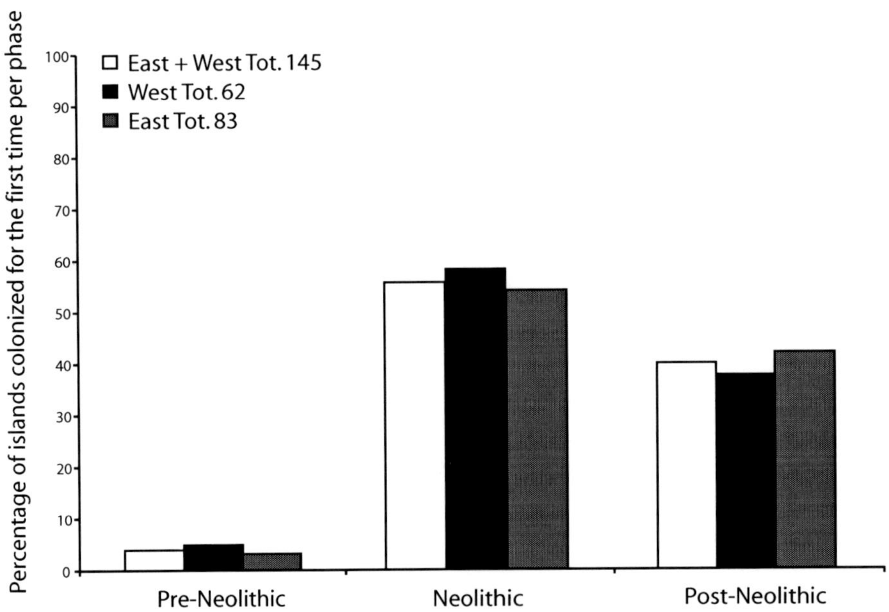

Figure 7.17: Non-cumulative plot showing percentages of islands colonized before, during, and after the Neolithic.

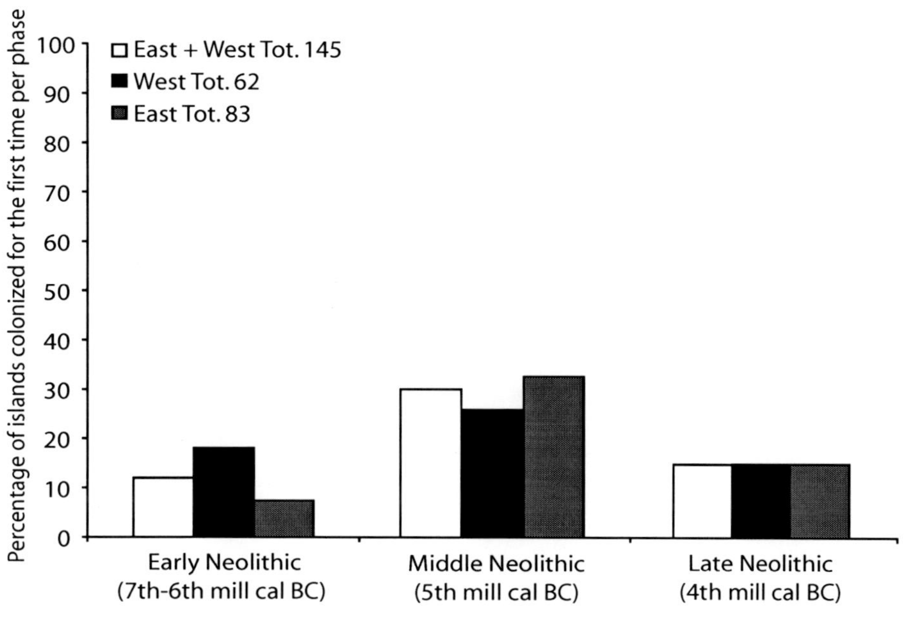

Figure 7.18: Non-cumulative plot showing percentages of islands colonized during the Neolithic.

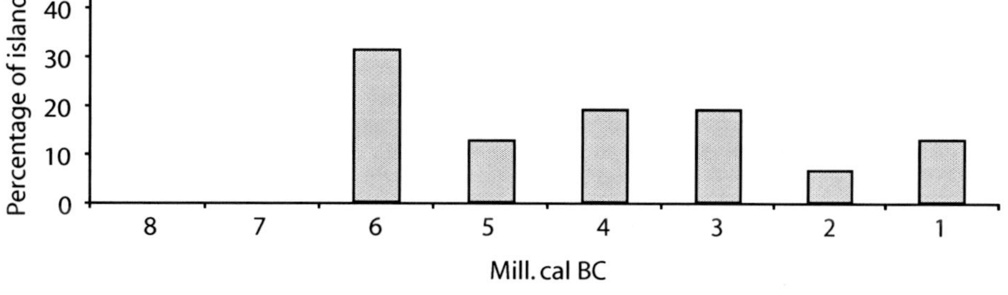

Figure 7.19: Islands off Sicily—non-cumulative plot (rates of colonization per millennium)

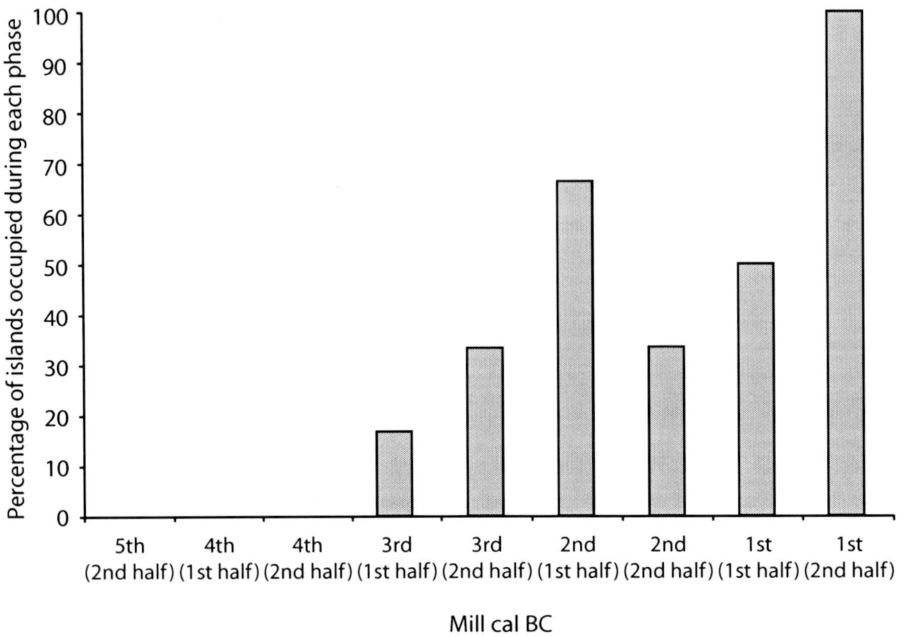

Figure 7.20: Spanish Islands—occupation plot.

they eventually left does not correspond to failure, since it may never have been their intention to settle permanently. Environmental analysis has in fact demonstrated that the rock-shelter of Akrotiri-*Aetokremnos* was used year-round, though it is unclear how long this occupation lasted overall (Simmons 1999: 208; note, however, Binford's [2000] doubts). This raises the question of what constitutes 'successful colonization', especially in the case of mobile communities.

Place-focused residence or repeated visitation of an island (and not just permanent residence) should be taken to represent 'colonization', albeit of a different kind. Changes in the archaeological record can represent distinct phases leading to permanent settlement (when substantiated by a long-term record), but variability can also point to different activities or types of island colonization. Accepting that there can be different types of colonization (each with its own prerequisites, aspirations, and manifestations) provides us with a framework for studying abandonment and recolonization more effectively. Since abandonment data may illustrate what conditions resulted in the demise of island-focused activities, they can also shed light on what may have prompted them in the first place. However, far from merely taking abandonment as a failed colonization experiment, we should view it as part of an integral strategy for using landscapes.

A common-sense characterization of abandonment refers to the absence of people where previously they had existed. However, even a short review will illustrate that this concept is not as straightforward as it might seem at first sight. Schiffer (1976: 30) originally defined abandonment as 'a type of cultural deposition process belonging to the S-A type', i.e., a kind of transformation from the systemic ('in use'), to the archaeological realm ('not in use'). Graves and Longacre pointed out that 'the process of abandonment may involve a number of complexly related events including *movement* by individuals, households, and larger social groups, and *changes* in birth and death rates; these processes may act differentially upon groups that comprise a community' (1982: 201, emphasis added). These definitions may not cover all aspects of abandonment, but they provide a starting point. Does abandonment refer to the extinction or the movement of people and/or settlements? Can it be extended to any kind of cultural activity? Does abandonment always entail interruption or is it rather a transformation (e.g., within an island cluster, a process such as settlement nucleation) and therefore does it have an element of continuity?

The following examples illustrate the points made so far. The Aleutian Islands are among 'the most isolated islands in the world', but they have been occupied for at least the last 8,000 years (McCartney and Veltre 1999: 507). Aleuts lived in extreme environmental conditions and relied on an exclusively marine diet owing to the lack of terrestrial fauna. In spite of these extreme conditions, the islands supported a large population during late prehistoric times, estimated at 12,000–15,000 people (McCartney and Veltre 1999: 503, 512). The Aleuts developed a complex strategy for coping with the harsh environment, the rich coastal economy sustaining population in the long term (*ibid*). McCartney and Veltre point out that because the archipelago is extensive (ca. 100 islands extending over a 1,600 km-long area), local changes, whether sudden or gradual, would have affected one or more villages (e.g., resulting in their abandonment) but not the entire chain, thus overall longevity could be ensured (1999: 512).

In the southern Pacific there is a longer tradition of studying abandonment than in the Mediterranean. Two categories are particularly useful for this study: the 'Mystery' Islands and the 'Outlier' Islands. As discussed, the Mystery islands display evidence for multiple colonizations and 'substantial delay between colonization and establishment', which 'was perhaps never achieved' (Graves and Addison 1995: 389). Irwin noted that the Pacific Mystery Islands are all less accessible from their neighbours than islands that remained occupied, so that when navigation declined (possibly as a result of the settlement of larger islands that were self-sufficient, e.g., Hawai'i and New Zealand) they were abandoned (Irwin 1992: 177). According to Irwin, these were not 'successful stand-alone islands' (1992: 178), and their abandonment seems to coincide neatly with the decline of voyaging (1992: 179).

The so-called 'Outlier' Islands are located outside the Polynesian triangle and, while displaying cultural differences from Melanesian and Micronesian islands to which they are closer, they are inhabited by communities with Polynesian cultural traits (Irwin 1992: 183). Irwin explains that these are no longer considered 'relic populations left behind during west to east migration through Polynesia', but rather it is believed that the islands were recolonized upon return. Although they are mostly small islands and some are atolls, they are close enough to Micronesia and Melanesia not to be abandoned (Irwin 1992: 185).

A review of several studies (in Dawson 2005), including those mentioned above, identified the following recurrent factors in the abandonment of land, both in general terms and specifically on islands: configuration (area, distance, location), geology, agricultural land, biodiversity, resources, rainfall, water sources, as well as human perception of a territory's potential for ensuring a community's survival and other cultural variables affecting human behaviour (e.g., in the case of islands, the 'pull' effect of a mainland or larger island). A selection of such factors will be used for in-depth study in the following section.

Data analysis—part 2

Gaps in the archaeological record cannot be automatically assumed to represent 'abandonment', and thus only islands for which a good archaeological record exists (either intensive field survey data or a fully excavated multi-period site) are included in the following investigation: Kythera (south of the Peloponnese, Greece); Melos, Kea, and Naxos (Cyclades, Greece); Cyprus (eastern Mediterranean); Palagruza and Hvar (east-central Adriatic Islands, Croatia); Ibiza and Formentera (the Pitiussae Islands, Spain); the Aeolian Islands (southern Tyrrhenian, Italy); Malta, Jerba, Pantelleria (southern Mediterranean); Palmarola (central Tyrrhenian, Italy); and the Tremiti Islands (western Adriatic, Italy) (table 7.3). The islands range in size from 0.3 sq km (Palagruza) to 9251 sq km (Cyprus). Seven are smaller than 10 sq km, which in the Pacific is considered to be the minimum acceptable size in terms of ensuring population survival (Keegan and Diamond 1987: 65). In the Mediterranean, it has been suggested that this threshold may be lower (or even irrelevant), because of reduced inter-island distances (Broodbank and Strasser 1991: 238).

The data are summarized in table 7.3, which contains information regarding the islands' size, maximum altitude, distance to nearest mainland, presence of water sources and mineral resources, annual rainfall, and population estimates in relation to island size and different settlement densities. Population estimates indicate that fewer than half the islands could support endogamous populations (ca. 300–500 people) (Wobst 1974; Adams and Kasakoff 1976; Jones 1976; MacCluer and Dyke 1976; Williamson and Sabath 1984; Broodbank 2000). When population estimates are translated into arable land needs (using figures from Broodbank [2000: 86] and Robb and Van Hove [2003: 246]), it emerges that most islands in the sample would have had sufficient agricultural land: a community of ca. 300–500 people would have required between 0.5–1.5 sq km of (not necessarily continuous) arable land, which most islands in the sample could offer. These figures indicate that commu-

Island	Size sq Km	Alt. m a.s.l.	Dist. km	Ground water	Rain mm	Min. resources	pop 0.5	pop 1	pop 2	pop 3
Palagruza	0.3	90	130	no	n/a	yes	<1	<1	<1	1
San Nicola	0.4	75	20	no	500	no	<1	<1	<1	1.2
Palmarola	1.4	262	32	no	n/a	yes	<1	1.4	2.8	4.2
San Domino	2	116	20	no	500	no	1	2	4	6
Panarea	3.4	421	42	no	500	no	1.7	3.4	6.8	10.2
Alicudi	5.2	675	87	no	500	no	2.6	5.2	10.4	15.6
Filicudi	9.5	774	47	no	500	no	4.75	9.5	19	37.8
Stromboli	12.6	924	56	no	500	no	6.3	12.6	24.6	37.8
Vulcano	21	500	22	no	500	yes	10.5	21	42	63
Salina	26.8	962	43	no	500	no	13.4	26.8	53.6	80.4
Lipari	37.6	602	30	no	500	yes	18.8	37.6	75.2	112.8
Formentera	82	202	95	no	370	no	41	82	164	246
Pantelleria	83	836	102	no	350	yes	41.5	83	166	249
Kea	130.6	568	12	yes	490	no	51.8	103.6	207.2	391.8
Melos	150.6	751	105	no	435	yes	75.3	150.6	301.2	451.8
Malta	246	253	85	yes	500	no	123	246	492	738
Kythera	280	507	15	yes	660	no	140	280	560	840
Hvar	312	626	4	no	800	no	156	312	624	936
Naxos	428	1000	132	yes	384	yes	214	428	856	1284
Jerba	568	40	2	no	200	no	284	568	1136	1704
Ibiza	572	475	92	yes	400	no	286	572	1144	1716
Mean	155.3	520.9	59	$y = 5/21$	480	$y = 7/21$	82	164	327	442
Cyprus	9251	1950	69	yes	500	yes	4626	9251	18502	27753

Note: alt: highest point; dist: distance to nearest mainland; pop.: population estimates at 0.5 people/sq km; 1p/sq km; 2 p/sq km; 3p/sq km

Table 7.3: Geographical characteristics and population estimates.

nities on most of these islands would have been able to survive, in view either of the island's actual size and/or of its proximity to other land. However, this was not always the case.

In some cases, occupational data have also been represented graphically as time-lines. Different periods of human use of the islands have been marked on a horizontal axis (the numbers indicate millennia BC and AD), through either a continuous line (=definite occupation) or a dotted line (=sporadic occupation). A gap represents definite abandonment, while a question mark indicates possible abandonment. Four types of analysis are carried out. The first aims to identify any general patterns of occupation and abandonment; the second, focusing on the development of four islands that are also obsidian sources, investigates to what extent the presence of resources affects island settlement; the third assesses whether or not small size and abandonment are related, using islands smaller than 10 sq km; and finally, the forth analyses the role played by distance and the 'pull' effect of the mainland, by focusing on distance from the nearest mainland (NM).

Overall patterns

Table 7.4 shows that, on average, occupation lasted longer than abandonment (approximately 4500 years vs. 1500 years per island). The average occupation period was in the order of just under 2000 years, while the average abandonment period lasted ca. 1000 years. The first occupation period for most of the islands was usually long (more than 1000 years), except in three cases (underlined): Formentera, Ibiza, and Cyprus. For Cyprus this is because the short initial occupation relates to the Akrotiri-*Aetokremnos* rock-shelter, but if that is not included, then the initial occupation period lasts much longer (3000 years), which is what one would perhaps expect on such a large island.

The data from table 7.5 indicate that on average prehistoric abandonment lasted ca. 800 years. This figure is supported by Butzer's (1996: 146) observation that settlement surveys in some Mediterranean regions have recorded occupational gaps lasting between 500–1000 years. During the first two millennia AD, the length of abandonment periods is much shorter (generally lasting ca. 300 years), while on average each island in the sample was abandoned for about a century during historical times. Table 7.5 also shows that islands colonized early (6th–4th mill. cal. BC) generally experienced initial occupation periods that lasted longer than those colonized later (3rd–2nd mill. cal. BC). Initial occupation periods on islands colonized later ranged between ca. 3000 (San Nicola) and 700 years (Formentera), or ca. 800 years on average per island; whereas occupation on islands colonized earlier varied between ca. 5000 (Lìpari) and 1000 years (Palagruza), or ca. 2250 years on average per island. On the other hand, abandonment periods vary in length (between 300 and 2500 years) throughout prehistory with no apparent patterning with regard to the period of initial occupation or to the islands' size and distance (with Cyprus, for example, experiencing longer abandonment periods than Lìpari). The fact that islands colonized later were abandoned sooner may reflect the fact that these tended to be smaller or less favourable to prolonged occupation than those occupied in earlier periods. The trend may also echo changes in the use of the islands over time—use that by the Bronze and Iron Ages had become increasingly specialized and thus tied in with socio-cultural processes of a more contingent nature than before. Importantly, it was at this time that maritime transportation became easier, thanks to the introduction of sail technology at the end of the 3rd millennium BC (Broodbank 1989: 327–9, 2000: 96).

Obsidian islands

I have also considered four islands that are also obsidian sources. They are, in order of increasing size, Palmarola, Lìpari, Pantelleria, and Melos. On average, occupation periods on the islands (ca. 2250 years) were much longer than abandonment periods (ca. 450 years) (table 7.6). This is true even if Palmarola is excluded, since its occupation record does not relate to permanent settlement but repeated visitation (ca. 1900 years vs. 240 years).

When viewed in pairs, it emerges that the two smaller islands (Lìpari and Palmarola) experienced slightly longer occupation (ca. 2800 years) and shorter abandonment periods (ca. 350 years), which may be related to the fact that they are the closest to their respective mainlands. Pantelleria and Melos, which are the largest, experienced slightly shorter periods of occupation than the smaller ones (ca. 2300 years) and twice as long abandonment periods (ca. 700 years). This is mainly due to the effect of Pantelleria's isolation, as individual abandonment figures in the table for the two islands indicate, but also to the high value attributed to Palmarola. If the island is excluded, it emerges that the two larger islands, Pantelleria and Melos,

Island	Size (sq. km)	Max. alt (m)	Dist NM (km)	Dist NL (km)	Ground water	Rain (mm/yr)	Min resources	Earliest col. mill BC	Earliest aban. mill BC	Tot. occupation (yrs)	Tot. abandonment (yrs)
Palagruza	0.3	90	130	0.2	No	n/a	Y	6	5	1000+700+2600	2500+1200
San Nicola	0.4	75	20	0.2	No	500	N	1	1 AD	3000	200
San Domino	2	116	20	0.2	No	500	N	6	4	2000+500	5000
Panarea	3.4	421	42	14	No	500	N	5	4	1500+1800+1500	500+1700
Alicudi	5.2	675	87	15.5	No	500	N	3	2	1000+500	3000
Filicudi	9.5	774	47	15.5	No	500	N	6	4	2000+1300+1500	1000+1700
Stromboli	12.6	924	56	18.25	No	500	N	3	2	1500+1500	2000
Salina	26.8	962	43	4.25	No	500	N	6	4	2000+1300+1500	1000+1700
Lipari	37.6	602	30	0.875	No	500	Y	6	1	5250+200+900	350+300
Formentera	82	202	95	3.7	No	370	N	2	2	700+1700+400	900+300
Pantelleria	83	836	102	70	No	350	Y	3	2	1000+2500	1000
Kea	130.6	568	12	9	Yes	492	N	4	3	1800+800+1500+700	300+400+300
Melos	150.6	751	105	2.5	No	435	Y	4	2	2900+1200+1300	400+200
Kythera	280	507	15	15	Yes	662	N	5	2	3900+1350+900	400+450
Hvar	312	626	4	2	No	800	N	6	2	4200+800+1900	700+400
Naxos	428	1000	132	7	Yes	384	Y	5	.	7200	-
Jerba	568	40	2	2	No	200	N	1	.	2400	-
Ibiza	572	475	92	3.7	Yes	400	N	2	2	700+2650	650
Average per island (without Cyprus)	155	521	59	10	Y=4/18	477	Y=5/18	.	.	4476 (ca. 1890 yrs/period)	1503 (ca. 1057 yrs/period)
Cyprus	9251	1950	69	69	Yes	500	Y	10	10	500+3000+6700	800+1000
Average (with Cyprus)	610	592	59	13	Y=5/19	479	Y=6/19	.	.	4763 (ca. 1984 yrs/period)	1518 (ca. 1047 yrs/period)

Note: alt: Highest point; dist NM: distance to nearest mainland; Dist NL: distance to nearest land (mainland or other island). N.B. Palmarola and Vulcano not included as not permanently settled.

Table 7.4: Assessing overall patterns.

Island	Earliest col. mill BC	Earliest aband. mill BC	Total prehistoric occupation (yrs)	Occupation bridging P and H	Total historic occupation (yrs)	Total prehistoric abandonment (yrs)	Abandonment bridging P and H.	Total historic abandonment yrs
Palagruza	6	5	1000+700	2600	.	2500+1200	.	.
San Nicola	1	1 AD	.	3000	.	.	.	200
San Domino	6	4	2000	.	500	.	5000	.
Panarea	5	4	1500+1800	.	1500	500	1700	.
Alicudi	3	2	1000	.	500	.	3000	.
Filicudi	6	4	2000+1300	.	1500	1000	1700	.
Stromboli	3	2	1500	.	1500	.	2000	.
Salina	6	4	2000+1300	.	1500	1000	1700	.
Lipari	6	1	5250	.	200+900	.	.	350+300
Formentera	2	2	700	1700	400	900	.	300
Pantelleria	3	2	1000	2500	.	.	1000	.
Kea	4	3	1800+800	1500	700	300+400	.	300
Melos	4	2	2900	1200	1300	400	.	200
Malta	6	.	.	7500	.	.	.	.
Kythera	5	2	3900	1350	900	400	.	450
Hvar	6	2	4200+800	.	1900	700	400	.
Naxos	5	.	7200	.	.	.	.	.
Jerba	1	.	.	2400	.	.	.	.
Ibiza	2	2	700	2650	.	650	.	.
Cyprus	10	10	500+3000	6700	.	800+1000	.	.
Mean[1]	.	.	2443	1655	665	588	825 (605[2])	105
Mean[3]	.	.	2035	3009	983	839	2062 (1642[2])	300

Notes: Palmarola and Vulcano not included as not permanently settled. [1]average by island; [2]without San Domino; [3]average by period.

Table 7.5: Assessing prehistoric (P) vs. historic (H) occupation and abandonment (approximate values).

Island	Size sq km	Max alt m	Dist NM km	Dist NL km	Ground water	Rain mm/yr	Min. res.	1st col. mill. BC	1st aband. mill. BC	Tot occupation years	Tot aband. years
Palmarola	1.4	262	32	11	No	n/a	Y	4	.	5000[1]	.
Lipari	37.6	602	30	0.9	No	500	Y	6	1	5250+200+900	350+350
Pantelleria	83	836	102	70	No	350	Y	3	2	1000+2500	1000
Melos	150.6	751	105	2.5	No	435	Y	4	2	2900+1200+1300	400+200
Average	68	613	67	21	No	428	Y	.	.	2250[2]	460[2]
										1900[3]	238[3]

Note: [1]sporadic occupation; [2]average is by period; [3]Without Palmarola.

Table 7.6: Obsidian islands. Geographical and occupation data.

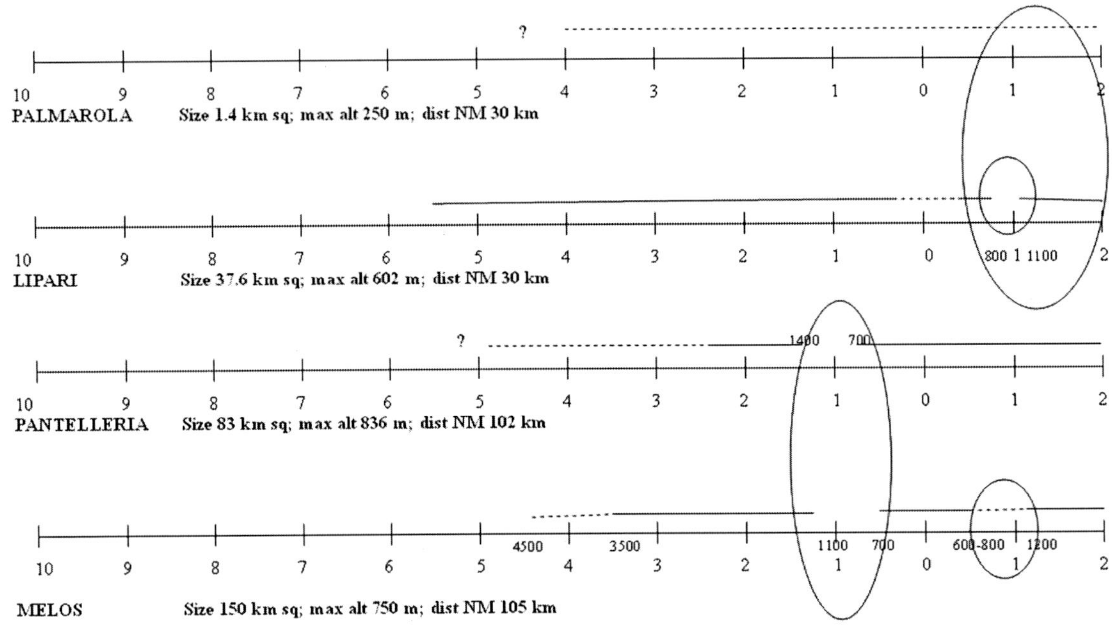

Figure 7.21: Timelines for the obsidian islands.

experienced longer occupation periods than the small ones. The abandonment periods of Lìpari and Melos are similar in length (350 and 400 years respectively).

Islands smaller than 10 sq km

The average occupation period for these islands is ca. 1700 years and the average abandonment ca. 1800 years (table 7.7). This means that, unlike what happens when all the islands are considered (table 7.5), occupation periods are on average shorter (ca. 1700 vs. 2000 yrs) and abandonment periods longer (ca. 1800 vs. 1000 yrs) on smaller islands. Abandonment periods lasted slightly longer than occupation on these islands, which is the opposite to the general trend noted for the islands overall. This may result from their size but also from a combination of other factors: although rainfall is adequate and a few have fertile soils, none of the small islands in question has water sources.

Comparing islands with distances to mainland less than and greater than 50 km

Fifty km was taken as representing a convenient break in the sample, but with 20 km being the distance a canoe can cover in one day (Broodbank 2000: 287), any journey beyond 50 km would involve planning and increase danger. Stròmboli and Alicudi have been included in this group, although they are more than 50 km away from the nearest mainland (Sicily). This is to take into account the 'stepping-stone effect': they are the only islands in the sample lying at less than a day away from nearby islands (Panarea and Filicudi respectively) that in turn are less than 20 km away from the nearest mainland. Although most of the islands excluded from this group lie close to other islands (generally much less than 20 km), apart from Pantelleria and Cyprus (ca. 70 km), all the islands more than 50 km from the mainland lie close to islands that in turn are distant (>20 km) from their nearest mainland.

On average, for the first group of islands, occupation periods lasted longer than abandonment (ca. 1700 vs. 1100 years). The same applies to the second group, but here occupation actually lasted longer (on average ca. 2600 years), and abandonment periods were shorter (ca. 650 years). On average, the islands in the second group are by far larger than those in the first group and most have water sources or mineral resources (or both). This combination of factors may have been more influential than distance

Island	Size sq km	Max Alt m	Dist NM Km	Dist NL km	Ground water	Rain mm/yr	Min. res.	1st col. Mill.	1st aband. Mill.	Tot occupation years	Total abandonment yrs
Islands smaller than 10 sq km											
Palagruza	0.3	90	130	0.2	No	n/a	Y	6 BC	5 BC	1000+700+2600	2500+1200
San Nicola	0.4	75	20	0.2	No	500	N	1 BC	1 AD	3000	200
Palmarola	1.4	262	32	11	No	n/a	Y	4 BC	.	5000[1]	.
San Domino	2	116	20	0.2	No	500	N	6 BC	4 BC	2000+500	5000
Panarea	3.4	421	42	14	No	500	Y	5 BC	4 BC	1500+1800+1500	500+1700
Alicudi	5.2	675	87	15.5	No	500	N	3 BC	2 BC	1000+500	3000
Filicudi	9.5	774	47	15.5	No	500	N	6 BC	4 BC	2000+1300+1500	1000+1700
Mean	3	345	54	8	No	500	Y=3/7			1727[2]	1867[2]
Islands with dist. NM>50 km											
Jerba	568	40	2	2	No	200	N	1 BC	.	2400	.
Hvar	312	626	4	2	No	800	N	6 BC	2 BC	4200+800+1900	700+400
Kea	130.6	568	12	9	Yes	492	N	4 BC	3 BC	1800+800+1500+700	300+400+300
Kythera	280	507	15	15	Yes	662	N	5 BC	2 BC	3900+1350+900	400+500
San Nicola	0.4	75	20	0.2	No	500	N	1 BC	1 AD	3000	200
San Domino	2	116	20	0.2	No	500	N	6 BC	4 BC	2000+500	5000
Lipari	37.6	602	30	0.9	No	500	Y	6 BC	1 BC	5250+200+900	350+300
Panarea	3.4	421	42	14	No	500	N	5 BC	4 BC	1500+1800+1500	500+1700
Salina	26.8	962	43	4.3	No	500	N	6 BC	4 BC	2000+1300+1500	1000+1700
Filicudi	9.5	774	47	15.5	No	500	N	6 BC	4 BC	2000+1300+1500	1000+1700
Stromboli[3]	12.6	924	56	18.3	No	500	Y	3 BC	2 BC	1500+1500	2000
Alicudi[3]	5.2	675	87	15.5	No	500	N	3 BC	2 BC	1000+500	3000
Mean	116	524	32	8		513	Y=2/12			1700[2]	1129[2]
Islands with dist. NM>50 km											
Cyprus	9251	1950	69	69	Yes	500	Y	10 BC	10 BC	500+3000+6700	800+1000
Malta	246	253	85	10	Yes	500	N	6 BC	.	7500	.
Ibiza	572	475	92	3.7	Yes	400	N	2 BC	2 BC	700+2650	650
Formentera	82	202	95	3.7	No	370	N	2 BC	2 BC	700+1700+400	900+300
Pantelleria	83	836	102	70	No	350	Y	3 BC	2 BC	1000+2500	1000
Melos	150.6	751	105	2.5	No	435	Y	4 BC	2 BC	2900+1200+1300	400+200
Naxos	428	1000	132	7	Yes	384	Y	5 BC	.	7200	.
Mean	1545	781	97	24		420	Y=4/7			2663[3]	656[3]

Note: [1]sporadic occupation; [2]average is by period; [3]islands included as < 20 km away from islands that are < 50 km away from nearest mainland.

Table 7.7: Geographical data.

alone, and thus explain why these islands experienced longer occupation and shorter abandonment periods.

Conclusions

A reconsideration of island settlement and of the different meanings of colonization and abandonment in different contexts is timely. The long-term material record from the Mediterranean islands is ideal to carry out such reassessment, particularly in the light of the recent developments in island archaeology (e.g., Fitzpatrick 2004), and the potential for comparison with other areas (both insular and non-insular). As a general point, environmental factors *per se* were not an obstacle to island life in the Mediterranean islands considered in this study. Resources played a more prominent role in the Mediterranean than distance and size in determining overall occupation and abandonment periods, with the presence of obsidian considerably reducing the length of abandonment periods experienced. On the other hand, islands smaller than 10 sq km, suffered slightly longer periods of abandonment, which could be related, albeit not exclusively, to their size. It seems that colonizers made the most of what they had and islanders developed successful survival strategies by creating social networks. On the other hand, excessive involvement in networks and in non-reciprocal relations may also have had negative effects, exposing islanders to fluctuations in the networks themselves. Satellite islands would have smaller fluctuating populations than islands that were more centrally (physically or socially) located within networks: e.g., Henderson in relation to Pitcairn or Mangareva (Irwin 1992: 179; Weisler 1995: 380), Antikythera in relation to Kythera (Conolly pers. comm.), Salina to Lìpari; Ibiza to Mallorca, and Formentera to Ibiza (Dawson 2005).

Overall, it is evident that a number of different survival strategies were available to island communities facing difficulties. Gosden's definition of colonization as a 'rearrangement of time and space as people re-order themselves and their world' (1993: 24; cf. Broodbank 2000: 110) captures these complexities well. In the long run, certain survival strategies may even have been implemented as a pre-emptive measure, if difficulties could be anticipated and alternatives, whether real or perceived, were available. Risk can be solved in different ways, and in that respect, abandonment and recolonization should be seen as a last resort. However, the abandonment of several Mediterranean islands in the recent past and in the present (Baggioni and Hache 2000) demonstrates that people abandon islands long before survival itself is at stake. With obvious caveats, this may offer a parallel to what happened in prehistory, when islands, including some large ones, were repeatedly abandoned and recolonized (before, during, and particularly after the Neolithic). Braudel wrote that 'the fate of many Mediterranean islands was a precarious, restricted, and threatened life' (1972: 154). This is certainly true for the historical period, which was often bleak for the islands, which suffered famines, epidemics, and pirate incursions. However, we should reject the hypothesis of 'insular fatality': 'islands in the Mediterranean are not subject to deterministic fates just because of their island status' (King and Kolodny 2001: 257). Rather, their abandonment and in some cases recolonization can be seen as a peripatetic and perhaps opportunistic means of survival: quite simply, if abandonment did not work as a strategy, in terms of moving on physically and/or mentally, other solutions would have been selected instead.

Although it may appear counter-intuitive, abandonment contributed to Mediterranean island life being continuous when viewed overall. Island abandonment is not intrinsically 'different' from any other kind of regional abandonment, since islands can undergo different types of abandonment, which find parallels in mainland situations, and which, in turn, are related to different kinds of activities. With regard to the nature of islands and insularity, and whether the colonization of islands is different from that of any other territory, evidence from the Mediterranean suggests that it is, since generally speaking Mediterranean islands were settled later than mainland territories, although this may be more a reflection of the type of evidence that researchers have been seeking and of the meaning given to 'colonization'. Rather, island life is best gauged on a relative rather than an absolute scale of enquiry, which emphasizes that concepts such as 'successful' colonization, 'established', or 'long-term' occupation should be assessed on a case-by-case basis against general patterns.

References cited

Adams, J.W. and Kasakoff, A.B. (1976). Factors underlying endogamous group size. In C.A. Smith (ed.) *Regional analysis II: Social Systems,* pp. 149–73. London and New York: Academic Press.

Ammerman, A.J., Flourentzos, P. McCartney, C., Noller, J. and Sorabji, D. (2006). Two new early sites on Cyprus. *Report of the Department of Antiquities, Cyprus*: 1–21.

Anderson, A. (2001). No meat on that beautiful shore: the prehistoric abandonment of subtropical Polynesian islands. *International Journal of Osteoarchaeology* 11: 14–23.

Baggioni, M.J. and Hache, J–D. (eds.) (2000). *Quel Statut pour les Îles d'Europe? (What Status for Europe's Islands?) Islands Commission of the Conference of Peripheral Maritime Regions of Europe.* Editions L'Harmattan. ISBN: 2-7384-9250-9. http://www.eurisles.com/statut_iles/EN/cadre.htm [Accessed 2003-10-06]

Bass, B. (1998). Early Neolithic offshore accounts: remote islands, maritime exploitations, and the trans-Adriatic cultural network. *Journal of Mediterranean Archaeology* 11: 165–190.

Bellwood, P. (1978). *Archaeological Research in the Cook Islands.* Honolulu: Bernice Pauahi Bishop Museum.

Binford, L. (2000). Review of Simmons. *American Antiquity* 65: 771.

Bowdler, S. (1995). Offshore Islands and maritime explorations in Australian prehistory. *Antiquity* 69: 945–958.

Braudel, F. (1972). The Mediterranean and the Mediterranean World in the Age of Philip II. London: Collins.

Broodbank, C. (1989). The longboat and society in the Cyclades in the Keros–Syros culture. American Journal of Archaeology 93: 319–37.

Broodbank, C. (1999). Colonization and configuration in the insular Neolithic of the Aegean. In Halstead, P. (ed.) Neolithic Society in Greece, pp. 15–41. Sheffield: Sheffield Studies in Aegean Archaeology.

Broodbank, C. (2000). *An Island Archaeology of the Early Cyclades.* Cambridge: Cambridge University Press.

Broodbank, C. (2006). The origins and early development of Mediterranean Maritime Activity. *Journal of Mediterranean Archaeology* 19/2: 199–230.

Broodbank, C. and Strasser, T.F. (1991). Migrant farmers and the Neolithic colonization of Crete. *Antiquity* 65: 233–45.

Butzer, K.W. (1996). Ecology in the Long View: Settlement histories, agrosystemic strategies, and ecological performance. *Journal of Field Archaeology* 23: 141–150.

Čečuk, B. and Radic, D. (1995). *Vela pilja: Pretpovijest Otoka Korule. Katlog Izlobe.* Dubrovnik: Knevev Dvor.

Cherry, J.F. (1981). Pattern and process in the earliest colonization of the Mediterranean islands. *Proceedings of the Prehistoric Society* 47: 41–68.

Cherry, J.F. (1990). The first colonization of the Mediterranean islands: a review of recent research. *Journal of Mediterranean Archaeology* 3 (2): 145–221.

Cherry, J.F. (2004). Mediterranean island prehistory: what's different and what's new? In S.C. Fitzpatrick (ed.) *Voyages of Discovery: The Archaeology of Islands*, pp. 233–250. Westport: Praeger Publishers.

Davis, J.L., Tzonou-Herbst, I. and Wolpert, A.D. (2001). Addendum: 1992–1999. In Cullen (ed.) *Aegean Prehistory. A Review*, pp. 77–94. Boston: American Journal of Archaeology Supplement 1.

Davis, S. (1984). Khirokitia and its mammal remains: a Neolithic Noah's Ark. In A. Le Brun (ed.) *Fouilles récentes à Khirokitia (Chypre) 1977–1981*, pp. 147–62. Paris ADPF. Éditions Recherche sur les Civilizations.

Dawson, H.S. (2005). *Island Colonization and Abandonment in Mediterranean Prehistory.* Unpublished Phd Thesis. Institute of Archaeology, University College London.

Evans, J.D. (1973). Islands as laboratories for the study of culture process. In A.C. Renfrew (ed.) *The Explanation of Culture Change: Models in Prehistory*, pp. 517–20. London: Duckworth.

Evans, J.D. (1977). Island archaeology in the Mediterranean: problems and opportunities. *World Archaeology* 9: 12–26.

Fitzpatrick, S.M. (2004). Synthesizing island archaeology. In Fitzpatrick (ed.) *Voyages of Discovery: The Archaeology of Islands,* pp. 3–20. *Westport: Praeger Publishers.*

Gaffney V., Kirigin B., Petri M., and Vujnovi N. (1997). *The Adriatic Islands Project. Contact, Commerce and Colonialism 6000BC–AD 600. Volume 1: The archaeological heritage of Hvar, Croatia.* Oxford: BAR International Reports.

Gosden, C. (1993). *Social Being and Time.* Oxford: Blackwell.

Graves, M.W. and Addison, D.J. (1995). The Polynesian Settlement of the Hawaiian Archipelago: integrating models and methods in archaeological interpretation. *World Archaeology* 26: 380–399.

Graves, M.W. and Longacre, W.A. (1982). Aggregation and abandonment at Grasshopper Pueblo, Arizona. *Journal of Field Archaeology* 9: 193–206.

Grove, A.T. and Rackham, O. (2001). *The Nature of Mediterranean Europe: An Ecological History.* Yale: Yale University Press.

Guilaine, J., Briois, F., Coularou, J., Carrere, I. (1995). L'établissement néolithique de Shillourokambos (Parekklisha, Chypre). Premieres résultats. *Report of the Department of Antiquities, Cyprus* 1995 (Nicosia): 11–36.

Guilaine, J., Briois, F., Coularou, J., Vigne, J.–D., Carrere, I. (1996). Shillourokambos et les débuts du Néolithique à Chypre. *Espacio, Tiempo y Forma,* sér. 1, *Prehistoria y Arqueologia* 9: 159–171.

Honea, K. (1975). Prehistoric remains on the island of Kythnos. *American Journal of Archaeology* 79: 277–9.

Irwin, G. (1992). *The Prehistoric Exploration and Colonisation of the Pacific.* Cambridge: Cambridge University Press.

Jones, R. (1976). Tasmania: aquatic machines and off-shore islands, in G. de G. Sieveking, I.H. Longworth and K. E. Wilson (ed.) *Problems in Economic and Social Archaeology.* pp. 235–263. London: Duckworth.

King, R. and Kolodny, E. (2001). Marginal or nodal? Towards a geography of Mediterranean islands. In R.King, P. de Mas and J.M. Beckin (eds.) *Geography, Environment and Development in the Mediterranean.* pp. 237–260. Brighton and Portland: Sussex Academic Press.

MacArthur, R.H. and Wilson, E.O. (1967). *The Theory of Island Biogeography.* Princeton, NJ: Princeton University Press.

McCartney, A.P. and Veltre, D.W. (1999). Aleutian island prehistory: living in insular extremes. *World Archaeology* 30/3: 503–515.

MacCluer, J.W and Dyke, B. (1976). Minimum size of endogamous populations. *Social Biology* 23/1:1–12.

Patton, M. (1996). *Islands in Time: Island Sociogeography and Mediterranean Prehistory.* London: Routledge.

Peltenburg, E. (2003). *The colonization and settlement of Cyprus: investigations at Kissonerga–Mylouthkia, 1976–1996.* Sävedalen: Åström.

Perlès, C. (1987). *Les industries lithiques tailées de Franchthi (Argolide, Grèce). I. Présentation générale et industries paléolithiques (Excavations at Franchthi Cave, Greece, Fascicle 5).* Bloomington: Indiana University Press.

Ramis, D., Alcover, A.A., Coll, J., and Trias, M. (2002). The chronology of the first settlement of the Balearic Islands. *Journal of Mediterranean Archaeology* 15 (1): 3–24.

Robb, J. (2001). Island identities: ritual, travel and the creation of difference in Neolithic Malta. *European Journal of Archaeology* 4 (2): 175–201.

Robb, J. and Van Hove, D. (2003). Gardening, foraging and herding: Neolithic land use and social territories in southern Italy. *Antiquity* 77: 241–254.

Rockman, M. (2003). Knowledge and learning in the archaeology of colonization. M. Rockman and J. Steele (eds.) *Colonization of Unfamiliar Landscapes. The Archaeology of Adaptation,* pp. 3–24. Routledge, London.

Rockman, M. and Steele, J. (2003). Editors' introduction. In M. Rockman and J. Steele (eds.) *Colonization of Unfamiliar Landscapes. The Archaeology of Adaptation,* pp. xix–xxii. Routledge, London.

Sampson, A. (1996). Excavations at the cave of the Cyclope on Youra, Alonessos. In E. Alram–Stern (ed.) *Die Ä gäische Früzheit, 2. Serie. Forschungsbericht 1975–1993, 1:*

Das Neolithikum in Griechenland mit Ausnahme von Kreta und Zypern, pp. 507–20. Vienna: Austrian Academy of Sciences.

Sampson, A. (1998). The Neolithic and Mesolithic occupation of the cave of the Cyclope, Youra Alonessos, Greece. *Annual of the British School at Athens* 93: 1–22.

Schiffer, M.B. (1976). *Behavioural Archaeology*. New York: Academic Press.

Simmons, A.H. (1999). *Faunal extinction in an island society: pygmy hippopotamus hunters of Cyprus*. New York: Plenum.

Spriggs, M. and Anderson, A. (1993). Late colonization of East Polynesia. *Antiquity* 67:200-17.

Stoddart, S. (1999). Long-term cultural dynamics of an island community: Malta 5500 BC–2000 AD. In R.H. Tykot, J. Morter and J.E. Robb (eds.) *Social Dynamics of the Prehistoric Central Mediterranean*, pp. 137–148. Volume 3 Accordia Specialist Studies on the Mediterranean. Accordia Research Institute, University of London.

Strasser, T.F. (2003). The subtleties of the seas: thoughts on Mediterranean island biogeography. *Mediterranean Archaeology and Archaeometry* 3/2: 137–148.

Trantalidou, K. (2004). Glimpses on the Aegean communities during the first millennia of the Holocene period: the zooarchaeological point of view. *Horizon. A Colloquium on the prehistory of the Cyclades. 25–28 March 2004*. Volume II. Cambridge: MacDonald Institute for Archaeological Research. University of Cambridge.

Trump, D. (2002). *Malta: Prehistory and Temples*. Valletta, Malta: Midsea Books.

Vayda, A. and Rappaport, R.A. (1963). Island cultures. In F.R. Fosberg (ed.) *Man's Place in the Island Ecosystem*, pp. 133–44. Honolulu, HI: Bishop Museum.

Vigne, J.-D. (1989). Le peuplement paléolithique des î les: le débat s'ouvre en Sardaigne. *Les nouvelles de l'archéologie* 35: 39–42.

Vigne, J.-D. (1996). Contribution des peuplements de vertébras insulaires à la connaissance de la navigation préhistorique en Méditerranée. In *Pour qui la Méditerranée au 21eme siecle? Navigation, echanges et environment en Méditerranée*, pp. 65–76. Actes du Colloque Scientifique, Montepellier France 11–12 Avril 1996.

Waldren, W.H. and Ensenyat, J.A. (eds.) (2002). *World Islands in Prehistory. International Insular Investigations. V Deia International Conference of Prehistory*. 301–311. BAR International Series 1095.

Weisler, M.I. (1995). Henderson Island prehistory: colonization and extinction on a remote Polynesian island. *Biological Journal of the Linnean Society* 56: 377–404.

Weisler, M.I. (1996). Taking the mystery out of the Polynesian 'mystery' islands: a case study from Mangareva and the Pitcairn group. In J.M. Davidson, G. Irwin, B.F. Leach, A. Pawley, and D. Brown (eds.) *Oceanic Culture History: Essays in Honour of Roger Green*, pp. 615–29. Dunedin: New Zealand Journal of Archaeology Special Publication.

Williamson, I. and Sabath, M.D. (1982). Island populations, land area, and climate: a case study of the Marshall Islands. *Human Ecology* 10: 71–84.

Wobst, H.M. (1974). Boundary conditions for Palaeolithic social systems: a simulation approach. *American Antiquity* 39: 147–78.

Chapter 8

What may be learnt about the archaeology of islands from archaeologically derived models of the exploration of Polynesia, 1966–2001?

Doug Sutton, University of Waikato

Introduction

Polynesian archaeology is one regional specialization in the world-wide practice of archaeological investigations of islands, oceans and seas. It is timely to consider how Polynesian archaeology fits within that newly-articulated framework of theoretical and methodological advances concerned with islands. To do this, I examine the history of archaeologically-derived models of the exploration of Polynesia developed since the invention of radiocarbon dating.

The focus on exploration models has been chosen for four reasons. First, they are more ubiquitous than any other topic pursued by Pacific archaeologists in the recent past. Therefore they provide the best window on change over time in the ways in which Pacific prehistory is made. Second, models are chosen because they are at least as difficult to build as anything else Pacific archaeologists construct: the stratigraphical evidence for the designation of *earliest* is vaguest, the traces of anthropogenic environmental impacts are most ambiguous, and speculation is necessary to fill the yawning gaps between data points, which can have untoward influences. For those reasons the development of models of exploration test collective wit and perspicacity to an exceptional extent. Third, they occupy an esteemed place in the panoply of models produced by archaeologists. They are at once most publicized, most exalted and fiercely contested. Fourth and finally, those models are chosen as the focus of this work because they are of a kind necessary to almost all archaeological sequences. This makes them part of the currency and topicality which Pacific archaeology exchanges with other parts of the world of scholarship. They comprise one of the principal bases upon which the ability of Pacific archaeologists to speak to issues of global importance is considered by archaeologists elsewhere, many of whom are also building models of exploration and population movements.

A succession of Polynesian exploration models has built up over the last 40 years. Those considered here were published by Sinoto (1966), Kirch (1984), Irwin (1992), Spriggs and Anderson (1993), and Kirch and Green (2001). Detailed consideration of these models leads to a summary of their differences and similarities and the theoretical content of each model as expressed in the vocabularies used by its author or authors, and recognition of substantial variation in the use of evidential support.

Doing these things makes it possible to consider how points of difference between successive schemes could be parsimoniously resolved, rather than remaining in the guise in which they are have been presented—i.e., as elements in alternative narrative accounts which exist in opposition to one another because of fundamental differences. Implementation of the five suggested resolutions presented below would strengthen the contribution which Polynesian research is making to 'archaeology of islands, oceans and seas', and lead to further development of the investigative capability of islands archaeology insofar as those resolutions influenced islands archaeology elsewhere (e.g., see Terrell 2004).

Similarities

The models are similar to one another in the degree to which they share three features. The first is the use of arrows to represent a succession of point-to-point 'voyages' in Sinoto's terms or 'explorations' in Irwin's language which connected widely separated places and resulted in first settlement of formerly pristine island groups. In doing so they continue a long tradition in Pacific scholarship, going back to the arch-diffusionists. There are a couple of reasons for thinking that the pointed arrow may have outlived its usefulness.

The first is that since Sinoto's Voyage 1 was discredited and Biggs (1972) critically reviewed archaeological use of historical linguistics, exploration is seen as most likely to have occurred wherever possible on a nearest-neighbour basis rather than via a small number of 'heroic' long distance journeys. Second, the directional arrow was standard fare in illustrations of the prehistory of continents but is no longer widely used in either North America or Europe, although it has currency still in Australia (e.g., O'Connor and Chapell 2003: figure 1). There are alternative approaches by which exploration can be rigorously modelled and accurately represented. For instance, Hazelwood and Steele (2004) use a very different approach in considering whether intrinsic limitations are brought to the modelling of population dispersal when these are based on either variation in radiocarbon dates of early sites or variable distributions of discarded artefacts. They suggest that models must be built to test the veracity of each of those approaches, applying a statistical model based on population-level description of the demographic characteristics of the expanding populations involved in the European Neolithic transition and the late glacial colonization of the Americas population of the New World. This leads them to the view that narrow constraints limit the usefulness of radiocarbon and artifactual approaches to the mapping and timing of population movement.

The second shared feature is that the models express a framework of inferred facts dealing with four topics: source of first settlers, order of first settlements, date of each first settlement, and the intervals between them. This is more an assumed form rather than a necessary one. It mirrors the dendritic form in which linguistic histories and biological distances are often expressed but may not provide the clearest possible reflection of the realities of knowledge about exploration.

The third shared feature is a short vocabulary of process descriptors used in all the models. The meanings attributed to its key terms vary from model-to-model. I consider that further in the next section.

Differences

The models differ from one another in: the chronologies they offer; their uses of process descriptors; what is taken into account as evidence, and; how different categories are treated in relation to one another. I consider these issues in the following four subsections.

Chronologies

Inferred first settlement dates in the models under review are such that the values for:

1. Samoa and Tonga have been to 3500 BP and returned to Sinoto's value
2. Tuvalu and Tokelau are added by virtue of recent research
3. the Marquesas have been to 2100–2200 BP and returned to Sinoto's value
4. Easter Island appears to be the most volatile. However, Kirch and Green's estimate is the only one outside the range of the other three
5. Hawaii appears to be volatile too, but again all values except the Kirch and Green estimate overlap
6. the Society Islands are receding in time, though primarily on the basis of inference as to what might be found there
7. Mangareva was important in the most recent model
8. Kermadecs is uncertain due to limited fieldwork
9. New Zealand is becoming more recent
10. the Chathams is the most recent.

Turning to the intervals between first settlement events it is evident that only two of Sinoto's pairs of consecutive first settlement dates were large enough to be separated by the radiocarbon assays available at the time. In effect he was arguing for the archaeologically instantaneous first settlement of Society Islands, Marquesas, Easter Island and Hawaii.[1]

Kirch's (1984) model includes no fewer than four pauses, some are archaeologically attested while others are supported either primarily or

[1]Davidson (1984: figure 21ff) made this point on the basis of her review of sources relating to Polynesian voyaging which were published between 1957 and 1979 (see also Davidson 1993).

entirely by reconstructions of the order in which language subgroupings are thought to have separated. At critical points, linguistic analysis is used to infer the order of separations and archaeological data are then used to establish the absolute date in time at which or by which a separation occurred. Linguistic inferences are used to repair weakness in archaeological data in the service of historical anthropology. Irwin's model represents Polynesian first settlement as a continuous process which occurred at an accelerating rate until faced with the discovery of New Zealand at which point it paused. Kirch and Green (2001) smooth most of the pauses out of Kirch's (1984) staccato model and bring the ages of first settlements of western Polynesia, the Marquesas and Hawaii to within archaeological range of Sinoto's values. They abbreviate the Easter Island sequence and lengthen the past of the Society Islands significantly. Sinoto's suggestion that the Marquesas acted as a primary dispersal centre is long gone (Allen 2004). Attribution of a similar role to central eastern Polynesia is questionable because recognition of the continuous nature of the exploratory process undermines the need for 'centers' from which dispersal emanated. In this respect the Polynesian sequence parallels several continental models in which punctuated movements of anatomically modern humans (AMHs) between nodes have been disproved as research has intensified, more fully revealing continuous movement.

Pauses are very important in the inferred chronology of Polynesian exploration. They have reduced in number and migrated eastwards since 1970. As Irwin (1981: 482) noted in 1981,

> 10 years ago a pause model was set in Western Polynesia itself... It is now known that the whole area was more 'equally' early than first thought. The pause model did not work... But now the same old kind of model has simply shifted to the border between Western and Eastern Polynesia.

In 2000 Irwin suggested that the West Polynesian pause may have been confused with a pause at the Andesite Line, seen in West Polynesia and in between, in Micronesia (Irwin 200). Two pauses remained in Irwin's (1992) *Prehistoric Exploration and Colonisation of the Pacific.* Continuing the eastward shift of such pauses, one was between tropical East Polynesia and New Zealand, which took 500 years. The other is between New Zealand and the Chathams to which Irwin attributes a further five centuries.

These pauses too are under question. The first is no longer accepted by a narrow majority of members of the New Zealand Archaeological Association who attended the recent conference in Russell (Prickett 2002). In addition, as Davidson (1984: 24) and Sutton (1980) earlier noted, the radiocarbon dates upon which Irwin (1992: 107–108) and McFadgen and Yaldwyn (1984) depend for the pause prior to the first settlement of the Chathams do not relate to evidence of first settlement there. This suggests that further changes will be forthcoming, and that they may continue a trend towards less pause, more pace, greater continuity.

Use of process descriptors

The shared vocabulary of process descriptors is problematic, not in its content but in its muddled usage. For instance, Kirch discriminated between dispersal and colonization in his book on the *Evolution of the Polynesian Chiefdoms* (1984: 71–86), but used discovery, colonization and settlement as more or less equivalent terms, in his influential essay on *Rethinking East Polynesian Prehistory* (1986). In 1992 Geoff Irwin clarified the meanings of key terms very significantly. 'Exploration' was seen as being based on systematic methods and strategies and not referred to as dispersal, avoiding any sense of 'scatter'. 'First settlement' could be proven by the earliest well-attested evidence of human presence, whether archaeological or palaeoenvironmental, and was sometimes inferred in the absence of empirical evidence. 'Second' or 'later settlement' also occurred, but was not referred to as 'secondary', which freed such events of any connotation of lesser importance. 'Colonization' occurred when occupation was established. 'Post-colonization changes' included abandonment of islands, decline in voyaging, cultural and linguistic differentiation, and other sociocultural changes which have been seen as the outcomes of cultural adaptation by Kirch and Green (2001). Spriggs and Anderson (1993) refer to colonization as an 'event' which has occurred when early initial, or first settlement has occurred.[2] Kirch and Green (2001: 79–81) use the terms 'discovery' and 'colonization' to refer to distinct events, though their use of 'first settlement' combines both in the sense that first settlers are seen as discoverers who established settlements, thus colonising a place. They use the term 'settlement' as a gloss for both 'discovery' and 'col-

[2] See Spriggs and Anderson (1993) use of 'colonized' as a synonym of 'first settled' in their paraphrase of Sutton (1987)

onization'. Elsewhere they use 'discovery' and 'colonization' together, with the clear implication that each refers to a different phenomenon. In one passage they state that New Zealand was "discovered, explored, and eventually colonized around AD 1000–1200" (Kirch and Green 2001). Reading their other usages, summarized here, leaves one uncertain as to whether one, two or three discrete events were involved, and whether these are thought of as simultaneous, concurrent or consecutive.

What counts as evidence

Sources of evidence are variously used; all by some, sometimes, but neither always nor consistently, while far fewer sources of evidence are used by others. Explicit reasons are sometimes offered to explain inclusions or exclusions.

For instance Sinoto (1966) used typological studies of artefacts which had been excavated and more or less accurately radiocarbon dated by the standards of the time at which he wrote, though very little of that dating would pass as accurate enough now. Kirch (1984, 1986) used archaeological evidence, linguistic reconstructions, and palaeoenvironment data indicating human induced ecological change in his critique of Sinoto's work. Irwin (1992) argued against the use of linguistic phylogenetics, preferring instead the use of archaeological data which informed and was anticipated on the basis of his reconstruction of the logic underpinning Pacific navigation. Spriggs and Anderson (1993: 200) state that they use "the results of radiocarbon dating applied to pollen cores, sediment columns and archaeological samples from Hawaii and the rest of East Polynesia". Actually, they do less than that as most of the archaeological radiocarbon dates are rejected and most of the relevant palynological and sedimentological dates are unmentioned in their paper. Kirch and Green (2001) attempted a correlation of linguistic and archaeological evidence in a major project intended to advance historical anthropology, and based on "a triangulation method in which historical linguistics, archaeology, comparative ethnology and biological anthropology independently contribute their data and assessments to the common objective of historical reconstruction" (Kirch and Green 2001: 42). They also use palaeoenvironmental information, sometimes but not consistently.

All of this illustrates that there are fundamental issues at play. One is the degree to which palaeoenvironmental evidence is actually being used in building exploration scenarios. The other question is of a general form and vastly more important than the first: how can categories of evidence which are very different in terms of epistemology be used together to form an aggregate whole which is more reliable or more robust as historical reconstruction than any of its constituent elements would be if used as the sole basis of reconstruction? Reviews of Kirch and Green (2001) raised this question in critiques of the use of historical linguistics in triangulation method but that is only part of the quandary, because archaeologists typically use multiple, divergent sources of evidence so the combinatorial issue applies to much more than the triangulatory use of the protolexeme and semantic reconstructions. I comment further on this point later in the paper.

The need for resolutions

The purposes of Polynesian archaeology can be glossed as follows: to map movement of peoples into and throughout Remote Oceania, to do with sufficient accuracy to enable the processes of change in human biology, culture, society and ecology to be better understood, and to form by these means the basis of scholarly enquiry into causes of change, diversity and relative homogeneity seen in the Pacific and elsewhere in the world's oceans and seas.[3]

Detailed examination of the exploration models show that the inferred facts which are the focus of this paper have varied a lot but changed little over the last forty years, leaving an overall impression of unstable churn rather than directional change in historical reconstruction based on increased precision. This is exceptional in contemporary science.

Means by which those resolutions can be achieved

I suggest things could be improved through change in five areas:

1. adoption of an islands framework
2. development of higher-resolution, continuous chronological sequences on Polynesian islands
3. the more rigorous use of process descriptors and the development of inferential time-place systematics as necessary parts of the investigative process

[3]This statement of purpose is influenced by papers by Terrell (2004), Anderson (2004) and Renfrew (2004), and proceeds from the idea that islands, can when considered carefully, make things visible.

4. use of high-resolution techniques in stratigraphical research, including cultural palynology
5. integration of excavation and environmental strategies so both are optimized.

First, the case for adoption of an islands framework is straightforward. Archaeology originated in Europe and the underpinnings of world archaeology were developed there and then in North America. Pacific archaeology inherited assumptions that came from continental contexts, the most general may date back to Thomsen and the origins of archaeology, others to Kroeber in the Americas and Kossina in Europe, some to the cultural evolutionists of the 1950s (see for example Hardin et al., 1960, and therein a remarkable foreword by Leslie A. White). But, as Jim Allen (2003) has noted recently, there can be basic differences between cultural and communication processes on continents and those which occur on islands in oceans and seas. He emphasizes the need to base assumptions about cultural process and interactions on the most directly relevant situations and to inform those assumptions with knowledge of the realities of island ecology, history and anthropology.

Second, there continues to be a pressing need for the development of fully accurate chronologies for the cultural sequences on each of as many islands and archipelagos as possible. It is important to note that current approaches to sampling for whole site dating, methods used for the combination of dates from within one site and between different, though interrelated stratigraphies are under question (Jones 2002; Jones and Nicholls 2002; Nicholls and Jones 2001). Taken together, these recent analyses indicate that for New Zealand the earliest date of occupation is currently misstated, while the duration and periodicity of occupation of key sites has been incorrectly calculated and is, as a result, quite wrong, and that full redress of these problems is a long-term undertaking (Higham and Jones 2004). This has obvious implications for the broader Pacific and suggests the need for new field and laboratory-statistical dating protocols, strengthening if not superseding those in use at present.

With those in place there could be increased emphasis on continuous cultural sequences of the long duration, and necessarily more consideration of synchronic cultural occurrences as they relate to neighbours and other contemporaries, antecedents and descendants.[4] There could also be an emphasis on the use of well-dated sequences as the single most important bench-mark alongside which natural and cultural changes could realistically be sized and scaled. In particular, the use of such independent chronology to clarify the strengths of historical linguistics as itself, rather than as a form of enquiry which, as Kirch and Green have urged, should not substitute for archaeological investigations in the Pacific.

Third, the use of process descriptors should be revised as advocated by many over the last twenty years (Davidson 1984: 222–225; Kirch 1984: figure 27*ff*; Irwin 2001; Graves and Addison 1995: 381). It is recognized that the discovery, colonization and settlement of Polynesian islands, and others, can be consecutive events with significant ecological, demographic and cultural implications (see for discussions of these issues in disparate islands research Broodbank 2000; Jones 2002; Vesteinsson et al. 2002; Whitley 2001). The key point of debate is whether archaeological methods can discriminate between discovery, colonization and settlement. In the Aegean, North Atlantic, including the Faroes, in the Falklands and the Caribbean archipelagos, those discriminations are routinely attempted in current research (references above and papers in Fitzpatrick 2004; Burney et al. 1993). In the Pacific, opinions are mixed but moving positively. For example, Anderson's position has recently adopted the Graves and Addison (1995) scheme in considering the early phase of the sequence in Fiji. I am optimistic that evidence of discovery can be satisfactorily identified.

The blurring of process descriptors in Polynesian archaeology, described earlier in this essay, did not help the development of interest in discriminating between discovery, colonization and settlement. On the contrary, it has apparently had the effect of allowing the continuation of lower resolution research in Polynesia than occurs elsewhere in islands archaeology. It would thus be useful to move the use of process descriptors from passive and muddled to clear, consistent and inferential. This would involve the use of an analytical vocabulary of descriptors based on the key issues under investigation, and aligned with the lived realities of the history of Polynesia. The terms discovery, exploration, colonization and settlement appear have broad support. In future, they could be used widely and consistently. In the general case, attribution of a site to one of these descriptor categories would be justified only by the presence of evi-

[4]This remark is, in part, a reaction to an historical tendency towards the treatment of Lapita as a discrete culture, of brief duration, strangely disembodied from other busy, interactive Pacific cultures and societies.

dence directly related to the process content of the descriptor in question. For instance, a site attributed to discovery could be expected to show evidence of pristine environment, initial anthropogenic impact, earliest reliable dates, limited use of more inaccessible local resources and artefact forms and materials from a source region.

Consistent space-time systematics could then develop through the analysis of well-dated portable and non-portable artefact assemblages which were discovered in strata attributed to specific descriptors. This would help to repair the present situation in which there are no agreed time-place systematics in Polynesian archaeology. Adopting the approach suggested here would make the use of time-place systematics an active part of the Pacific-wide investigative process.

Fourth, there is a pressing need for the use of high-resolution techniques in stratigraphical research, including cultural palynology in Polynesian research. For instance, there is at present a considerable mix of methods in use in palynological studies relevant to identification of the earliest human presence. Narrowing the range of methods used, to produce consistently high-resolution research would be advantageous.

Fifth, in their work work in northern Europe and the North Atlantic, Edwards and Mithen have sought the integration of excavation and environmental strategies so both are optimized (Edwards and Mithen 1995). Following that lead, this paper calls for a reintegration of archaeological chronometrics and cultural palynology into the core of archaeological methods, following their successful application in island contexts elsewhere in the world. This can be achieved simply in two key areas: (i) the inclusion of well-established palaeoenvironmental change when it is specifically enough dated to actually contribute to the objectives of archaeological research.[5] Obviously there will be concern at the widening gaps which exist between some palynological methods used in New Zealand archaeology and those now in place in the most acute investigations of the dispersal of Neolithic populations through Europe and across the North Atlantic. (ii) Further collaborative research with dating scientists to establish the general use of optimal sampling and statistical methods is needed, which should repair a schism between the two which was identified recently by Allen (2003: 38).

[5]This supports Anderson's (1995: 116) objection to the possibility that palynology might become 'archaeology by proxy'.

One invaluable source of insights into cultural processes is the library of specific understandings which has accumulated from anthropological observations of island life. So there may be very good reasons for making sure that interpretations of archaeological, linguistic and environmental data and whatever else is used archaeologically time to time are grounded in what are believed to be the most directly relevant understandings. Similarly, investigations should be scaled and dated on the basis of those insights. More often than not they will be from islands in oceans and seas, not the Pacific alone but the others also, and less often from continents.

Conclusion

Particular conclusions reached in the body of the paper are as follows:

- There is a high degree of churn in dates of discovery or, to make the same point differently, little by way of sustained directional change in the archaeological timing of Polynesian exploration.
- In the absence of an agreed time axis and the presence of diverse goals and interpretative methods, there are no generally agreed archaeological time-place systematics which apply across Pacific archaeology.
- Instead, several semi-discrete, variously founded and differently intended schema are in simultaneous use.
- Under those conditions there is plurality and dissonance in the current conceptual framework.
- However, these conditions may not prevail. This is because of the momentum in the archaeology of islands as a new, distinctive research practice.
- In particular, it supports the recognition of islands as distinctive, highly valuable research settings (Burney 1997; Terrell 2004) and the further development of research methods capable of enabling islands to more clearly 'make things visible'.
- A focus on islands would see a shift of emphasis from culture area approaches, and the content and boundary condition issues they evoke, to a more generalized search for understandings of cultures that inhabited oceans and seas.

References cited

Anderson, A.J. (1995). Current approaches in east Polynesian colonization research. *Journal of the Polynesian Society* 104: 110–133.

Anderson, Atholl. (2004). Initial human dispersal in Remote Oceania: Pattern and explanation. In C. Sand (ed.), *Pacific archaeology: Assessments and prospects. Proceedings of the International Conference for the 50th Anniversary of the first Lapita excavation, Koné-Nouméa 2002*. Le Cahiers de L'Archéologie en Nouvelle-Calédonie 15, pp. 71–84. Nouméa: Museum of New Caledonia.

Anderson, A.J. (2004). Islands of ambivalence. In S. Fitzpatrick (ed.), *Voyages of Discovery. The Archaeology of Islands*, pp. 251–274. London: Praeger.

Allen, J. (2003). Discovering the Pleistocene in island Melanesia. In C. Sand (ed.), *Pacific Archaeology: Assessments and Prospects.* Proceedings of the International Conference for the 50th Anniversary of the first Lapita excavation, Koné-Nouméa 2002. Le Cahiers de L'Archéologie en Nouvelle-Calédonie 15, pp. 33–42. Nouméa: Museum of New Caledonia.

Allen, M. (2004). Revisiting and revising Marquesan culture history: new archaeological investigations at Anaho Bay, Nuku Hiva Island. *Journal of the Polynesian Society* 113(2): 143–196

Biggs, B. (1972). The implications of linguistic subgrouping with special reference to Polynesia. In R.C. Green and M. Kelly (eds.), *Studies in Oceanic Culture History* 3: 143–152. Pacific Anthropological Records 13. Honolulu: Bishop Museum Press.

Broodbank, C. (2000). *An Island Archaeology of the Early Cyclades.* Cambridge: Cambridge University Press.

Burney, D. (1997). Tropical Islands as palaeoecological laboratories: gauging the consequences of human arrival. *Human Ecology* 25: 437–457.

Burney, D., L.P. Burney, R.D.E. MacPhee (1993). Holocene charcoal stratigraphy from Laguna Tortuguero, Puerto Rico, and the timing of human arrival on the island. *Journal of Archaeological Science* 21: 273–281.

Davidson, J.M. (1984). *The Prehistory of New Zealand.* Auckland: Longman Paul.

Davidson, J.M. (1993). Issues in New Zealand prehistory since 1954. In M. Spriggs, D.E., Yen, W. Ambrose, R. Jones, A. Thorne, and Ann Andrews (eds.), *A Community of Culture. The People and Prehistory of the Pacific*, pp. 239–258. Canberra: The Australian National University.

Edwards, K.J. and S. Mithen (1995). The colonization of the Hebridean Islands of western Scotland: evidence from the palynological and archaeological records. *World Archaeology* 26(3): 348–365.

Fitzpatrick, S.M. (ed.), (2004). *Voyages of Discovery. The Archaeology of Islands.* London: Praeger.

Graves, M.W., and D.J. Addison (1995). The Polynesian settlement of the Hawaiian archipelago: integrating models and interpretation in archaeological interpretation. *World Archaeology* 26(3): 380–399.

Haberle, S. (1994). Anthropogenic indicators in pollen diagrams: problems and prospects for Late Quaternary palynology in New Guinea. In J.G. Hather (ed.), *Tropical Archaeobotany*, pp. 172–201. London: Routledge.

Harding, T.G., D. Kaplan, M.D. Sahlins and E.R. Service (1960). *Evolution and Culture.* Ann Arbor: The University of Michigan Press.

Hazelwood, L. and J. Steele (2004). Spatial dynamics of human dispersals: constraints on modelling and archaeological validation. *Journal of Archaeological Science* 31(6): 669–679.

Higham, T. and M. Jones (2004). Chronology and Settlement. In L. Furey and S. Holdaway (eds.), *Change Through Time*, pp. 215–250. Volume 26, New Zealand Archaeological Association. Auckland: New Zealand Archaeological Association.

Irwin, G.J. (1981). How Lapita lost its pots: the question of continuity in the colonisation of Polynesia. *Journal of the Polynesian Society* 90: 481–494.

Irwin, G.J. (1992). *Prehistoric Exploration and Colonisation of the Pacific.* Cambridge: Cambridge University Press.

Irwin, G.J. (2000). No man is an island: The importance of context in the study of the colonization and settlement of the Pacific Islands. In Atholl Anderson and Tim Murray (eds.), *Australian Archaeologist: Collected Papers in Honour of Jim Allen*, pp. 393–411. The Australian National University. Canberra: Coombs Academic Publishing.

Jones, M.D. (2002). *A Brief History of Time.* Unpublished Ph.D. Dissertation, Anthropology, University of Auckland.

Jones, M.D. and G. Nicholls (2002). New radiocarbon calibration software. *Radiocarbon* 44(3): 663–674.

Kirch, P.V. (1984). *The Evolution of the Polynesian Chiefdoms.* Cambridge: Cambridge University Press.

Kirch, P.V. (1986). Rethinking east Polynesian prehistory. *Journal of the Polynesian Society*, 95: 9–40.

Kirch, P.V. and R.C. Green (2001). *Hawaiki, Ancestoral Polynesia. An Essay in Historical Anthropology.* Cambridge: Cambridge University Press.

McFadgen, B.G. and J.C. Yaldwyn. (1984). Holocene sand dunes on Enderby Island, Auckland Islands. *New Zealand Journal of Geology and Geophysics* 27: 27–33.

Nicholl, G. and M.D. Jones (2001). Radiocarbon dating with temporal order constraints. *Journal of the Royal Statistical Society, Series C-Applied Statistics* 50: 503–521.

O'Connor, S. and J. Chappell (2003). Colonization and coastal subsistence in Australia and Papua New Guinea: different timing, different modes? In C. Sand (ed.), *Pacific archaeology: Assessments and prospects. Proceedings of the International Conference for the 50th Anniversary of the first Lapita excavation, Konè-Nouméa 2002.* Le Cahiers de L'Archèologie en Nouvelle-Calèdonie 15, pp. 17–32. Noumèa: Museum of New Caledonia.

Prickett, N.J. (2002). First settlement date and early rats: an opinion poll. *Archaeology in New Zealand* 45: 288–292.

Renfrew, C. (2004). Islands out of time? Towards an analytical framework. In S.M. Fitzpatrick (ed.), *Voyages of Discovery. The Archaeology of Islands*, pp. 295–304. London: Praeger.

Sinoto, Y.H. (1966). A tentative prehistoric cultural sequence in the northern Marquesas Islands, French Polynesia. *Journal of the Polynesian Society* 75: 287–303.

Spriggs, M. and A.J. Anderson (1993). Late colonization of east Polynesia. *Antiquity* 67: 200–217.

Sutton, D.G. (1980). The culture history of the Chatham Islands. *Journal of the Polynesian Society* 89: 67–93.

Terrell, J.E. (2004). Island model of reticulate evolution: The "Ancient Lagoons" Hypothesis. In S.M. Fitzpatrick (ed.), *Voyages of Discovery. The Archaeology of Islands*, pp. 203–223. London: Praeger.

Vesteinsson, O., T.H. McGovern and C. Keller (2002). Enduring impacts: social and environmental aspects of Viking age settlement in Iceland and Greenland. *Archaeologica Islandica* 2: 98–136.

Whitley, J. (2001). *The Archaeology of Ancient Greece.* Cambridge: Cambridge University Press.

Chapter 9

Farmers, fishers and whalemen: the colonization landscapes of Lord Howe Island, Tasman Sea, Australia

Kimberley Owens, Australian National University

Introduction

I here present preliminary results and interpretations from several months' fieldwork and extensive historic research undertaken on early settlement sites on the Australian offshore island of Lord Howe.[1] Following preliminary work by other researchers focused on confirming the antiquity of the island's human occupation, a more detailed investigation of the historic record and its revelations about the existing settlement landscape was desired to enable an informative and comparable case study of historically recorded human colonization on a Pacific island. Further, I present a study of the nature of island landscapes, the influence of Pacific trade and exchange networks and the development of communities in varying scales of isolation to situate Lord Howe Island in its own historic and geographic context.

Lord Howe Island

The Lord Howe Island group lies in the north Tasman Sea, 770 km northeast of Sydney (figure 9.1) and the main island and its 27 offshore islets are the eroded remains of a shield volcano that rose from the ocean floor approximately 6.9 million years ago. The group has a total area of 1540 hectares, with Lord Howe (LHI) accounting for 1455 hectares, making it the only island large enough to sustain human settlement. This settlement currently covers about 398 hectares, which is occupied by approximately 295 permanent residents, whose main industries include tourism and palm cultivation (Australian Bureau of Statistics 2001; Hutton 1998). The island's southern end is dominated by Mounts Gower and Lidgbird, which rise 875 m and 777 m above sea level respectively and account for half of the island's area (figure 9.2). These mountains, along with warm ocean currents flowing down from the Queensland coast have created a microclimate on the island that has fostered the development of many endemic species and subspecies of flora and fauna. This diversity includes the world's most southerly coral reef and, despite the extinction of several species of land birds and insects during 170 years of human occupation, LHI's ecology has remained largely intact. A permanent park preserve covering 70% of the island was proclaimed in 1953 and in 1982 the Lord Howe Group was inscribed on the World Heritage List on the basis of its unique natural beauty and ecological importance and thus management efforts have focused on maintaining the balance between human settlement and the island ecosystem (Edgecombe 1987; Hutton 1998).

LHI was sighted and named by Lt. Ball of the *Supply*, on the 17th of February 1788, while en route from Port Jackson to Norfolk Island. On return from Norfolk Island in early March, the *Supply* landed a company of people to claim the island for the crown and conduct preliminary explorations (Nicholls 1952). Primary geographical features were named and vegetation and fauna described including the abundance of turtle, tame birds and fish, some of which were taken back to the starving and scurvy ridden throng at Port Jackson (Edgecombe 1987). This initial bounty prompted the *Supply's* return in May 1788, and concurrent visits by several First Fleet transports en route to Norfolk Island and Canton.

Unlike the quickly observed evidence of human occupation on neighbouring Norfolk Island such as remnant banana plantations and stone tools, LHI did not reveal any indications of

[1] This forms a component of my PhD work being undertaken in the Research School of Pacific and Asian Studies at the Australian National University.

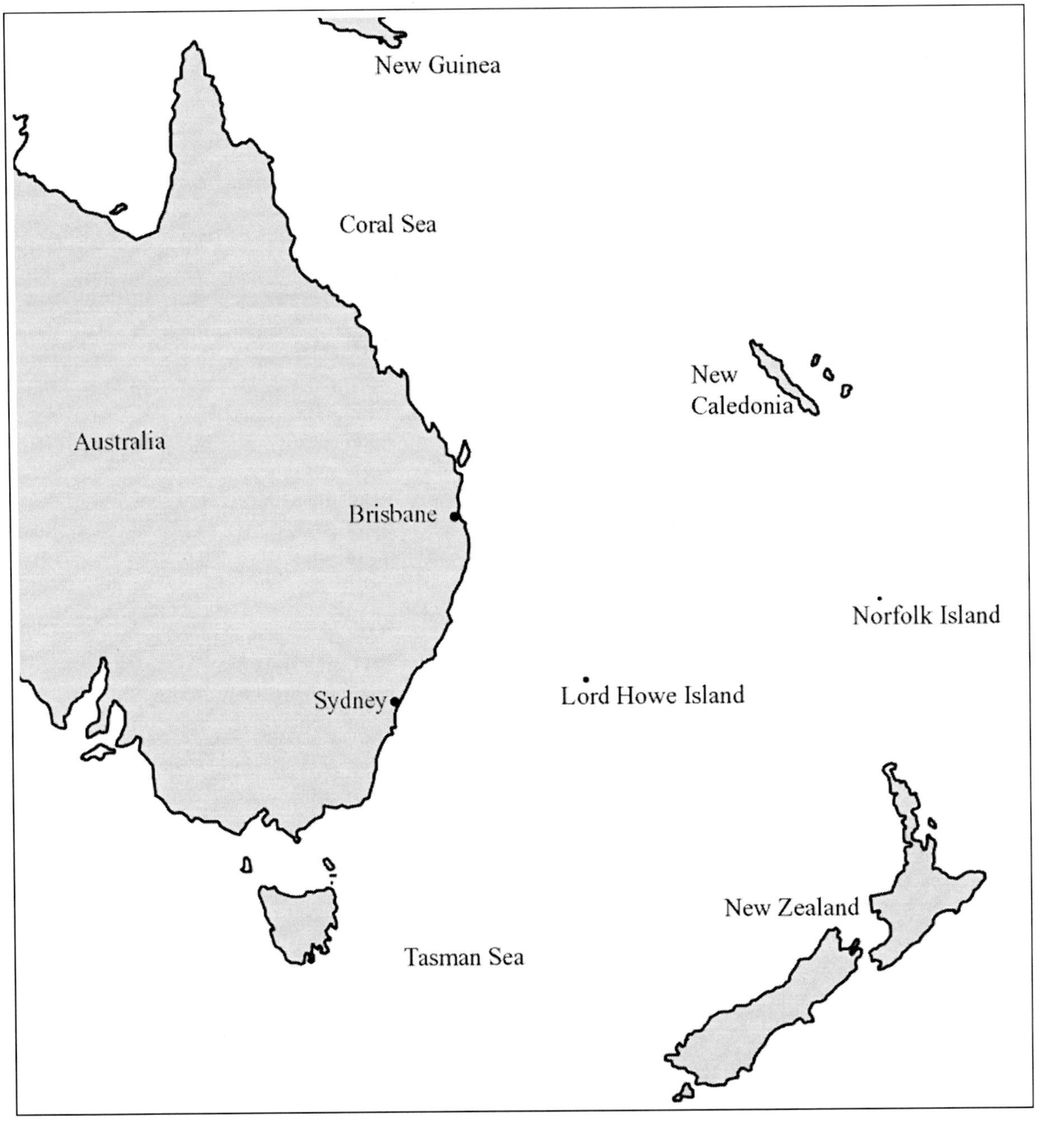

Figure 9.1: Lord Howe Island in relation to Australia, New Zealand and Norfolk Island.

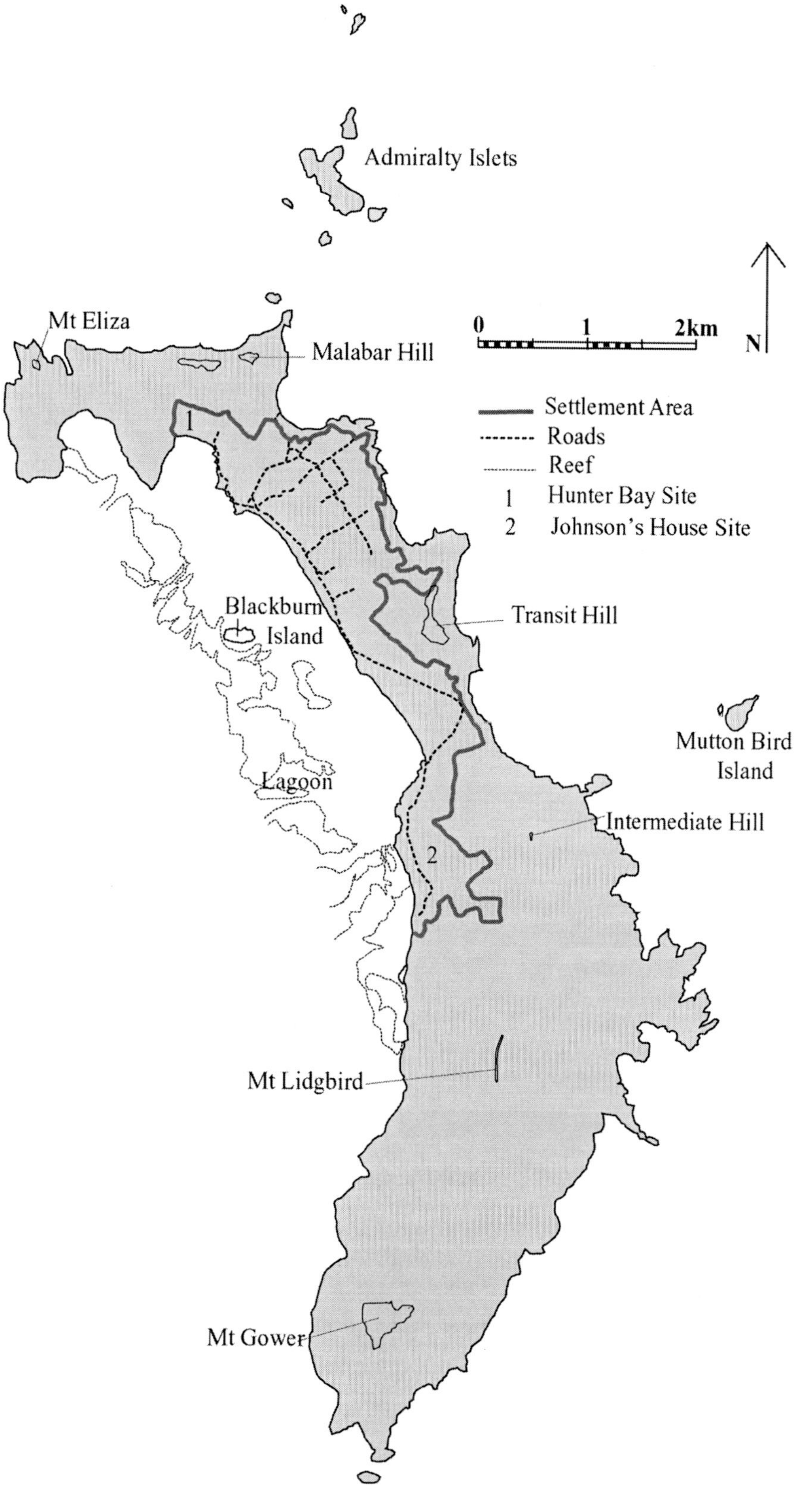

Figure 9.2: Lord Howe Island showing settlement extent and site locations.

prior human settlement during these early explorations. The incredible abundance and tameness of the island's marine and avi fauna, including game sized and flightless land birds lent weight to the idea of it being an untouched paradise:

> Great numbers of gannets, very large and fat, were walking with less fear and concern than geese in a farm yard; and they were taken by hand, with much more ease. (Gilbert nd:10)
>
> Partridges likewise, in great plenty, ran along the ground, very fat, and exceedingly well tasted. Several of those I knocked down, and their legs being broken, I placed them near me as I sat under a tree. The pain they suffered caused them to make a doleful cry, which brought five or six dozen of the same kind to them, and by that means I was able to take nearly the whole of them. I might not otherwise have procured so many; for, although they were by no means shy, yet they ran very fast when chased. No traces whatever of any human being were to be observed in this island, except those made by the people of the Supply when they were here before. (Gilbert nd:11)

Following several days of successful foraging, hunting and fishing, the *Supply* returned to Port Jackson without any of the highly prized turtle, and Governor Phillip concluded that due to the reported lack of fresh water or good anchorage that LHI was of little consequence for the future good of the colony (Finch and Finch 1967; Nicholls 1952). Despite the Governor's conclusion, colonial ships periodically stopped at LHI during voyages en route to Norfolk Island, and whaling and trade vessels began visiting for water, game and fire wood. These visits appear to have been relatively frequent as pigs and goats were released to provide regular supplies of meat and at least two species of land bird described by the first fleet ships had probably been hunted to near extinction prior to 1834 (Edgecombe 1987; Hutton 1998; Nicholls 1952).

In 1834 seven adults and two children were landed on LHI with the express purpose of establishing a trade depot to provide supplies to whaling and trade vessels (Hutton 1998). This small group quickly established a small settlement, including gardens and thus supplied produce and meat to passing ships in a fairly short time. This set the stage for LHI's future development, which initially was dependant on trade from whaling vessels hunting the 'middle grounds' between Sydney and the New Zealand coast (Nicholls 1952). LHI is almost in the middle of these grounds and during the peak year 1865 approximately 70 ships visited LHI, with up to 12 vessels in harbour at one time (Edgecombe 1987). Over time, a variety of crops were grown with varying success including: potatoes, onions, carrots, maize, pumpkins, cabbages, taro, arrowroot, ginger, coffee, papaws, figs, oranges, lemons, bananas, guavas, peaches, passionfruit, grapes, and sugar cane. Pigs and goats were raised from feral stocks on the island, local game birds and fish were gathered and eventually European fowl, cattle and some horses were introduced. Many of crops were grown primarily for islander use, but staples such as potatoes, pumpkin, onions, local 'cabbage palm', fish, pork, chevon and poultry were traded in exchange for goods such as tea, sugar, flour, clothing and other items that could not be produced locally (Edgecombe 1987; Nicholls 1952).

During the 1870s the whaling industry declined sharply with the wider introduction of petroleum products, and the LHI community became isolated from the world. Three years passed at one stage without a ship calling, and the settlement had to be entirely self sufficient, with no comforts such as tea or flour to lift the spirits of people who sat and watched their surplus produce rot for lack of traders (Edgecombe 1987; Nicholls 1952). With the eventual return of trade, islanders still lacked the means to acquire goods, as they had no money and fewer visitors to barter with. The need for a new industry was quickly realized, and it was soon found that LHI's unique climate had produced a palm tree, the Kentia, that was tolerant to cool climates and low sunlight, making it an ideal export to Europe as an indoor plant (Hutton 1998). The palm seed industry formed the basis of the island's economy from 1878 onwards and subsequently LHI developed other ventures including the export of fruit, vegetable and flower seeds, whole frozen fish and the growth of tourism. Today, tourism, meteorological work, environmental management and Kentia palm cultivation provide employment for the majority of islanders (Hutton 1998; Nicholls 1952).

The community on the island grew slowly but steadily, despite several settler groups selling up or leaving their improvements within the first 12 years. New settlers were often retired mariners wanting to start families, and other wandering individuals who ended up on LHI and decided to stay. Despite its size the LHI community was quite cosmopolitan with individuals from South

Africa, Portugal, the Gilbert Islands (Kiribati), America, Britain, New Zealand and Australia including Maori and Aboriginal individuals, and recent census statistics indicate similar variety in the present population (Australian Bureau of Statistics 2001; Finch and Finch 1967; Nicholls 1952). Perhaps half of the LHI's population have familial connections on the island back to at least the 1880s, and many properties have seen continued occupation by one family for several generations (Edgecombe 1987; Nicholls 1952). During early settlement, islanders handled administrative issues such as land tenure, education, health care, registration of births and marriages and law and order on a democratic, ad hoc basis. Occasional visits by magistrates, registrars and other officials to settle particular matters and report on various issues to parliament occurred sporadically, but until the establishment of the Lord Howe Island Board of Control in 1911, official government involvement in island affairs was limited (Edgecombe 1987; Nichols 1953). Various alterations have been made to island administration since the formation of the board, culminating in the development of the *Lord Howe Island Act 1953*, and its amendment in 1982. One of the main outcomes of the act granted perpetual leases to long-term island residents, which ended 120 years of virtual 'squatting' by island inhabitants, and formalized the processes of further development in the community (Edgecombe 1987; Hutton 1998; Nicholls 1952). Most recently the construction of an airstrip has ensured regular cargo and passenger services to the island, and in turn has fostered the growth and success of tourism on the island. Lord Howe Island's World Heritage Area status has been the main focus of tourist marketing and helps ensure tourism, settlement size and development are sustainable and the community demographic is maintained.

Research background

Since 1834, LHI has seen 170 years of continuous occupation, during which the majority of habitable land has been settled in some form. Throughout this time, as indicated by the first explorations, no suggestion of prior human occupation has ever been uncovered. Historically known faunal introductions such as mice in the 1860s and rats in 1918 and the recorded consequences for native fauna and flora have further confirmed the lack of prior human occupation. However this historic lack of evidence was not deemed conclusive, as similar scenarios on a number of other remote, Pacific islands later revealed evidence of occupation and subsequent abandonment (Anderson 2003; Anderson and White 2001). Early pollen work conducted in 1977 by the University of NSW and 1996 coring investigations by the University of Wollongong and the Australian National University obtained similar results. Samples from several sedimentary sites contained charcoal and pollen evidence of vegetation clearance occurring only during modern European occupation, and this coupled with historic evidence of the late introduction of murids and subsequent faunal extinctions, lends weight to the proposition of LHI having no prehistoric occupation (Anderson 2003; Dodson 1982). To date, archaeological investigation of LHI has been limited to a preliminary heritage listing study in 1984 (Birmingham, Kelly and Tanner 1984) and the coring investigations in 1978, 1982 and 1996 (Anderson 2003; Dodson 1982). This leads to questions relating to LHI that have implications not only for the history of the island but the nature of island colonization by European communities and the impact on a Pacific environment, and a general analogue for colonization processes. Anderson (2003) elaborates:

> [LHI] is an uncommonly visible landscape of initial human colonization. It is important in its own right as a clear and contained example of the archaeology of European colonization, but also holds out the prospect of a valuable analogue of Pacific prehistory. The basic questions that are usually asked of prehistoric colonization about the locations of the first settlements, the nature of their subsistence patterns, the rate and directions of human impact on the environment, and the early direction of economic and demographic change, may be more amenable to archaeological investigation on Lord Howe than on any other remote Pacific island and provide, thereby a valuable comparative case. (101)

Further, there are other issues to be investigated relating to the islands past, from locally relevant issues such as the development of the settlement, processes of change and heritage values, to broader studies with implications for Australian and Pacific studies, and archaeology in general. These include the development of communities in isolation and frontier theory, the nature of island landscapes and networks of maritime interaction, and processes of island colonization and settlement. Each of these in turn inform on more specific themes such as environmental impact and change, processes of social, technological and subsistence adaptation and the influence of various maritime industries in the Pacific. This project has set out to determine the nature of the settlement landscapes present on LHI from 1834 to 1904, a period which has been deemed to be the early settlement period and the end of its more marked isolation with the advent of regular steamer services in the early 20th century. Specifically the project is aimed at addressing a number of issues relating to the nature of and development and spread of early settlement on the island, environmental impacts and the development (or not) of islander adaptations to the environmental, geographical, temporal and political contexts of LHI. These issues are being addressed through a number of standard methods including historic research; community consultation and collection of oral histories; surface survey and ground penetrating radar investigations; test and area excavations; GIS map generation of settlement phases; and assemblage analyses including dating, sourcing, use-wear, species identification and selection.

During preliminary historic research and reconnaissance visits to the island six sites of potential interest where identified for further surface and excavation investigations. These sites were comprized of a mix of domestic and domestic/agricultural sites, were of varying ages within the first phase of settlement, had all be subsequently abandoned, were subject to minimal disturbance but were still accessible and all had partially known histories with important questions which could only be addressed by archaeological investigation. Preliminary results from two of four excavated sites have revealed some interesting possibilities regarding the lifeways on LHI from the first settler group to the early 20th century.

Two phase colonization of Lord Howe Island

Although the stated period for this project is from 1834 to 1904, in actuality the processes of human colonization on LHI commenced earlier and extended beyond the arbitrary year markers indicated. From the first explorations in 1788 onwards, LHI's stunned and previously naïve wildlife had increasing encounters with humans and their influence. The effects of sporadic but intensive predation by visiting mariners on land and sea birds, and habitat disturbance by feral goats and pigs, were being strongly felt by at least two species of land bird at permanent settlement in 1834. Accounts are scarce and somewhat contradictory but it appears likely that the White Gallinule (*Notornis alba*) and the White Throated Pigeon (*Columba vitiensis godmanae*) were endangered by this time, if not extinct (Hutton 1990). Phase two of colonization commenced at the establishment of the first permanent settlement on the island, from which time successive bird and invertebrate extinctions continued periodically, however the most devastating loss of species occurred after the stated study period. The LHI group remained completely rat free until the running aground of the SS *Makambo* in 1918— within two years the rats (most likely *Rattus rattus*) were numbered in the thousands and their impact on wildlife and crops was being sorely felt (Nicholls 1952). Of the fifteen land-bird species noted by the first fleeters, nine are now extinct, five species only becoming extinct within five years after rats came ashore (Hutton 1990; Recher 1977). At least one well-documented insect was also decimated; a flightless stick insect, the phasmid or 'land lobster' (*Dryococelus australis*) endemic to the LHI Group was completely wiped out on Lord Howe, and is now found only on Ball's Pyramid, a basalt sea stack south of the main island.

First settlers?

The first people to settle on LHI arrived in June 1834 and were in the employ of an Australian whaling and merchant firm to set up a supply station for passing whaling and trade vessels (Nichols 2005; Nicholls 1952). Three British New Zealanders, Ashdown, Bishop and Chapman, their three Maori wives, an adult brother of one of the women and two children settled in Hunter Bay (now referred to as Old Settlement Beach), and cultivated gardens of potatoes, carrots, maize, pumpkin and taro. Six months after their arrival, a government surveyor was sent to the island and he recorded the location of five huts in Hunter Bay and of one small garden further south. The garden location is now under the modern airstrip, however it was thought possible to relocate the site of at least some of the huts recorded in 1835 (Nichols 2005; Nicholls 1952). Later records indicate subsequent settler groups also occupying the bay, and suggest that one later group occupied the same improvements as Ashdown, Bishop and Chapman, at least for a period. The improvements and provisioning business were purchased in 1841 by a new incoming business partnership, for the substantial sum of £350, indicating that the improvements were of some significance (Nicholls 1952). Subsequent maps up until 1853 show continued activity in the bay of varying scales but the relationship of these to the original Ashdown, Bishop and Chapman holdings is not clear (Denham 1853; Foulis 1851). Surface indications along the foreshore of Hunter Bay comprized mounds and hollows of interest, which were confirmed in part by ground penetrating radar surveys, which revealed the two most prominent mounds contained defined, consistent horizons of regular dimension. Shovel tests confirmed that both horizons consisted of the same material, and given their similarity of dimension and depth shown on the radar, likely to be of the same nature.

Area excavations conducted on one of the mounds revealed a substantial, roughly rectangular stone and clay-render hut foundation measuring approximately four by six metres with a north–south lengthwise orientation and western facing entryway, denoted by a section of more regularly sized and tightly packed stone (figure 9.3). An outer course of stone spaced between ten to twenty centimetres from the edge of the foundation, interspersed with medium sized post holes, would have served to keep walls made of palm thatch panels flush against the foundation edge and aid water drainage into the dune sand. Two large, deep, stone wedged, centralized postholes and a small intervening depression suggest a central beam from which the roof was pitched to form a simple but sturdy single room hut, of a style similar to that shown in later photographs on LHI and relatively common in Polynesia and elsewhere (see figure 9.4) (Anderson, pers. com.; Bowie-Wilson 1882). Charcoaled remains of small to medium sized lengths of wood interspersed with smaller, lineal fragments lying on parts of the floor surface further indicate the likely building fabric and suggest a fire either directly before or soon after abandonment as the structure was at least partially standing when burnt. Located centrally in the east facing wall is a large, stone flagged area measuring approximately 160 centimetres by 120 centimetres, containing ash, charcoal and animal bone fragments, indicating some form of hearth structure for cooking and perhaps some heating despite LHI's moderate climate. Adjacent to this hearth, three large, hand-cut calcarenite blocks were recovered, which are possible remnants of some form of chimney structure for the hearth. For this hearth to be a functional cooking space it is likely that it contained larger quantities of cut stone and pieces of metal for hooks, spits or grills and perhaps sheet metal as part of a flue or chimney. The lack of expected quantities of these materials suggests the possibility of scavenging from the site after abandonment for use in other structures, a behaviour that is particularly common in isolated communities. Materials that are in short supply and can't be reproduced locally such as metal, glass, ceramic and textiles or those which require intensive labour to replace such as hand cut local stone or large timbers are often reused until the effort of recovering or adapting a material to a new use outweighs that to replace it outright or the material disintegrates (Davies 2001; Lawrence 2000).

At least two discard zones were found adjacent to the foundation; one occurring one and a half metres to the south between the house and the shore which contained a variety of glass, ceramic, metal and faunal material; and a second occurring directly around the south-west corner of the foundation and was comprized entirely of pipi shells (*Plebidonax deltoids*) with the exception of one glass artefact, a portion of bottle base which was possibly used as a makeshift knife. The majority of material was recovered from these two zones, with some occasional finds on the living floor, and largely comprizedd what would be an expected assemblage from a 19th century Australian-British household situated near a marine environment. Expected types of ceramic, glass, pipes, buttons and consumption of pig (*Sus domesticus*) were obvious, as well as exploitation of marine resources including strawberry cockle (*Fragum unedo*), black nerite (*Nerita atramentosa*), and 'mutton bird'—either the flesh-footed shearwater (*Puffinus carneipes*) and/or the wedge-tailed shearwater (*Puffinus pacificus*). So far one large portion of skull from a southern elephant seal (*Mirounga leonia*) has been identified from the middle to lower sections of undisturbed discard, as well as a humerus from a large sea-bird, possibly a species of albatross or booby, however more detailed identification and quantification of faunal materials is yet to be undertaken. An endemic species of land snail (*Placostylus bivaricosus*) found in discard deposits is also a possible food source, but as it is a land based species that wasn't found exclusively in the discard deposits, testing for size selection is needed to verify its role. Strangely, almost no fish remains were recovered at all from this site, which is notable as LHI has always been known for excellent fishing of both reef and pelagic species, and the first fleet visitors found the fish were so plentiful as to almost literally jump into the boats (Gilbert nd). Differential preservation may be a contributing factor, but a separate processing/discard site for fish also a possibility. Preliminary observations

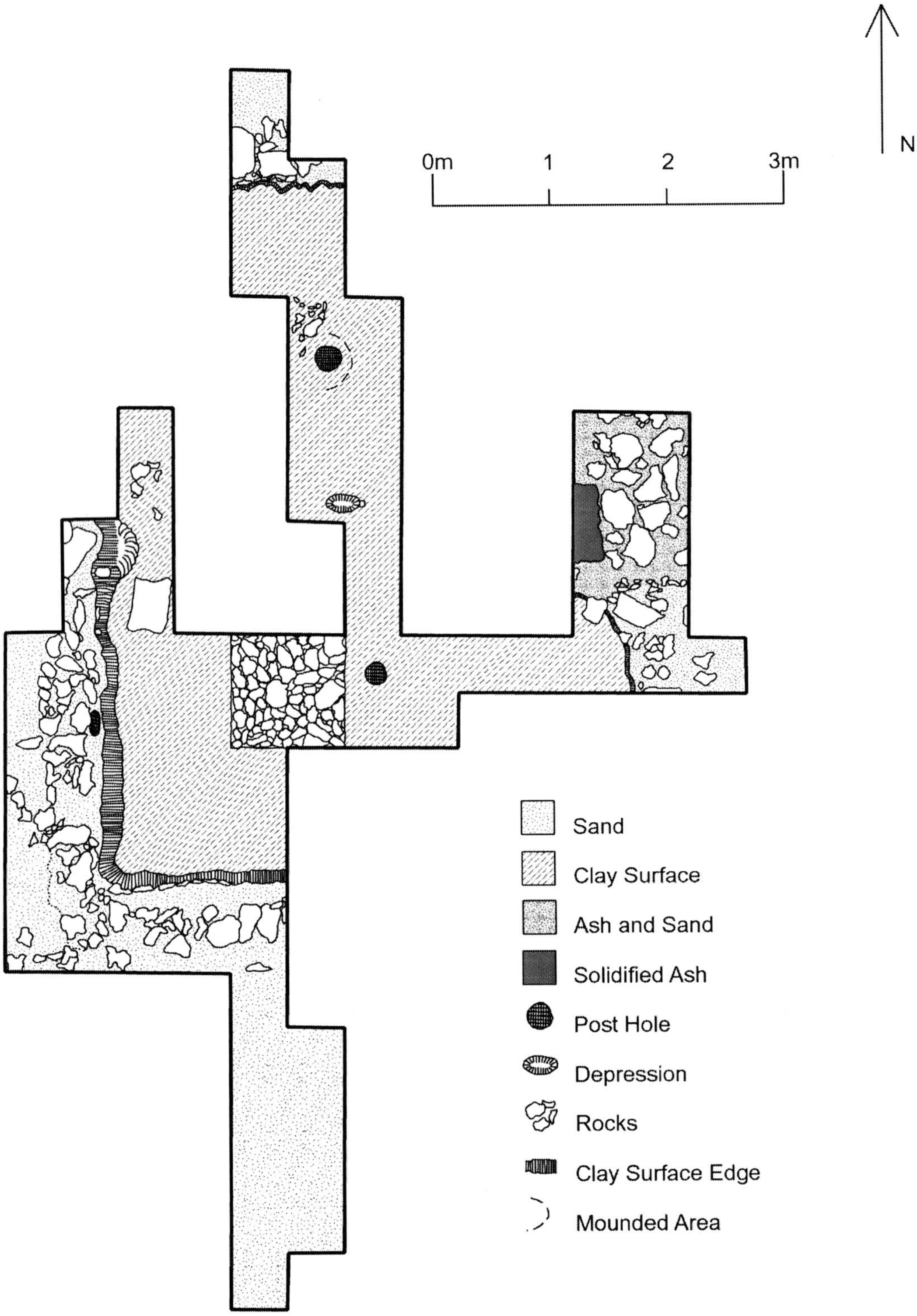

Figure 9.3: Plan of hut foundation excavated at Hunter Bay.

Figure 9.4: 1882 picture of the Johnson's thatched house, a typical example of the traditional Lord Howe Island homes until the early 20th century (Bowie-Wilson 1882).

of numbers of pipi in this site and current islander assertions that such numbers would be impossible to collect today indicate that there has been a definite impact over time on at least some shellfish communities. Prior to the historic clearing of native foreshore vegetation in Hunter Bay for pasture, there would have been large colonies of shearwaters nesting in the sand floor of palm forests extending throughout the bay flat. Remnant palm forest and shearwater colonies now exist at the eastern margin of the bay and adjacent headland and given that their numbers have been effected by generations of human, cat and rat predation, the previous population of Hunter Bay colonies would have been of considerable size (Hutton pers. com.).

Although it is not absolutely certain at this stage if this dwelling is a part of the Ashdown, Bishop and Chapman occupation, the history shows it predates at least 1853, and given the substantial but simple nature of the dwelling and the large amount paid by the new business partnership that took over in 1841, there is a high possibility that this foundation, and the horizon detected under the adjacent mound are the remains of two of their dwellings. The location of these mounds, while occurring at the same end of Hunter Bay, is different to the five huts shown on the 1835 map. However, there is a high possibility that those shown on White's map were temporary dwellings that were built immediately upon settlement, particularly as White visited only six months after their arrival (Nicholls 1952; White 1835). The new settlers may well have waited for a period to gauge the viability of the provisioning business before investing labour and materials into building more substantial holdings. Further, although there were three couples on the island, only two had children, with the one couple having two children on arrival at LHI, and the Ashdowns having all four of their children while on LHI, their first being born two years after their arrival (Nichols 2005). The expansion of families may also have been a catalyst for building more substantial houses, and may explain the apparent existence of only two foundations, as families with children may have moved to larger quarters while the childless couple and Maori man remained in their initial dwellings recorded by White. Fine-grained dating of ceramic and glass assemblages may resolve the issue of the dwellings age, but given the common nature of ceramics and glass initially observed and the small period of differentiation between known occupations (20 years maximum) this may not be possible. The discovery of one glass artefact in the middle to lower layers of the discard zone may be the best indicator that the initial builders were indeed Ashdown, Bishop or Chapman. The presence of one small piece of clear bottle glass that appears to be an exhausted core or scraper may be indicative of occupation by Polynesians or at least European settlers who have had extended contact with stone tool manufacturers; currently the nature of the glass artefact is yet to be confirmed by more detailed manufacture and edgewear analysis.

Even if this site is not part of the holding of the very first colonists on LHI, it is most certainly the earliest remaining evidence of settlement on the island. Searches for the very first huts recorded by White yielded nothing, and given known and recently observed slippage and silting events, no trace is likely to remain. The absence of mouse (*Mus musculus*) remains in this site further indicates that it was built, abandoned and likely partly buried by the time of their introduction from Norfolk Island around 1868 (Etheridge 1889). The faunal and cultural materials recovered show that the people living in this hut were inclined to regularly utilize a range of native foods available to them including different types of shellfish, seabirds, feral game and probably opportunistic finds such as the southern elephant seal. Such a broad use of native foods suggests both a smaller, less established and more isolated group of people and an environment that has economic quantities of faunal and floral resources remaining.

Later settlement

By 1853 the population of LHI was 16 men women and children and was host to approximately 60 whaling and trade vessels a year (Denham 1853; Foulis 1851; Nicholls 1952). In 1855 a crewman on the American whaler *Will 'o the Wisp*, emancipated slave Perry Johnson, left ship to settle on LHI. Five years later Perry and fiancèe Sarah returned from Sydney, married on the island and settled on a large portion of land at the southern end of the Island in an area known as Soldier's Creek. Farming a large portion of land and hand cutting an extensive network of drainage ditches, the Johnsons remained on their land until their deaths in 1915 and 1918, leaving their holdings to their neighbours as they had no children (Nicholls 1952; Rabone 1940). Throughout their nearly 60 year occupation, the Johnson's took in a number of individuals and families while new accommodations were constructed, and on a couple of occasions small portions of their large holding were subdivided and occupied by other families (Nicholls 1952; Rabone 1940). In 1882 a visiting government commission camped on the Johnson's land and photographs taken for the commissioners visit depict the Johnson's home and two adjacent buildings (figure 9.4). These photographs along with local place names, community knowledge and historic accounts make it possible to locate the site of the Johnson's residence to within about 15 sq m.

Following inconclusive radar surveys, test pits were dug where the house site was likely to be, and although material was recovered from these that are consistent with underfloor deposit, no evidence of structural features were evident to confirm this. A test pit located slightly downhill revealed a small hand cut calcarenite block, and subsequent investigation revealed six calcarenite blocks of similar size and spaced at regular intervals suggesting they are possible footings for a small structure. Subsequent pits were excavated and on first observation contained a higher concentration of hardware material, along with an array of domestic and industrial artefacts such as ceramics, glass, buttons, beads, pipes and faunal materials in addition to a range of nails and a tacks, wire and general scrap metal. The more industrial inclusions in the assemblage in addition to the small area delineated by the blocks suggest this might be an underfloor deposit from an auxiliary building such as a work shed that had only a partial timber floor in one section of the building. The easily observed species range of faunal materials recovered from this site are similar to those observed in the earlier site at Hunter Bay; pig, mutton bird, pipi, and black nerite, but with the addition of goat, mouse and possibly cow. Of particular note is again the almost complete lack of fish remains, despite the recovery of at least two iron fish-hooks. The native land snails are also present in this site, but as there is no definite discard zone identified from these excavations, their presence is even more ambiguous at this point. The glass, ceramics, buttons, beads and pipes recovered from this site are again consistent with those expected from a 19th century Australian-British household.

Historic accounts indicate that islanders were harvesting mutton bird chicks and eggs well into the 20th century, and continued to hunt feral goats and pigs until at least the end of World War Two (Hutton 1998; Nichols 1952). There were likely to be local shearwater colonies available in the Soldier's Creek area prior to pasture clearance, but the extent and longevity of such colonies is unknown. Remnant nesting colonies exist on small islets at the very southern end of the lagoon beach system, and islanders have been travelling by boat to the nearby Admiralty Islets to gather seabirds since at least the 1880s (Clarson 1882). Localized marine gathering from Johnson's Beach would have been less profitable than that in the mud flats of Hunter Bay and nearby Neds Beach, but nevertheless the much smaller observed amounts of shell found in the Soldier's Creek site may be an indication of depleted shellfish resources, dietary preference (with the more varied crops and livestock available in the more established settlement) or may be entirely a product of the Soldier's Creek site not having a discrete discard area sampled. Size selection may be able to resolve this issue, as with the question of land snails as a food resource. Currently no obvious evidence of the influence of tenants with children has been observed nor any overt indicators of the Johnson's ethnicity, Perry being African-American and Sarah a South African Hottentot (Nicholls 1952; Rabone 1940). At first observation there also appears to be fewer indications of recycling of materials as might be expected in smaller and more isolated communities, but this is again to be confirmed by more detailed analysis.

From both the history and the assemblage it is obvious that the Soldier's Creek site is a part of the more established settlement landscape than that at Hunter Bay. The lack of any indication of a clay foundation or an even more basic hard dirt floor, makes it likely that the main residence was constructed using wood or stone piers with cut joists and floor boards. This construction would require more intimate knowledge of local timbers and their locations, saws, pits and sufficient labour, which would be a much more realistic undertaking for a settlement of sixteen rather than one of nine. The more restricted range of native foods is perhaps indicative of impacts of earlier settler predation and also wider choice from a more traditional range of introduced European and American livestock available from the early 1860s (Cloete 1869).

Although the results presented thus far are preliminary and based largely on field observations and historically known facts, there appears to be tantalising hints of an emerging pattern of colonization of LHI. Initial settlement in a small, contained resource rich area; early, heavier reliance on easily gathered native foods and easily raised introduced crops and livestock; eventual diversification of introduced resources as local resources become depleted and/or as settlements ability to invest human resources and take risks with more difficult agriculture and animal husbandry improves; settlement of more marginal land possible and/or required with economic and population growth and so on. Following more detailed analysis of materials and GIS mapping of settlement location and resource catchments should help clarify these emerging patterns and allow comparisons with other colonization examples from the Pacific and beyond.

Is Lord Howe really an island?

In terms of biogeographical concerns such as the land and resource limits, environmental sensitivities and marine reliant food, transport and economy, Lord Howe is very much an island. However, the LHI community is one that has a rather unusual place, for while it is indeed a nicely recorded example of human colonization of a Pacific island, it is one which is not removed from its own historical and political context. In some ways, due to LHI being colonized during the 'age of sail', access to neighbouring landmasses, economic and cultural development opportunities with international visitors and the scale of trade and exchange networks are easier, more frequent and larger than those available to earlier colonists of the Pacific. Paradoxically, due in part to a freak of colonial government indifference and the free settler, squatter nature of the settlement, LHI was largely isolated from the rest of the Australian colonies both politically and economically for approximately the first 80 years (Nicholls 1952) while at the same time being part of an international economic network spanning both the Pacific and Atlantic. As Foulis wrote in 1851:

> Howe's Island has for many years been a place of resort for whalers to procure wood, water, and fresh provisions, to enable them to prosecute their voyage without the necessity of going into Port. There were generally from 60 to 80 vessels in the course of the year that touched there for the above purpose, and it not unfrequently happened that we had English news from American vessels some weeks before the same was known in this Colony. (7)

Further, occurring closer to the current time, available technologies and the advent and changes to formalized local governance have influenced hunting and gathering efficiency, social and cultural developments, as well as apprehend some processes of colonization which have not had adequate time depth to manifest fully. Conservation and management efforts on LHI in recent years have served to halt further settlement expansion, faunal and floral introductions, hunting and gathering of most local resources, and have restricted fishing and settlement population size. Rat, mouse, cat, pig and goat eradication programs and captive breeding and monitoring of endangered species have also altered the expected patterns over time. In general most if not all patterns of colonization are expected to be similar, but perhaps in the example of LHI, some processes have been accelerated while others have been arrested due various factors described above. Regardless, LHI remains a interesting and important opportunity to explore issues of similarity of island colonizations, while also providing a well documented example of the influence of historic, geographic and political contexts and the varying scales of 'island' that can apply to one speck in a vast ocean.

Acknowledgements

Many thanks to the Lord Howe Island community and their council and parks staff; the many keen locals who participated in the project especially Daphne Nichols and Cindy Shick; and the Lord Howe Island Museum Committee. Much appreciation to my hard working field volunteers and radar operator Paul Brugman and especially to Professor Atholl Anderson for his supervision and sorely needed guidance in the howling wind and rain.

References cited

Anderson, A. (2003). Investigating early settlement on Lord Howe Island. *Australian Archaeology* 57:98–102.

Anderson, A. and White, P. (2001). Approaching the prehistory of Norfolk Island. In A. Anderson and P. White (eds.), *The Prehistoric Archaeology of Norfolk Island, Southwest Pacific*, pp. 1–9. Sydney: Australian Museum.

Australian Bureau of Statistics. (2001). Census 2001: Lord Howe Island Snapshot. `http://www.abs.gov.au`. Accessed 23/6/2003.

Birmingham, J., Kelly, M. and Tanner, H. (1984). Subsidiary report for Pannell, Kerr, Forster Services Proprietary. Lord Howe Island Regional Environmental Plan—Tourism Study Final Report to the Lord Howe Island Board.

Bowie Wilson, J. (1882). *Report on the Present State and Future Prospects of Lord Howe Island*. Sydney: Thomas Richards, Government Printer.

Clarson, W. (1882). *The Island of Lord Howe: the Madeira of the Pacific*. Sydney: Gibbs, Shallard and Co.

Davies, P. (2001). Isolation and Integration: The Archaeology and History of an Otways Forest Community. Unpublished PhD Thesis. School of Historical and European Studies, Faculty of Humanities and Social Sciences, La Trobe University. Melbourne.

Denham, H.M. (1853). Remarks upon Lord Howe Island by Captain Denham, R.N., F.R.S. communicated at the request of His Excellency Sir Charles Augustus Fitz Roy, Governor General of the Australian Colonies. Correspondence No 12, Enclosure No 1. In New South Wales Government (ed.), 1853 *Correspondence Relative to Proposed New Penal Settlement at Lord Howe Island*, pp. 12–13. New South Wales Parliamentary Accounts and Papers.

Dodson, J.R. (1982). Modern pollen rain and recent vegetation history on Lord Howe Island: evidence of human impact. *Review of Palaeobotany and Palynology* 38:1–21.

Edgecombe, J. (1987). *Lord Howe Island: World Heritage Area*. Balmain: Australian Environmental Publications.

Etheridge, R. Jnr. (1889). The general zoology of Lord Howe Island. In R. Etheridge, Jnr. (ed.), *Lord Howe Island: Its Zoology, Geology and Physical Characters. Memoirs of the Australian Museum* 2:3–42.

Finch, A. and Finch, V. (1967). *Lord Howe Island*. Adelaide: Rigby Ltd.

Foulis, J. (1851). Statement of circumstances in reference to Lord Howe's Island, situated off the coast of New South Wales, between the Heads of Port Jackson and Norfolk Island. Correspondence No 5. In New South Wales Government (ed.), 1853 *Correspondence Relative to Proposed New Penal Settlement at Lord Howe Island*, pp. 5–7. New South Wales Parliamentary Accounts and Papers.

Gilbert, T. (nd) *Voyage from New South Wales to Canton, In the Year 1788, with Views of the Islands Discovered*. London: George Stafford.

Hill, E.S. (1870). *Lord Howe Island: Official Visit by the Water Police Magistrate and the Director of the Botanic Gardens, Sydney; Together with a Description of the Island by Edward S. Hill.* Sydney: Thomas Richards, Government Printer.

Hutton, I. (1998). *The Australian Geographic Book of Lord Howe Island.* Terry Hills: Australian Geographic Pty Ltd.

Hutton, I. (1990). *Birds of Lord Howe Island: Past and Present.* Coffs Harbour: Ian Hutton.

Lawrence, S. (2000). *Dolly's Creek: An Archaeology of a Victorian Goldfields Community.* Carlton: Melbourne University Press.

Nichols, D. (2005). *Lord Howe Island Rising.* In Press.

Nicholls, M. (1952). *A History of Lord Howe Island.* Hobart: Mercury Press.

Rabone, H.R. (1940). *Lord Howe Island: Its Discovery and Early Associations 1788 to 1888.* Sydney: Australia.

Recher, H.F. (1977). Colonization and extinction. In N. Smith (ed.), *Lord Howe Island,* pp. 23–26. Sydney: The Australian Museum.

White, H.J. (1835). Report on Lord Howe's Island. Correspondence no. 3, enclosure no. 3. In New South Wales Government (ed.), 1853 *Proposed New Penal Settlement at Lord Howe Island: a collection of papers submitted by the Governor General to the Legislative Assembly,* pp. 3–4. New South Wales Parliamentary Accounts and Papers.

Chapter 10

Prehistoric sea-faring: Bronze Age sewn plank sea craft from the Humber Estuary, England, UK and their role in an island economy

Malcolm Lillie, University of Hull

Introduction

Research into the sewn plank boats of the Humber Estuary, and elsewhere in England and Wales (figure 10.1) represents in excess of sixty-four years of investigation on the part of Ted Wright and other researchers (e.g., Wright and Wright 1939; Wright 1990; Wright et al. 2001; Gifford and Gifford 2004a and b; Lillie 2005; amongst others). Obviously, the periodization of these craft was enhanced considerably by the advent of radiocarbon dating in the late 1940s but, despite this, the process of ascribing an accurate age ultimately relies on new techniques of analysis, including the application of AMS dating (Wright et al. 2001).

The protracted nature of this research has led to other finds being considered to be of earlier date, for example, Caldicot (Nayling and Caseldine 1997:27), which was dated to 3439±19 BP (UB-3472), and calibrated at 2σ to 1874–1689 BC. In addition, the find from Kilnsea (Van de Noort et al. 1999) produced a calibrated age of 1870–1670 cal BC (OxA-6939 3455±35 BP and OxA-6940 3340±BP). Despite assertions by Van de Noort et al. that Kilnsea represented the earliest example of these craft, alongside the clear overlaps with the Caldicot example (1999), the re-dating of Ferriby 1, 2 and 3 has further refined their chronology. The new dates for Ferriby 1 calibrate to 1880–1680 cal BC, Ferriby 2 calibrates to 1940–1720 cal BC, and finally, Ferriby 3 is calibrated to 2030–1780 cal BC, making Ferriby 3 the earliest in the sequence.

The Ferriby boats represent the earliest examples of sewn plank boats from England and Wales yet discovered. Of some interest in this debate over dating is the fact that the dates from the Kilnsea find fully overlap with the new dates for Ferriby 1. Wright (1990:20) has noted that during the survey and partial excavation of this particular vessel, fragments of planking were lost to the estuary between 1940–41. In addition, excavation in 1943–44 of a further 1 m of the planking of Ferriby 1 of the northern outer plank and the keel plank occurred after '... their southern equivalents [had] already been washed away' (1990:22). The derived context of the Kilnsea plank fragment (Van de Noort et al. 1999) could conceivably reflect the re-deposition of those parts of the Ferriby 1 that were dislodged during these original investigations and, as such, the entire context of Kinsea is brought into question, a consideration that is assessed in greater detail below.

Clearly, in light of the recent re-dating work, and the earlier periodization indicated, the entire social context of these craft requires revision. Perhaps more significant in this context is the re-assessment of the excavation archive for the Ferriby ships as undertaken by Ted Wright, as this work is fundamental to our recognition of the sea-going potential of these craft. To date, despite the increasing number of finds of these craft, or parts of them, the significance of more complete finds like the Ferriby and Dover boats (Wright 1990; Clark 2004) is that they allow us to gain important insights into construction techniques and the potential scale of plank boats. This forms the basis for considered attempts at reconstruction.

In the case of the Ferriby examples, Ferriby 1 was the more complete, being ca. 16 metres long, 2 metres wide and constructed using massive oak planks. It has been suggested that their size alone justifies designating them as ships in their chronological context (Coates *pers. comm.* and 2004:21). Reconstructed, these craft would have weighed ca. 4 tonnes unladen, and they

Figure 10.1: Location of main findspots for Ferriby-type craft and elements from these, along with location of source of jet at Whitby. 1. Whitby, 2. North Ferriby, 3. Kilnsea, 4. Brigg, 5. Dover, 6. Goldcliff, 7. Caldicot.

would be able to carry about 7 tonnes in a seaworthy manner. Alternatively, they would have been able carry ca. 30 passengers while being propelled by up to 18 men working paddles or a smaller number working oars, or by sail (though there is as yet no uncontroversial archaeological evidence of sail in NW Europe before about 500 BC—Coates, pers. comm., 2001). Of equal importance are the social implications of such craft, a theme which will be expanded upon below.

Certain aspects of the Ferriby craft require assessment within the context of their possible function in their prehistoric setting. In relation to the evidence for oars, and a sea-going capability, recent research has provided important information, which is fundamental to both topics. In the case of the evidence for oars, John Coates (forthcoming a) has reconsidered the report in Wright (1990:154) relating to the find of a 'paddle' at Ferriby. The work of Coates suggests that on the basis of the actual dimensions of the Ferriby example, Wright's original interpretation was erroneous. The find at Ferriby is considered to be more realistically interpreted as an oar, as 'the blade has the same bending strength as the necks of modern oars, whose shafts increase in diameter towards the thole or pivot point where bending is greatest' (*op. cit*). On reconsideration it is suggested that the blade from Ferriby could have been suitable for an oar of between 2.5–3.5 m length. Ted Wright (1990) was conservative in his original interpretation of the Ferriby 'paddle' due to the fact that finds of prehistoric oars in northwest Europe were unknown at the time of his writing.

When reconsidering the original excavation archive, Wright (1990:18, 90) was careful to point out the fact that he was incorrect in his original interpretation of the Ferriby craft as having a flat hull. In actual fact, the original depth soundings taken when assessing the length of Ferriby 1 in the ground, confirmed that the hull was either curved or 'rockered'. This observation is of fundamental importance, as the existence of a rockered keel is helpful if these craft were to have the hypothesized sea-going capacity.

Coates (forthcoming b) notes that one of the more limiting aspects of research into the reconstruction of the Ferriby craft and theories relating to their sea-going capabilities has been the insistence of other specialists that the original flat bottomed interpretation of the Ferriby craft continues to be used when discussing their significance (e.g., McGrail 1987, 2001). More recently Gifford and Gifford (2004b:501) have noted that in addition to Ted Wright's depth soundings, the photographic and drawn archive for Ferriby 2 confirm that this vessel was also rockered. This additional observation lends considerable weight to the suggestion of a sea-going capability for these craft.

Reconstructing Bronze Age vessels

Wright (1990) has suggested that the findspot of the Ferriby craft represents a boat-yard, or equivalent, where these vessels were both constructed and repaired. In this context, it perhaps worth noting that when considering the origin of the massive oaks used in the construction of the Ferriby boats, two possible sources can be inferred. Palaeoenvironmental investigations in the Vale of York, to the northwest of Ferriby (Lillie and Gearey 1999) have shown that the Vale would have been heavily wooded during the Bronze Age. Similarly, in the Trent valley, a river that flows into the Humber, massive oak trees of Bronze Age date were recovered during gravel extraction (Lillie and Grattan 1995a, 1995b).

In addition, a chronologically later, though equally significant find from the Humber region is the Iron Age Hasholme log boat, found in the Foulness River, a tributary of the Humber (Millett and McGrail 1987). This vessel is dated to ca. 300 BC, and was 12.78 m long by 1.4 m in width. Whilst craft such as the Hasholme boat would have been more suited to riverine transportation, the significant point to note in this context is the fact that the boat was constructed from a single oak tree. Oaks of a size suitable for use in the construction of the Bronze Age Ferriby boats were clearly still available in the Iron Age.

The evidence above leads us into a consideration of recent attempts at reconstruction, and the following represents a summary of the outline presented in Lillie (2005). This summary is based on the work of Edwin and Joyce Gifford and John Coates, all of whom have provided the current author with helpful advice and access to their research on the Ferriby craft (Coates 2004; Gifford and Gifford 2004a, 2004b).

It should be noted that when undertaking a reconstruction project a number of considerations have to be addressed, and factors such as the differences inherent in the construction of a scale model require assessment as different characteristics will vary in their responses to scaling down in reconstruction (Gifford and Gifford 2002). However, the precise impacts relating to scaling down may be less significant than the sum of the responses to displacement, sail area, speed, strength and seaworthiness, when considered in terms of the information gained during tests of the scale model.

When assessing displacement, the half-scale model will have a displacement equivalent to one-eighth of the full-scale ship. As the half-scale model will have a waterline length that is half that of the full-scale ship, the half-scale craft will have a speed that is the full-scale speed divided by $\sqrt{2}$ (Coates *pers. comm.* 2001). Propulsive force would require, under sail, a half-scale sail or, alternatively, one-eighth the assumed crew paddling or rowing in order to replicate propulsive force. In relation to strength, two figures are produced as the strength of the hull structure will be one eighth that of the full-scale craft, but the bending moments from the displacement will be one-sixteenth those of the full-scale ship, so the model hull structure will have an equivalent strength of twice that of the ship. The last parameter considered here, that of seaworthiness, is less easy to measure, but in essence if the half-scale reconstruction can function well in a 1 m sea, then the full-scale version will be able to withstand a 2 m sea (Gifford and Coates *pers. comm.* 2001). The above parameters were assessed during sea trials of the half-scale model in 2002, and will be considered further.

In light of Wright's (1990) observations relating to the probability that Ferriby 1 had a rockered keel, and those of Gifford and Gifford (2004a, 2004b) in respect of Ferriby 2, it is now generally assumed that these craft were sea-going, or that they had the potential to be. However, other examples of sewn plank craft differ in design features. For instance, the Dover and Brigg vessels are both flat bottomed, initially suggesting that it was unlikely that they would have had any significant sea-going capacity. However, recent re-examination of the Dover boat (Marsden 2004:18–19), has suggested that a combined river, and limited coastal use, would be a reasonable interpretation for this crafts function, and in this context the 'significance' is clearly open to debate and is dependent upon context.

The finds at Kilnsea (Van de Noort *et al.* 1999), Goldcliff (Bell et al. 2000; McGrail 2000) and Caldicot (Parry and McGrail 1991a, 1991b; Parry and McGrail 1994, McGrail 1997) are all fragments of sewn plank boats (as noted above, the provenance of the Kilnsea fragment may be open to debate). However, the finds discussed above are in no way comparable to the examples from Ferriby (Wright 1990), Brigg (McGrail 1994) and Dover (Clark 2004) in terms of scale, and the information that they provide in relation to construction techniques. Despite this, it should be noted that a number of the finds discussed are located in close proximity to the major estuaries of the Humber and Severn, perhaps providing some indication of their use in these environments. In essence though, it would appear that the finds from Ferriby are the only examples (found to date) that might be considered as having any significant sea-going capabilities.

As a consequence of the above observations, it is clear that Ferriby 1 is of considerable importance for a number of reasons. First, this vessel appears to be technically the most advanced form of sewn plank boat yet recovered. Second, its new dating (along with Ferriby 2 and 3) moves it towards the earliest part of the Bronze Age—it is probably unlikely that such a level of technical competence occurred overnight (Coates *pers. comm.* 2002). The process of development must suggest that the genesis of these vessels lies at the Neolithic–Early Bronze Age boundary as a minimum estimate. Consequently, and third, the significant socio-political changes that characterize this period revolve, in part, around the control of prestige items and the rise of the individual (discussed below). Therefore, the determination of the sea-going capabilities of these craft is central to the debate relating to their significance in relation to late Neolithic and early Bronze Age socio-economic and probably political developments across this transition.

The following discussion represents a summary of the 2002 papers by Gifford and Gifford (2004a), and Coates (2004), both of which were given at the Dover Boat conference. During the initial preparation for the construction of the half-scale replica two problems were encountered. The first was that in order to construct the vessel it was necessary to bend and carve green oak. Fortunately Richard Darrah obtained an oak suitable for this purpose. The second problem was the fact that yew withies were used as stitches in the construction of the Bronze Age vessels. Clearly, for practical purposes, an alternative for use in the half-scale reconstruction was needed. In this respect a recently developed pre-stretched polyester halliard rope was used. Finally, with respect to the sealing of planks with moss, an alternative caulking agent in the form of mastic compound was decided upon. After preliminary preparation of the planks in Shropshire, these were moved to Hampshire and stored in a water-tank until required. On the basis of the above discussion and the identification of Ferriby as a 'ship yard' (Wright 1990), plus the considerable size of the oak planks used in the original construction, it is likely that the oaks used were obtained from the immediate vicinity.

When attempting the reconstruction, Gifford and Coates realized that in order to replicate the rocker (or curve) in the keel, it would be necessary to bend the two parts of the keel (Gifford and Gifford 2004a; Lillie 2005). The lack of evidence for the use of fire in this process led to the hypothesis that heated water may have been used in this stage of the original construction process. Consequently, for ease of reproduction, the keel plank parts were enclosed in a PVC sleeve which had steam fed in to it from a wallpaper stripper. Weights were placed on the planks to ensure curvature, and after four hours of steaming at 70°C, the 60 mm planks were bent to shape.

In the original craft, the system of longitudinal carved cleats and transverse timbers used in construction served to align the bottom planks during assembly and stitching. In the reconstruction, the transverse timbers were placed at twice the original spacing in order to ensure that the sheer strength of the stitching would be to scale. An explanation for the absence of stitches at the boxed butt joint in the keel plank amidships (figure 10.2) was provided by John Coates who observed that the long end overhangs and fine lines both fore and aft would have caused the ends of this ship to droop. This effect, known as hogging, would serve to put the bottom planks in compression and the top strakes in tension. In addition, this characteristic would result in the force on the stitching occurring in only one direction, thus lessening the wear on them. Some difficulties were encountered with the three-dimensional carving of the plank hood ends, and with their final fitting to the keel plank. Despite this difficulty, it was apparent that this technique was a significant development in the construction of hull shape through planking as opposed to the hollowing out of boats from logs in prehistory.

The remaining details of the reconstruction project are outlined in Gifford and Gifford (2004a and b), and Lillie (2005). However, given the above discussion of the technical knowledge necessary for the construction of these vessels it is apparent that sewn plank ships would have represented a significant advance in northwest European boat building in the later Neolithic and earlier Bronze Age.

The sea-going capabilities of the reconstructed Ferriby ships were tested during trials in 2002 and more recently in January of 2005 (figure 10.3). In general the craft handled well, and the use of nine people in the reconstructed craft during the initial sea trials confirmed an estimate of seventy-two for the full-scale vessel. A speed equivalent to 6 knots with 16 rowers was established, and under sail a speed equivalent to Coates' origin calculations of 6–7 knots was achieved (Gifford and Gifford 2004b: 503). All of the above observations reinforce the suggestion that if we accept Wright's (1990) identification of keel curvature (or rockering), then the Ferriby craft would have been significant sea-going vessels at the end of the Neolithic period in British (and Northwest European) prehistory.

Discussion

Given the above observations, it is perhaps unsurprising that there has been some renewed debate relating to the social context of sea-faring in the Bronze Age, particularly given the evidence from the Humber region (e.g., Van de Noort 2004; Chapman and Gearey 2004). However, the 'social' context of these studies is in fact more of a focus on the landscape context, viewed from perceptions of the practical and ritual elements relating to the finds locations of Kilnsea and Ferriby, in the Humber region. This focus, in my option (Lillie 2005), is too myopic and fails to fully realize the potential implications of the dating and sea-going capabilities of the Ferriby ships in relation to the major Neolithic and Bronze Age socio-economic and political developments occurring in Yorkshire and Derbyshire (*cf.* Wright *et al.* 2001).

In brief, when discussing the social context of sea-faring, Van de Noort (2004) acknowledges the fact that long distance trade and exchange would have occurred during the Neolithic and earlier Bronze Age, indeed the archaeological record is incontrovertible in this respect. However, Van de Noort (2004: 407) only considers the integration of sewn-plank boats within the elite exchange networks as 'plausible', while also considering the involvement of political and religious leaders to be essential.

Despite these observations, the thrust of Van de Noort's paper appears to focus more on the actual act of sea-faring and the landscape context of the ship finds from the Humber Estuary as opposed to the more fundamental significance of these craft in relation to the rise of the individual in Yorkshire and Derbyshire during the Neolithic (Lillie 2005). This point is not made to denigrate the importance of sea-faring and landscape context, more to highlight the potential limitations of the work discussed.

One further intriguing area of debate considered by Van de Noort (2004:412) is the idea that the sea may have been perceived as a liminal

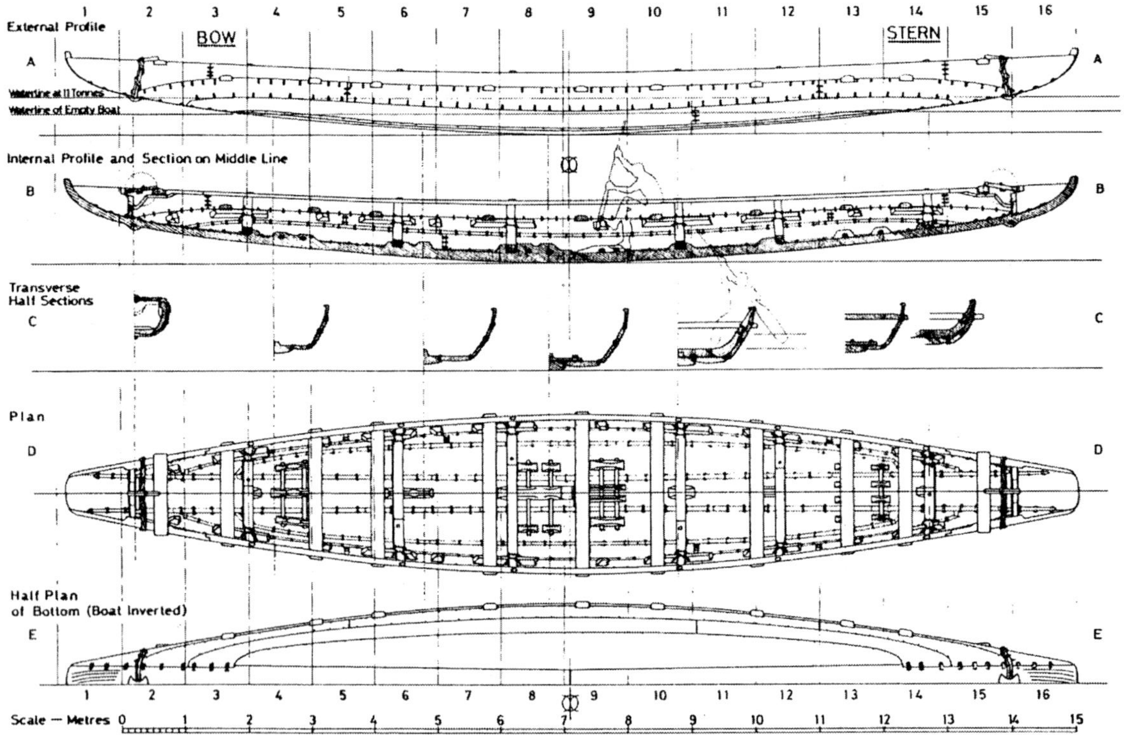

Figure 10.2: General plan of hypothetical reconstruction of a complete sewn plank boat (after Coates in Wright 1990).

Figure 10.3: Launching of the Ferriby $\frac{1}{2}$-scale reconstruction in January of 2005 (photograph ©Malcolm Lillie 2005).

boundary (*cf.* Van Gennep 1960). Van de Noort suggests that this liminality heightened the significance of the prestige items involved in the elite trade and exchange networks (2004). However, and as will be noted below, the source of jet on the east coast of England at Whitby, and its inclusion in numerous elite grave contexts throughout Britain, may throw some doubt on this assertion, or indicate a need for a more considered theoretical basis for the significance of sea-faring. This suggestion is made on the basis of the fact that jet has clear significance in the Neolithic as a prestige item (e.g., Whitegrounds, Yorkshire—discussed below), despite the fact that it did not need to be integrated into Neolithic-Bronze Age trading networks through exchange via sea voyages.

In their response to Van de Noort, Chapman and Gearey (2004) reconsider the landscape context of the Ferriby and Kilnsea finds. These authors suggest that *contra* Van de Noort (2004), the Ferriby boatyard could be argued to have been imbued with ritual meaning, whilst Kilnsea could have served a practical function in relation to navigation, alongside any ritual context associated with the embarking upon and the actual act of seafaring. The fundamental basis of these suggestions is not unequivocal, not least because the actual context of Kilnsea is brought into question by the discussions earlier in this paper (e.g., Wright 1990:20–22), and as noted below, by the authors themselves.

Chapman and Gearey (2004:453) express doubts about the context of the Kilnsea find, suggesting that it may have 'been transported by fluvial processes some distance before its final deposition', a point noted by Lillie at the WARP conference in Olympia in 2003, and by Wright himself in 1990. Interestingly, and despite their own arguments relating to 'visibility' in the landscape (Chapman and Gearey 2002:316), these authors fail to adequately consider the fact that the ritual monuments at Kilnsea may not have been readily visible as navigation aids to the sailors arriving at this location after crossing the North Sea due to palaeovegetation distributions. As such it is apparent that their own 'stud[y] of past landscapes [is therefore rendered] incomplete' (2002:316), and their own discourse could be considered to be 'at best confusing and at worst contradictory' a failing that they themselves have accused other researchers of (Chapman and Gearey 2002:317).

In this context, palaeoenvironmental studies from the Holderness region (Dinnin 1995:39) have shown that the evidence suggests that during the Neolithic and Bronze Age periods, clearances were of limited extent, with little evidence for arable activity, and regeneration of species such as lime and elm occurring in pollen diagrams from this region. One final point regarding Kilnsea, as made by Chapman and Gearey (2000:453), is that 'the suggestion that the Kilnsea site was a landing place for cross-estuarine travel during prehistory is not currently supported by any archaeological evidence'. Surely then, the question that must be asked is why do Chapman and Gearey (2004:455-6) dedicate so much time to questioning the validity of Van de Noort's work when they apparently deny the 'reality' of the Kilnsea landscape in relation to sea-faring?

The above discussion highlights the myopic and limited nature of discussions relating to the ritual and landscape aspects of the Humber finds when considered against the potential socio-political and economic aspects of these craft if they had a sea-going capability. There are considerable changes occurring at the later Neolithic to Bronze Age transition in the Yorkshire and Derbyshire regions, and elsewhere in the UK. Given these socio-political shifts, it is suggested here that the fundamental limitation of the discussions of Van de Noort (2004) and Chapman and Gearey (2004) is that these authors fail to acknowledge the fact that these represent some of the earliest expressions of the rise of the individual and individual power in Britain. In addition these occur in and around the Humber region in close proximity to the Ferriby shipyard. This paper has reinforced the suggestion that these changes are intimately associated with the earliest sea-going vessels yet identified in North-west Europe.

The changing socio-political articulations occur in the form of the large round barrows of Neolithic date that exist in these regions, for example, at Duggleby Howe, eastern Yorkshire (Kinnes 1979; Kinnes et al. 1983), where successive inhumation burials occur in a deep grave pit. At Whitegrounds, Yorkshire, an adult male was placed in the centre of a previously existing burial cairn with a jet slider and flint axe in association (Clarke et al. 1985:66). The bone material from this burial has been dated to 2570±90bc (Har-5587) which gives an age of ca. 3500–2900 cal BC at 2σ. These examples suggest that the importance of the individual, and concomitantly perhaps the control of goods and people, was rising in this region during the Neolithic. Intuitively, mechanisms that facilitate this control, such as sea craft capable of moving between Britain and the continent, would be exploited by such individuals.

During the Bronze Age period one important element of the prestige goods in circulation are the artefacts made from Whitby jet. These have been found at a range of sites in Scotland, such as Mound Stuart, Bute, Pitkennedy in Angus, Harehope in Peebleshire and Migdale in Sutherland. They have been found at other important sites such as Garton Slack in Yorkshire and Irthlingborough, Northamptonshire (Parker Pearson 1993). At the former site ochre, which is used as a polishing agent for jet, has been found. At Irthlingborough, one of the mounds produced significant evidence for structured grave deposits with an individual of some importance: an adult male was interred in a wooden burial chamber covered by the remains ca. 184 young male cattle. Whilst objects from eastern England occurred in the grave goods, an 'exotic' grave good in the form of an amber ring from the Baltic Sea region was also in association. Obviously, we cannot necessarily determine the mechanisms by which this latter item arrived into Britain, but the fact that the Ferriby-type craft were capable of crossing the North Sea at this time, may well provide some insights into one aspect of these exchange networks.

I am of the opinion that the discussions to date have failed to appreciate how significant the sewn plank sea-craft of the Humber may have been in relation to the rise of the individual and the expansion of elite trade and exchange networks during the Neolithic period. The early dating of the Ferriby ships at ca. 2000 BC must be considered a minimum age for these craft as no precursors to the advanced forms of Ferriby 1-3 have yet been recovered. The long-distance exchange of Whitby jet and the concomitant trade of prestige items must have placed the individuals who controlled the Ferriby-type vessels in an extremely strong socio-political and economic position. As such, the idea that the construction and everyday functioning of these craft was 'not embedded with any special meaning' (Van de Noort 2004:413), severely underestimates the likely significance of these craft.

Conclusion

In conclusion it should be noted that the author has developed his opinions in light of the reconsideration of the context and excavated archives of Ted Wright and the work of Gifford and Gifford (2004a, 2004b), and Coates (2004), along with discussions with these colleagues since 2001. The possible implications of a sea-going potential for the Ferriby craft are of considerable importance when considered in the context of later Neolithic and earlier Bronze Age socio-economic and political developments in the centuries around 2000 BC. It is possible that the individual interments occurring at sites such as Duggleby Howe and Whitegrounds, earlier in the Neolithic, persist as embedded social memories relating to the articulations of power by these significant social actors, through into the later Neolithic. The social significance of individual interment with prestige goods could conceivably be re-used and legitimized through social memory in the earlier Bronze Age. However, a general consensus has yet to be achieved, and indeed, even in the publication in which Gifford and Gifford (2004a) and Coates (2004) argue for a sea-going capacity (Clark 2004), McGrail still argues that 'seagoing was not their usual role' (2004: 59). Perhaps the final word will occur with the discovery, excavation and analysis of a Ferriby-type craft in the current context of a heightened awareness of the potential significance of sea-faring in the prehistory of Northwest Europe.

References cited

Bell, M., Caseldine, A., and H. Neumann. (2000). Prehistoric Intertidal Archaeology in the Welsh Severn Estuary. *Council for British Archaeology: Research Report* 120:74–82.

Chapman, H.P. and B.R. Gearey. (2002). Palaeoecology and the perception of prehistoric landscapes: some comments on visual approaches to phenomenology. *Antiquity* 74:316–19.

Chapman, H.P. and B.R. Gearey. (2004). The social context of seafaring in the Bronze Age revisited. *World Archaeology* 36(4):452–8.

Clarke, D.V., Cowie, T.G. and A. Foxon. (1985). *Symbols of Power at the Time of Stonehenge*. Edinburgh: National Museum of Antiquities of Scotland (HMSO).

Clark, P. (ed.) (2004). *The Dover Bronze Age Boat in Context: Society and Water Transport in Prehistoric Europe*. Oxford: Oxbow Books.

Coates, J. (2004). The legacy of Ted Wright. In Clarke, P. (ed.), *The Dover Bronze Age Boat in Context: Society and Water Transport in Prehistoric Europe*, pp. 20-30. Oxford: Oxbow Books.

Coates, J. (forthcoming a). Early seafaring in north west Europe: could planked vessels have played a significant part? *Mariners Mirror.*

Coates, J. (forthcoming b). The Bronze Age Ferriby boats: sea-going ships or estuary ferry boats? *International Journal of Nautical Archaeology.*

Dinnin, M. (1995). Introduction to the palaeoenvironmental survey. In R. Van de Noort and S. Ellis (eds.), *Wetland Heritage of Holderness: An archaeological survey*, pp. 27–48. Hull University: Humber Wetlands Project.

Gifford, E. and J. Gifford. (2002). *The use of half-scale model ships in archaeological research with particular reference to the Graveney, Sutton Hoo and Ferriby ships.* Paper presented at the Dover Boat Conference, Dover. 31st October–2nd November 2002.

Gifford, E. and J. Gifford. (2004a). The use of half-scale model ships in archaeological research with particular reference to the Graveney, Sutton Hoo and Ferriby ships. In P. Clarke (ed.), *The Dover Bronze Age Boat in Context: Society and Water Transport in Prehistoric Europe*, pp. 67–81. Oxford: Oxbow Books.

Gifford, E. and J. Gifford. (2004b). The Ferriby Ship Experiment. *Current Archaeology* 191:498–505.

Kinnes, I. (1979). *Round barrows and ring ditches in the British Neolithic.* London: British Museum Occasional Paper 7.

Kinnes, I., Schadla-Hall, T., Chadwick, P., and P. Dean. (1983). Duggleby Howe reconsidered. *Archaeological Journal* 140:83–108.

Lillie, M.C. and J.P. Grattan. (1995a). Geomorphological and palaeoenvironmental investigations in the Lower Trent Valley. *The East Midlands Geographer* 18(1):12–24.

Lillie, M.C. and J.P. Grattan. (1995b). On flood events and fluvial reworking: observations on terminology and the vagaries of geomorphological analyses in the Lower Trent Valley. *The East Midlands Geographer* 18(2):92–5.

Lillie, M.C. and B.R. Gearey. (1999). The palaeoenvironmental survey of the rivers Aire, Ouse, Wharfe and Derwent. In R. Van de Noort and S. Ellis (eds.), *Wetland Heritage of the Vale of York: An archaeological survey*, pp. 35–77. Hull University: Humber Wetlands Project.

Lillie, M.C. (2005). Deconstructing reconstruction: the Bronze Age sewn plank boats from North Ferriby, River Humber, England, UK and their context. *Journal of Wetland Archaeology* 5:107–19.

Marsden, P. (2004). Reconstructing the Dover boat. In P. Clarke (ed.), *The Dover Bronze Age Boat in Context: Society and Water Transport in Prehistoric Europe*, pp. 17–19. Oxford: Oxbow Books.

McGrail, S. (1987). *Ancient Boats in Northwest Europe: The Archaeology of Water Transport until AD 1500.* London: Longman.

McGrail, S. (1994). The Brigg raft: a flat bottomed boat. *International Journal of Nautical Archaeology* 23:283–97.

McGrail, S. (1997). The Boat Fragments. In N. Nayling and A. Caseldine (eds.), Excavations at Caldicot, Gwent: Bronze Age Palaeochannels in the Lower Nedern Valley. *Council for British Archaeology: Research Report* 108:210–7.

McGrail. S. (2000). The Boat Planks. In M. Bell, A. Caseldine, and H. Neumann (eds.), Prehistoric Intertidal Archaeology in the Welsh Severn Estuary. *Council for British Archaeology: Research Report* 120: 77.

McGrail, S. (2001). *Boats of the World from the Stone Age to Medieval times.* Oxford: University Press.

McGrail, S. (2004). North-west European seagoing boats before AD 400. In P. Clarke (ed.), *The Dover Bronze Age Boat in Context: Society and Water Transport in Prehistoric Europe*, pp. 51–66. Oxford: Oxbow Books.

Millet, M. and S. McGrail. (1987). The archaeology of the Hasholme boat. *Archaeological Journal* 144:1–68.

Nayling, N. and A. Caseldine. (1997). Excavations at Caldicot, Gwent: Bronze Age palaeochannels in the lower Nedern Valley. *Council for British Archaeology: Research Report* 108.

Parker Pearson, M. (1993). *Bronze Age Britain.* London: Batsford.

Parry, S. and S. McGrail. (1991a). Sewn Plank boat and a hard from Caldicot Castle Lake, Gwent. *Newswarp* 10:9–10.

Parry, S. and S. McGrail. (1991b). Prehistoric plank boat fragment and a hard from Caldicot Castle Lake, Gwent. *International Journal of Nautical Archaeology* 20:321–324.

Parry, S. and S. McGrail. (1994). A Bronze Age sewn boat fragment from Caldicot, Gwent, Wales. In C. Westerdahl (ed.), *Crossroads in Ancient Shipbuilding.* Oxford: Oxbow Monograph 40:21–8.

Van de Noort, R. (2004). An ancient seascape: the social context of seafaring in the early Bronze Age. *World Archaeology* 35(5):404–15.

Van de Noort, R., Middleton, R., Foxon, A., and A. Bayliss. (1999). The Kilnsea-boat, and some implications from the discovery of Englands oldest plank boat remains. *Antiquity* 73: 131–5.

Van Gennep, A. (1960). *The Rites of Passage.* Chicago: University Press.

Wright, E.V. and C.W. Wright. (1939). A submerged boat at North Ferriby. *Antiquity* 151: 349–54.

Wright, E.V. (1990). *The Ferriby Boats: Seacraft of the Bronze Age.* London: Routledge.

Wright, E.V., Hedges, R.E.M., Bayliss, A., and R. Van de Noort. (2001). New AMS radiocarbon dates for the North Ferriby boats. *Antiquity* 75:726–34.

Chapter 11

The sea is not land: comments on the archaeology of islands in the western Solomons

Peter Sheppard, University of Auckland

Richard Walter, University of Otago

Introduction

The Oxford English Dictionary defines insularity as "1. The state or condition of being surrounded by water. 2. The condition of living on an island; hence, narrowness of mind or feeling, contractedness of view." In a recent paper, *Islands out of time: towards a critique of island archaeology*, Paul Rainbird (1999) has made the point that this model of islands as isolates characterized by some form of distinctive relationship to the rest of the world is a western or continental view which biases archaeological study of these locations. More fundamentally the critique is of the islands as laboratories model where isolated islands could be used to study evolutionary development in a fashion analogous to that used by Darwin in the Galapagos.

Although the anthropological roots of such thinking date back at least to the 19th century (Kuklick 1996) it was revitalized in the post WWII resurgence of evolutionary approaches, chiefly in American anthropology. The cultural anthropologist Goodenough (1957) led the field in the Pacific arguing that the islands of Oceania would serve as natural laboratories where independent evolution could be observed in islands comparatively free from contact with other cultures. As Terrell (1997) notes this call was echoed by other anthropologist working in the Pacific and subsequently taken up as a research model (Goldman 1955; Sahlins 1957; Sahlins 1958; Vayda and Rappaport 1963). Archaeologists joined the effort with Terrell (1977; 1986) following the theoretical promise offered through the publication of *The Theory of Island Biogeography* (MacArthur and Wilson 1967) while Kirch (1975; 1979; 1980; 1984; 1986), amongst others, applied the cultural evolutionary/ecology approach current among those forms of processual archaeology, that derive in large measure from the work of Julian Steward (1955) and Leslie White (1949) and their student Marshall Sahlins (Sahlins and Service 1960). The work of the latter was naturally of particular interest to archaeologists in the Pacific. The clearest expression of the developed evolutionary approach to the sorting out of Pacific phylogenetics, as originally proposed by Goodenough, was ultimately, based on the model of Romney (1957) as developed by Kirch and Green (1987).

Although all of these authors celebrated the laboratory-like (controlled/ limited interaction) aspects of Oceanic islands in contrast to the less controlled interaction and complex histories of continental land masses, it would be hard to find any archaeologist or anthropologist working in the Pacific who truly believed in the "myth of the primitive isolate" (Terrell *et al.* 1997). It appears no island was so isolated people did not get to it and traditional histories related stories of voyages and migrations while ethnography and history documented at least intra-archipelago voyaging. Even Sahlins in his early work noted the importance of considering inter-island contact in explaining cultural variation (e.g., Sahlins 1958:174, 245). The question was not whether there was interaction but how much. The islands as isolates theme was reinforced early on by the work of Andrew Sharp (1956) in which he concluded Pacific colonization was conceivably carried out through accidental drift-voyaging. Subsequent research on traditional voyaging (Finney 1979) and experimental modelling (Irwin 1989; 1992) demonstrated that purposeful highly skilled long-distance voyaging was common in the Pacific and more adequately explained the archaeological data. Studies of the sourcing of adze stone and obsidian have subsequently confirmed interaction in Pacific prehistory (Best *et al.* 1992, Bird *et al.* 1981, Walter and Sheppard 1996, Weisler 1993; 1997). If there is a danger today, it is of erring on the other side of the debate and assuming that inter-island communications were easy and always common, the so-called mosaic or panmictic solution.

No matter what the voyaging technology the degree of island isolation (as measured through effort expended and danger encountered) must increase, minimally, with inter-island distance. For archaeologists working in the Solomons or coastal Papua New Guinea (PNG), interaction by sailing or paddling canoe along coasts or between inter-visible islands is a commonplace in a region where even small children can be seen paddling out to sea in dug-out canoes no bigger than a large baby buggy. The sea is a road not a barrier. However as Kirch and Green (2001:88) point out the world of Remote Oceania varies considerably in degrees of isolation and consequently on-going interaction. The islands of marginal east Polynesia are truly isolated while those of tropical central Polynesia are much less so, although by Solomon standards even the archipelagos in this area typically have very large inter-island distances. It seems clear therefore that 'insularity' is a factor which influences the archaeology of islands. The sea is not land. Simply replacing some apparent bias towards isolation with another one of interaction does not improve our models. What is needed is a more developed notion of how the properties of islands affect the cultural landscape and the archaeological record (Terrell 1986:123). In the following paper we look at how the cultural landscape and archaeological record are affected by the properties of the islands of the western Solomons. We argue that insularity or 'islandness' provides physical properties which can structure cultural systems and the archaeological record in ways not necessarily seen where 'islandness' is not salient.

The western Solomons

The geographic Solomon Islands are a set of wholly inter-visible islands (figure 11.1) stretching from Bougainville some 860 kilometers east to the tip of Makira which marks the easternmost point of Near Oceania beyond which we have a 360 km open sea crossing to the Reef Santa Cruz Group and the start of Remote Oceania (Green 1991). The maximum open sea crossing required to traverse the region between Buka in the northwest to Makira in the southeast is 50 km (between Vella Lavella or North New Georgia and Choiseul). This archipelago truly does form, with the Bismarck Archipelago to the north, what Irwin has described as a safe voyaging corridor (Irwin 1991) easily settled using simple maritime technology.

Although many of these islands are long they are generally narrow. With the exception of Bougainville none could be considered to have remote interiors. Beyond Bougainville only on Guadalcanal is the ocean more than 15 km from the island mid point. For most Solomon islanders prior to the 20th century the coast was within half a days walk. While today most people live on the coast, they often walk four or more kilometers inland during the day. Although the islands are narrow they generally only have a short coastal flat which then rises steeply across a series of longitudinal ridges generally to a maximum height of much less than 1500 m. Although the bush is rugged very little of it can be described as mountainous. In the western Solomons the only true high terrain is on the volcanic island of Kolombangara which rises to 1500 m and serves as a focal point in the centre of the New Georgia Group. Of particular interest are the very large lagoon systems which fringe many of the islands. These are a significant feature of the New Georgia Group where Marovo Lagoon at the eastern end of New Georgia is one of the largest inshore lagoon systems in the world.

As in most of northern island Melanesia the western Solomons is linguistically a very diverse region with over 25 languages spoken today and more in the past. Most languages belong to the very large Austronesian language family, however there are two (Bilua, Baniata) extant non-Austronesian languages. Many people in the region speak multiple languages and today most villages have numerous members from other language groups. Up until the start of the 20th century headhunting and inter-island warfare was practiced by all language groups in the western Solomons with the Roviana people considered to be amongst the most successful (Zelenietz 1979).

Interaction

As might be expected, given the nature of this archipelago, inter-island communication and travel was and is extremely common for economic and social reasons. This has ultimately formed a community of culture within the region. This is perhaps best demonstrated by an extract from the following traditional story which is common to the region. The story is

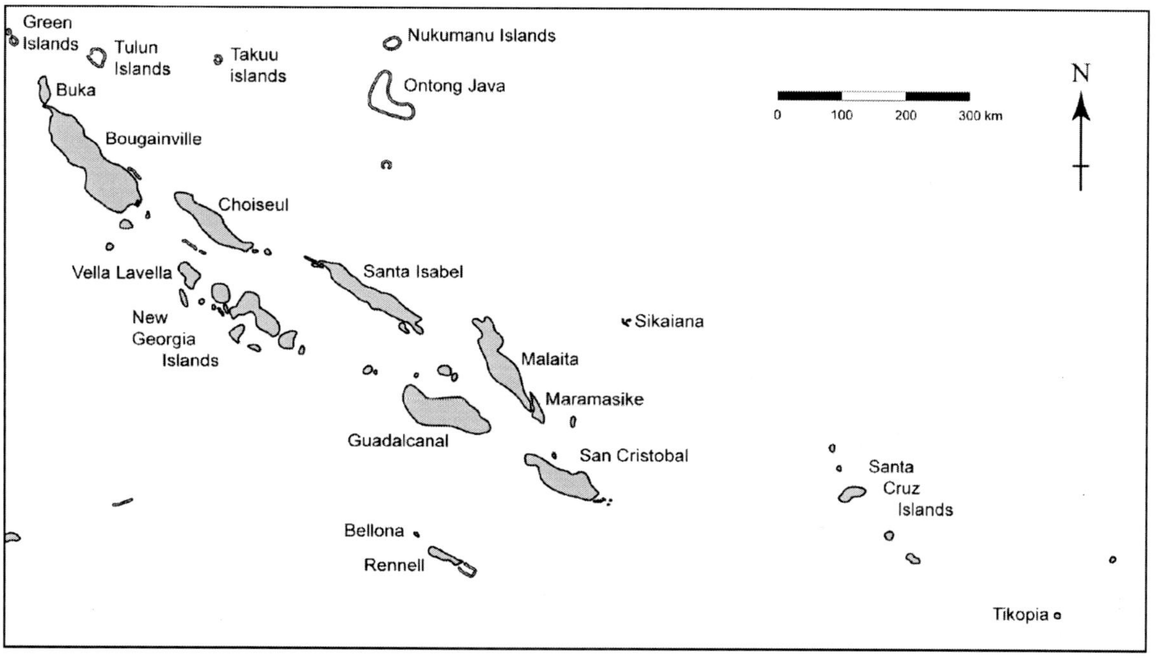

Figure 11.1: Map of the geographic Solomon Islands.

about the culture hero Tiola and his five brothers who came forth into the world from a cave on the slopes of Kolombangara. Whilst arguing among themselves Tiola's brother Heleveni comes up with a plan to kill their brother Veonona.

> ... Heleveni decided to do something so he invited Tiola, rat (*kuresu*), hermit crab (*komba*) and three sea birds, *belama, vangolo*, and *dekere*. He said you go down and get a canoe to go to Bugotu (eastern Santa Isabel). Go down and trade for *baha* and *tuva* [killing spells] and then come back not later than four days from now so we can kill Veonona and collect his head and put it on the shrine. That was Heleveni's secret plan with Tiola and the others. So they went, at which point Tiola turned into a dog. When they arrived at Bogotu they waited until dark and then went out and collected fire flies. They collected them and the spells. Then they decided to return to Nduke (Kolombangara). When they returned they were between Laru (Choiseul) and Nduke and they decided they would have a farting competition. So they suggested that *Belama* should go first. He did and they said it smelled like *rebirebi* dead fish because he ate flying fish. So then it was *Vangolo*'s turn and they said it smelled like *hipuhipu* (small schools of bait fish near the shore). When it was *Dekere*'s turn it smelled like *imabuku* (deep sea bait fish). Then it was *Kurezu* the rats turn. Then they said it smelt like rotten food. Next it was Tiola's turn and they said it smelt like human shit, because dogs eat shit. Finally it was *Komba's* turn. *Komba* moved the seat which he was sitting on and sat on the bottom of the canoe. The force of his fart split the canoe and it sank. The birds flew away. *Komba* sank down to the sea bed. Then Tiola started to swim but before he could the rat jumped and sat in Tiola's ear. The current took Tiola closer to Nduke but when the current changed it took him down towards Vella Lavella. When he was close to the shore the current changed again and carried him down through Gizo passage to Simbo. When he was near Simbo he started to swim towards shore. When he reached the shore he shook his body to dry himself and the rat *Kurezu* jumped down. When Tiola saw *Kurezu* was not wet he said: "I have drifted days and nights and am very wet. Why are you still dry?" So *Kurezu* said: "I was just riding between your ears and your face (*tili tala namu*)." *Tiola* was angry and chased *Kurezu*. He chased him up to the top of the hill on Simbo. There *Kurezu* went into a hole and went down so *Tiola* started to dig until he reached the cen-

tre of the earth. When he broke into the centre, a great wind came out and blew *Tiola* into the air. *Tiola* flew through the air until he fell on Nusa Roviana (at the eastern end of Roviana Lagoon on south-western New Georgia). When he fell people saw him and said: "We have found a stranger." He then spoke like a human and stayed there on Nusa Roviana with some people. (As told by Silas Oka of Patmos village in Roviana 1997.)

In this journey the culture hero Tiola travelled (figure 11.2) a total of 770 kilometers and covered most of the western Solomons. This is analogous to the journeys of the modern people who travelled by paddle canoe throughout the region in search of items that often included magic but also food, shell valuables, captives and heads (Hocart 1931). Most headhunting journeys in the western Solomons during the 19th century took place in 30-man war canoes (Waite 1990; Woodford 1909). These often travelled distances in excess of 150 kilometers (e.g., from Roviana to Choiseul or Santa Isabel) but occasionally exceeded 250 kilometers when Roviana warriors raided the western end of Guadalcanal (Bathgate 1985). Although travel over long distance was common it was not necessarily without danger from storms, currents and water spirits which could sink canoes or pull them off course. The famous *nuzunuzu* figure placed just above the water line at the bow of each war canoe is reputed to have been placed there to scare away evil sea spirits (Somerville 1897:271). The sea was also not a neutral or featureless body of water. Some areas were more difficult than others (e.g., the strait between Vella Lavella and Kolombangara traversed by Tiola) being characterized by strong currents and exposures to prevailing winds such as the southeast trades which can blow for long periods in mid-year and make sea travel very difficult even for people with large boats and outboard motors. Travel within lagoons is comparatively easy by paddle canoe and interaction within lagoons or across short straits such as exist in the western Solomons has always been common. However, even today open ocean crossings of more than 10 km are treated with respect and canoes often go down in bad weather.

The physical arrangement of islands with one another as well as the nature of their coasts, especially with regard to prevailing winds and currents, also strongly influences patterns of settlement. Although the use of paddle canoes makes many shores or shallow lagoons much more accessible than they are to larger sailing canoes or ships, some coasts are inhospitable to anyone attempting a landing. The difficulty of landing on rock-bound coasts with fringing reefs makes areas such as the south coast of New Georgia east of Roviana lagoon fairly inhospitable. This is probably reflected in the modern lack of settlement in that area. Coasts which are oriented toward other nearby coasts as opposed to facing long stretches of open sea would also appear, all else being equal, to be favoured as travellers have ready access to many other locations and people. Again where you have many near neighbours, settlement is favoured in those areas. In this case the density of coasts is a variable unique to islands. The modern distribution of villages on Vella Lavella (figure 11.3) would appear, in the absence of additional analysis, to be a case in point with a clear difference in settlement density between the east coast which looks in to the centre of the group (Kolombangara, Gizo, New Georgia), and the west and north coasts which look out to the open sea and distant PNG.

Given the fact, however, of long term interaction throughout the region, the question for archaeologists is whether this prohibits us from using biogeographical modelling techniques such as nearest neighbour analysis which predict cultural similarity arising from proximity. Figure 11.4 is a nearest neighbour analysis produced after the method employed by Terrell (1977; 1986). This figure would predict high degrees of similarity within the New Georgia Group and across to western Santa Isabel and eastern Choiseul but with a break at eastern Santa Isabel and western Choiseul which have minimal linkage to the rest of the New Georgia Group.

Figure 11.5 illustrates what is known of linguistic terms and forms associated with skull shrines within which are kept the skulls of important ancestors. The linguistic terms include the terms for sacred place, chief and a supernatural being associated (in Roviana) with chiefs. In Roviana these are *hope*, *bangara* and *mateana*. Figure 11.5 indicates that the best predictor of a term within the region is generally the neighbouring term. However, there is strong commonality within the region which is only broken at the points indicated by the nearest neighbour analysis. In western Choiseul human remains are cremated and kept in stone containers (*dolo*) and the terms found in the New Georgia Group are absent while the cultural pattern is most closely shared with eastern Bougainville where cremation is practiced. Although at the eastern end of Santa Isabel we have little data, the linguistic shift is marked (*padaghi*) and it seems that its cultural pattern is most closely shared

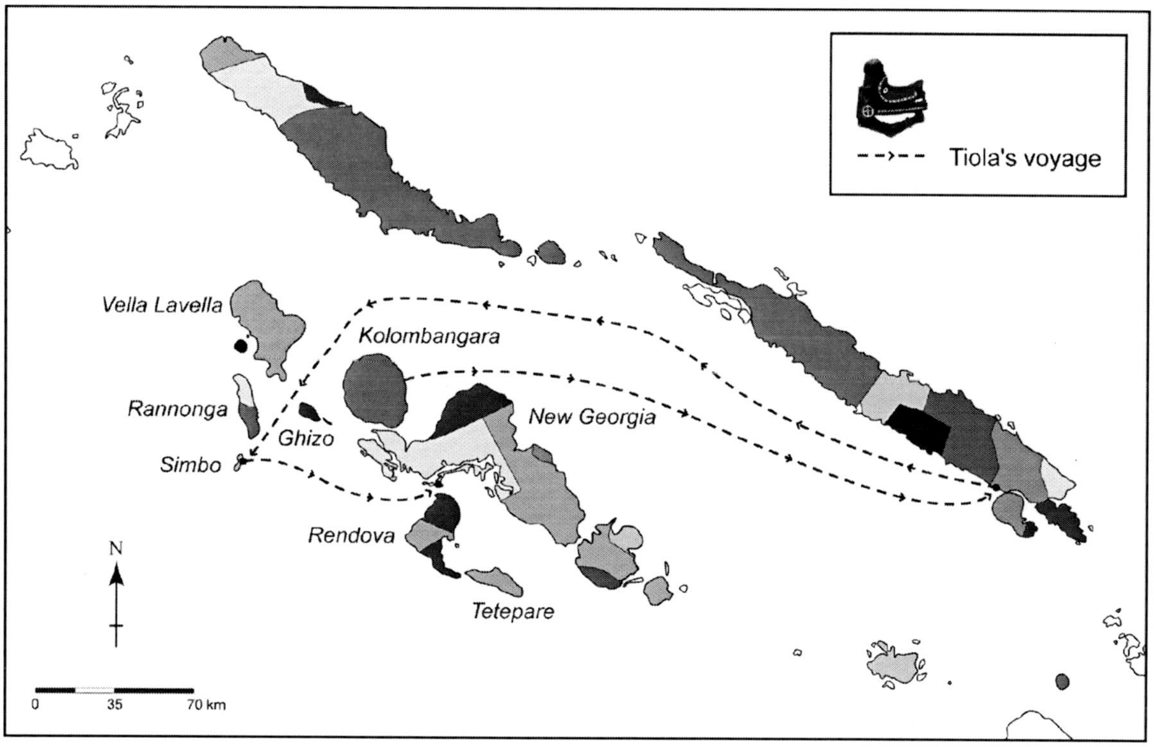

Figure 11.2: Tiola's voyage.

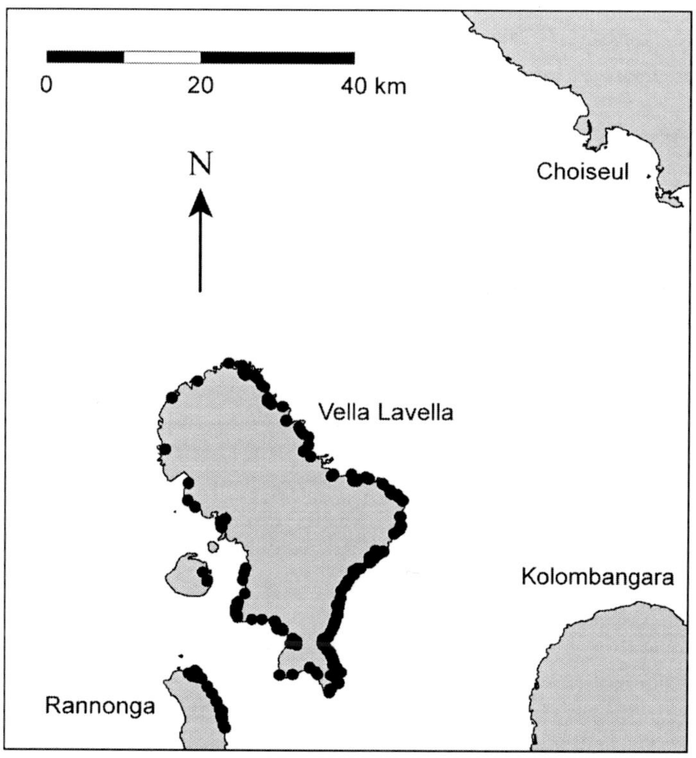

Figure 11.3: Distribution of modern villages on Vella Lavella.

with Malaita or the Florida Group to the east. This latter break corresponds to a major linguistic division; the Tryon-Hackman line which separates the languages of the Solomon Islands into two major linguistic groupings. It should also be noted that the people of Vella Lavella, who speak a non-Austronesian language, have apparently borrowed terms associated with the skull cult from their Austronesian speaking neighbours on Choiseul. Our research would suggest that the cultural complex involving headhunting, the skull cult, and shell valuable exchange, developed in Roviana at some point after 1500 AD (Aswani and Sheppard 2003, Sheppard *et al.* 2000; Walter and Sheppard 2001). Whether this points to its centre of origin, or whether it was part of an interactive regional development process is unclear, however, what is clear is that it spread throughout the region crossing all linguistic boundaries. Variation in the linguistic and cultural forms is definitely not confused or mixed—it is predicted by geographical proximity. If we wish to understand the last 500 years of the archaeological record in the New Georgia Group we must clearly pay attention to island geography which structures interaction and its effects. Just because people can interact over long distances does not mean that this will necessarily have a significant impact.

The type of interaction is, of course, an important variable in understanding the effects of interaction. From a long-term archaeological perspective it may simply be geographical propinquity—and therefore amount of interaction—which is of any explanatory importance. However, historically we know that the nature of interaction varied considerably. Today some people move over great distances within the Solomon Islands on a regular basis using light planes and ferries but many (perhaps most) people do not travel far from their own villages. As in most societies economics and family connections ultimately determine the extent of travel. Interaction is also determined by religious affiliation which remains one of the key bases of modern social networks. During the headhunting period interaction was apt to be short and violent, although very often considerable numbers of captives were returned to distant islands (McDougall 2000). The movement of captives will have influenced local culture but also created social ties between islands as the captives became assimilated into the local societies and married. Marriage to a captive could produce rights to land which are recognized to this day. In the case of sharing of cultural forms such as headhunting and shrines, which are associated with chiefly authority, it might well have been systematic interaction among chiefly classes which facilitated sharing of cultural features. The genealogies of Roviana chiefs indicate considerable intermarriage among the *mbangara* lineages. The famous chief Taebangara of Nusa Roviana was married to women from throughout the western Solomons and it seems probable that during that time (late 18th–early 19th century) Roviana influence dominated the region.

Islands as identities: 'New Georgia' is British

'Where are you from?' is a popular question in the Solomon Islands just as it is in most places. If the question is asked in the capital Honiara the answer may be 'the west' indicating the speaker is from the Western Province or essentially the New Georgia Group. Closer enquiry will likely elicit a language area (e.g., Roviana) as the question is now not so much geographical as social, for people are generally seeking to determine if they are 'one-talks' or people who speak the same language and therefore likely related. People tend, in our experience, not to identify with an island but with a language area. To our knowledge there is no Solomon Island name for New Georgia in its entirety. It is subdivided into language areas (e.g., Roviana, Marovo, Kusaghe, etc.). Often the names which are Solomon Island language terms now applied to some islands (e.g., Rannonga) were originally terms only applied to areas within the islands. Europeans eager to name mapped islands extended terms from a part to the whole. For Solomon Islanders there was little need, prior to colonialism, for a geographical unit bigger than a language area. Given the ease of travel within lagoons, and to nearby islands, language areas often extend across the water to adjacent islands rather than along the coasts for long distances. Therefore the Roviana region includes the mainland coast of southwest New Georgia, the barrier islands of Roviana Lagoon and the islands adjacent to Vonavona Lagoon. A similar pattern has been noted by Moss (2004) among the Tlingit whose social geography cross-cuts an extensive archipelago of islands along the coast of southeastern Alaska.

Islands as cultural structures

Islands are land masses with high ratios of coast to land area. This boundary between land and sea effectively defines an island both physically and culturally. The coastal transition marks an

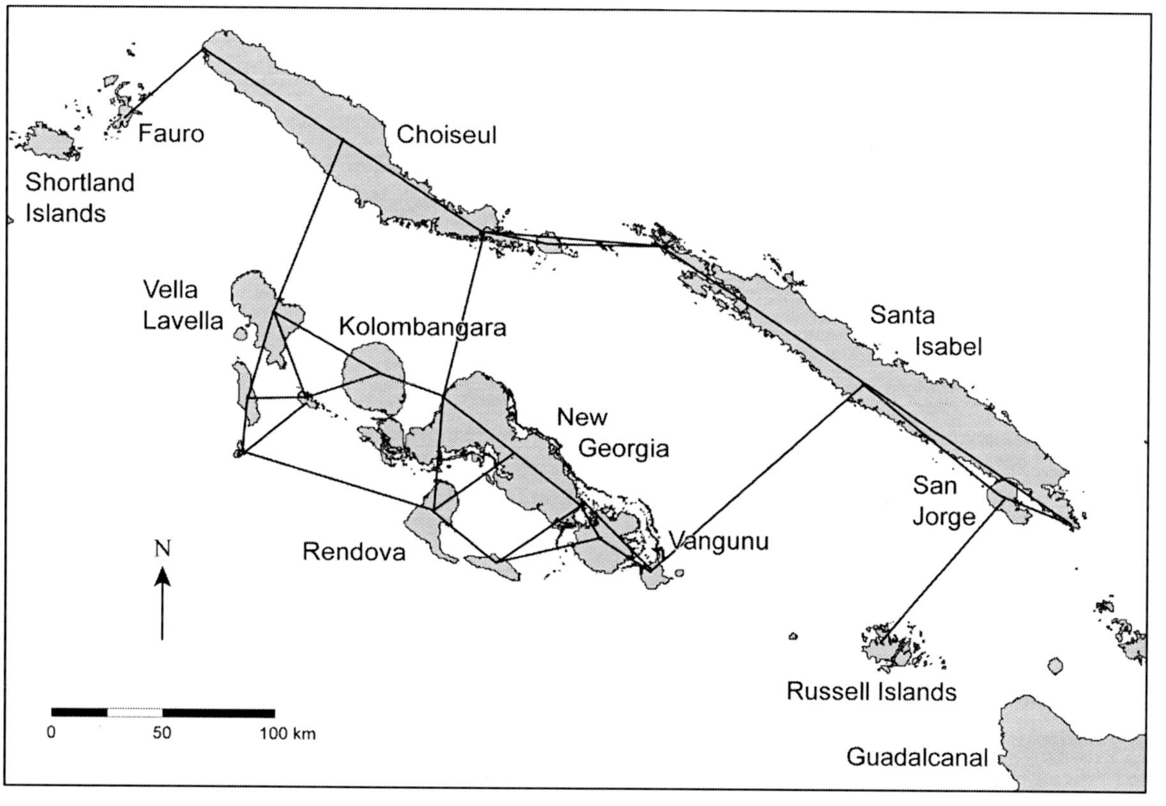

Figure 11.4: Third order nearest neighbour analysis (method after Terrell 1986: 130).

important ecological edge which generally makes the coast rich in economic terms. While this proximity to marine resources has obvious cultural importance; the dramatic physical transition to it from the land can also serve as a very important symbolic resource. Roviana Lagoon provides very rich marine resources to the modern inhabitants of its shores (Aswani 1997). The use rights to patches of reef are assigned to groups, as is common throughout the Solomons, however no-one attempts to restrict travel around the lagoon. It is only when you land that you suddenly cross the boundary from public to restricted access. Travelling to an island may be the easy part—surviving the landing may be much more problematic. It is probably safe to say that virtually all coastal land in the Solomons Islands is under some cultural control. Someone will claim rights to that land and feel they should be consulted about strangers landing on their coast. Such areas, are of course fairly easily demarcated as coasts provide their own obvious boundary markers. In the interior of the island land is not so finely managed, although large areas will be assigned to large descent groups. In the recent past, prior to the advent of logging and after the major population crash of the early 20th century, the interior for a time ceased to have much interest. It was a place of feared wandering spirits and generally only frequented by hunters chasing wild pigs. Today the interior has once again become important in the western Solomons as foreign logging companies move across the islands and the question of who controls what piece of interior is on most people's lips. Rights to logging royalties are today determined through demonstration of prior usage of the interior, as made manifest through knowledge of shrines, old villages and other cultural sites lying in the bush.

Although the physical and ecological natures of islands structure the economic bases of societies, these structures are often expressed symbolically through reference to island physiography. At the simplest level islands provide their own set of significant directions. These are typically toward or away from the coast or along the coast. Movement away from the coast is also generally accompanied by a gain in height therefore coastal is low and interior is high. The coast, which is often continuously cultivated, is also the place of culture while the interior is wild. In the past when interior settlement was more common the culture/nature divide was at the boundary between the bush and the gardens. Throughout the Solomons the bush/sea distinction has

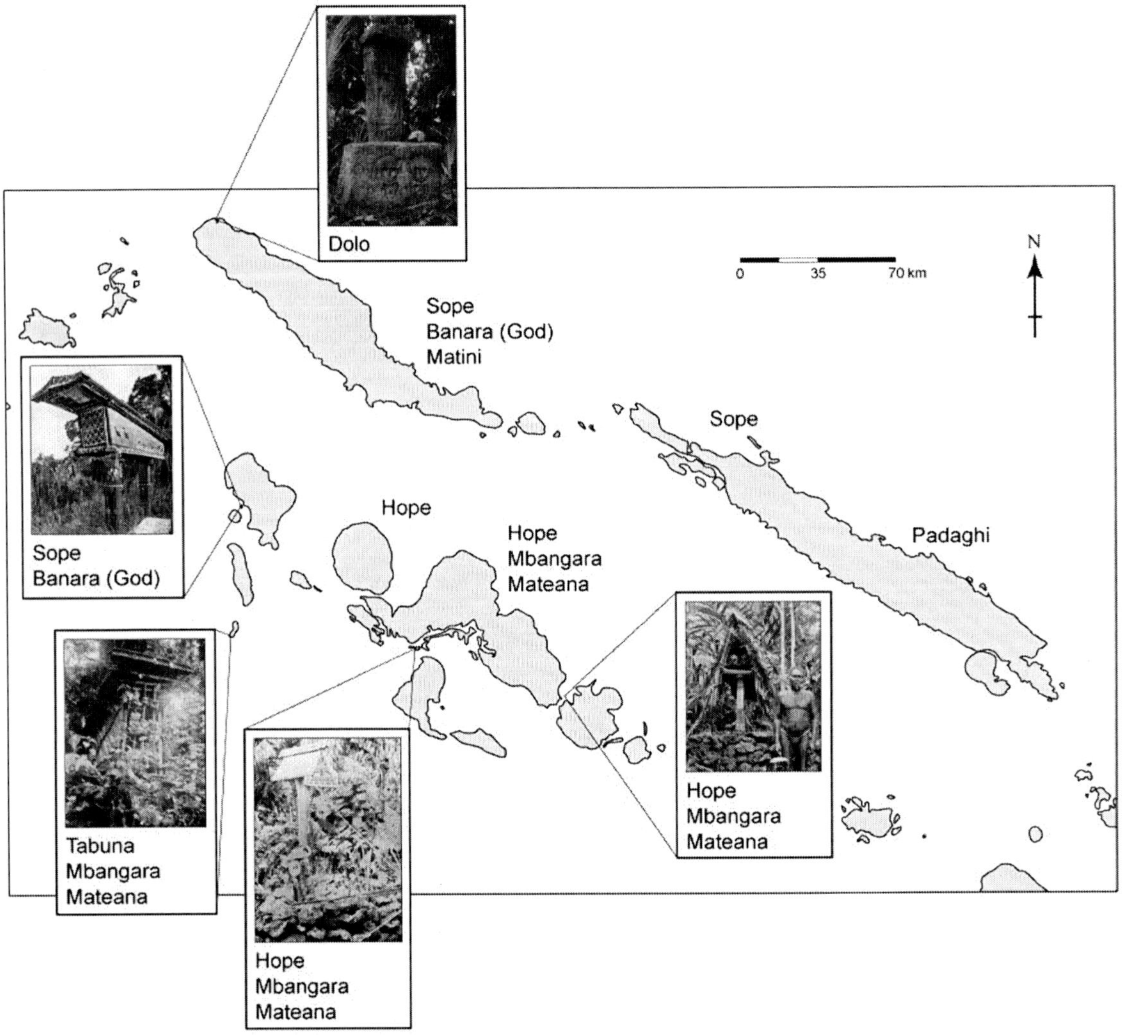

Figure 11.5: Distribution of shrine forms and linguistic terms in the western Solomons.

had cultural significance with people described as bush people or sea people (Roe 2000). Today, in the western Solomons almost everyone lives on the coast although archaeology and oral tradition clearly show that the interior was occupied prior to the 20th century. In Roviana, some groups of people claim to be descendants of 'bush tribes' and this is used to differentiate themselves from the Roviana sea people. Although all these people today speak Roviana, it is possible that the bush people originally spoke another language. Certainly being able to claim preferential use rights to interior land is useful in claiming logging royalties. However, all Roviana people ultimately claim descent in fact from people who lived in the interior behind modern Munda.

Oral tradition states that Roviana people descend from groups which aggregated at a place called Bao in the centre of North New Georgia and that they came down to the coast led by the chief Ididubangara. After arriving at the coast they fought with coastal peoples and displaced them from the Roviana Lagoon, finally establishing their base on Nusa Roviana (Sheppard *et al.* 2004, Sheppard *et al.* 2000; Walter and Sheppard 2001). This tradition of origin in the interior of islands followed by migration to the coast, often marked by a series of descending shrines or old village sites, is very common in the Solomon Islands and we have encountered it on a number of islands. Miller (1980) has suggested this phenomenon is a cognitive structure which relates social groups to the high interior with height associated with status in many Austronesian societies. We have argued elsewhere (Sheppard *et al.* 2004) that at least in Roviana the archaeology and chronology associated with features identified in the oral tradition strongly suggests the tradition is more than myth or common cognitive structure. Whatever the historical truth of the Roviana origin tradition it is certainly evident that many people in Roviana, as elsewhere in the Solomons, had relatives who came down from the interior at the end of the 19th century. Repeated flight into the interior, and return to the coast, have in general been an historical reality in the last 200 years in many areas of the western Solomons. The tension between coast and interior, a distance which in most instances is less than half a days walk, is a fundamental reality which is reflected both in the sea people/bush people distinction and the traditions of movement from the interior to the coast. The ultimate reason for this tension is perhaps revealed in the Roviana story which states that the chief Ididubangara wished to come down to the coast to get better access to shell valuables. This key structuring tension is ultimately founded on differential access to resources and mobility which are characteristic of most islands. It seems probable that this tension continues as an underlying driver of much culture history in the Solomon islands.

History piles up on the beach: islands as archaeological structures

The archaeological records of islands have unique properties. We may know nothing about the archaeology of an island, yet we can be confident that the earliest inhabitants arrived on the coast. When we step onto the beach the probability of standing in the footsteps of the earliest inhabitants is infinitely higher than if we were to descend into the bush of the interior. The density of human activity per unit area over time is clearly highest at the coast and declines as we move into the interior. In Roviana the record of the 20th century (before logging) is found solely on the coast where modern settlement is located. The area just back of the beach records all the significant historical events, modern houses, tourist resorts, government offices, WWII installations (wharves, airstrips, dumps, battle fields), mission stations, traders establishments, 19th century Roviana shrines, wharfs and in the inter-tidal zones ceramics dating back over 2000 years. At the coast it is possible, in many places, to stand in one spot and see over 2000 years of archaeological record at a glance. As we move into the interior the record becomes more scattered and in the dense jungle often impossible to find. The corollary of this effect, is of course, that destruction of the record is often greatest at the coast where, for example, major earthworks conducted during WWII resulted in the destruction of much of the earlier record, especially near Munda in Roviana. Near continuous use of the coastal flats and hills for gardens will have also disturbed much of the earlier record. In the interior the Pompeii effect may be much more pronounced, especially if we have a rapid abandonment followed by minimal subsequent use of the area. Finding the record in the dense interior jungle provides, however, another significant archaeological challenge. Disturbance and clearance on the coast allows for rapid survey and targeting of regions while in the interior today the creation of logging roads facilitates both access and exposes long stretches of road cut stratigraphy.

Conclusion: western and other ideas of 'islandness'

The islands of the western Solomons are obviously not isolates but neither are they just disjoint continents. Their 'islandness' provides them with physical characteristics and cultural potentialities which make them distinctive environments. In summary we have argued that:

- Islands are not isolates because of continued interaction. Thus settlement is structured by the physical characteristics, the relationships among the islands, the sea conditions and the type of interaction that occur in any given locality;
- Despite interaction over long distances, cultural similarity and difference correlates with geographical proximity—i.e., nearest neighbour analysis predicts variation;
- People in this archipelago do not relate so much to islands as to language areas which may extend over nearby islands to which access is easy;
- The coast/interior difference provides islands with structuring principals and tensions which may be used in framing core cultural schema;
- The archaeological record of islands is structured by differential access and resource potential creating higher probability of deposition on the coast.

It could be argued that the fundamental underlying features of islands which appear to make them distinctive environments, such as mobility/access and patterned differential resource richness, are simply features which may be found on continents (e.g., desert edges or oases). However the sea which surrounds islands provides a unique transport medium which during much of the Holocene was unrivalled—making islands and coasts unique locations. Islands as human environments have changed with the development of maritime technology and food production. With the advent of sailing technology and food production many of the islands of the Pacific became both accessible and inhabitable. Although western culture may have had a continental bias, out of which developed a notion of islands as isolated and exotic places which needed to be found, mapped and tamed, it seems probable that the people who lived on islands developed their own notions about landscapes and seascapes of this form. These notions in turn structured their behaviour and the resulting archeological record.

Acknowledgements

We would like to acknowledge the on-going support of the Royal Society of New Zealand Marsden fund which has funded our work on Vella Lavella and in Roviana. We also thank the many students and colleagues who have worked with us over the years. In particular we thank Kenneth Roga for his support and friendship. He has taught us something of the nature of islands while delivering us safely across many inter-island crossings in small canoes. Finally we thank the people of Roviana and Vella Lavella who have generously cared for us these many years. Leana hola and Ouri.

References cited

Aswani, S. (1997). *Customary Sea Tenure and Artisanal Fishing in the Roviana and Vonavona Lagoons, Solomon Islands: The Evolutionary Ecology of Marine Resource Utilization.* Unpublished PhD, University of Hawai'i.

Aswani, S. and P. Sheppard. (2003). The archaeology and ethnohistory of exchange in precolonial and colonial Roviana: gifts, commodities and inalienable Possessions. *Current Anthropology* 44 (supplement): S51–S78.

Bathgate, M.A. (1985). Movement Processes from precontact to contemporary times: the Ndi–Nggai, west Guadalcanal, Solomon Islands. In M. Chapman and R. Prothero (eds.), *Circulation in Population Movement: substance and concepts from the Melanesia case.* pp. 83–118. London: Routledge and Kegan Paul.

Best, S., P. Sheppard, R. Green and P. Parker. (1992). Necromancing the stone: archaeologists and adzes in Samoa. *Journal of the Polynesian Society* 101: 45–85.

Bird, J.R., W.R. Ambrose, L.H. Russell and M.D. Scott. (1981). *The Characterizations of Melanesian Obsidian Sources and Artefacts Using the Proton Induced Gamma–Ray Emission (PIGME) Technique.* Sydney: Australian Atomic Energy Commission.

Finney, B. (1979). Voyaging. In J.D. Jennings (ed.) *The prehistory of Polynesia.* pp. 323–

3551. Canberra: Australian National University.

Goldman, I. (1955). Status rivalry and cultural evolution in Polynesia. *American Anthropologist* 57: 680–96.

Goodenough, W. (1957). Oceania and the problem of controls in the study of cultural and human evolution. *Journal of the Polynesian Society* 66: 146–55.

Green, R. (1991). Near and Remote Oceania: disestablishing 'Melanesia' in culture history. In A. Pawley (ed.) *Man and a Half: Essays in Pacific Anthropology and Ethnobiology in Honour of Ralph Bulmer*, pp. 491–502. Auckland: The Polynesian Society.

Hocart, A.M. (1931). Warfare in Eddystone of the Solomon Islands. *Journal of the Royal Anthropological Institute of Great Britain and Ireland* 61: 301–24.

Irwin, G.J. (1989). Against, across and down the wind: a case for the systematic exploration of the Pacific Islands. *Journal of the Polynesian Society* 98(2): 167–206.

Irwin, G.J. (1991). Pleistocene voyaging and the settlement of Greater Australia and its Near Oceania neighbours. In J. Allen and C. Gosden (eds), *Report of the Lapita Homeland Project, Occasional Papers in Prehistory, No.20*, pp. 9–19. Canberra: Dept. of Prehistory, Research School of Pacific Studies, Australian National University

Irwin, G.J. (1992. *The Prehistoric Exploration and Colonisation of the Pacific.* Cambridge: Cambridge University Press.

Kirch, P.V. (1975). Cultural adaptation and ecology in western Polynesia: an ethnoarcheological study. Unpublished PhD thesis, Yale University.

Kirch, P.V. (1979). Archaeology and the evolution of Polynesian culture. *Archaeology* 32(5): 44–52.

Kirch, P.V. (1980). Polynesian prehistory: cultural adaptation in island ecosystems. *American Scientist* 68(Jan–Feb): 39–48.

Kirch, P.V. (1984). *The Evolution of Polynesian Chiefdoms.* Cambridge: Cambridge University Press.

Kirch, P.V. (ed.) (1986). *Island Societies: Archaeological Approaches to Evolution and Transformation.* Cambridge: Cambridge University Press.

Kirch, P.V. and R.C. Green. (1987). History, phylogeny, and evolution in Polynesia. *Current Anthropology* 28(4): 431–56.

Kirch, P.V. and R.C. Green. (2001). *Hawaiki, Ancestral Polynesia: An Essay in Historical Anthropology.* Cambridge: Cambridge University Press.

Kuklick, H. (1996). Islands in the Pacific: Darwinian biogeography and British anthropology. *American Ethnologist* 23(3): 611–38.

MacArthur, R. and E.O. Wilson. (1967). *The Theory of Island Biogeography.* Princeton: Princeton University Press.

McDougall, D. (2000). Paths of 'pinauzu': captivity and social reproduction in Ranongga. *Journal of the Polynesian Society* 109(1): 99–114.

Miller, D. (1980). Settlement and Diversity in the Solomon Islands. *Man* 15: 451–66.

Moss, M.L. (2004). Island societies are not always insular: Tlingit territories in the Alexander Archipelago and the adjacent Alaska mainland. In S.M. Fitzpatrick (ed.) *Voyages of Discovery: The Archaeology of Islands.* London: Praeger.

Rainbird, P. (1999). Islands out of time: towards a critique of island archaeology. *Journal of Mediterranean Archaeology* 12(2): 216–34.

Roe, D. (2000). Maritime, coastal and inland societies in island Melanesia: The bush–saltwater divide in Solomon Islands and Vanuatu. In S. O'Connor and P. Veth (eds), *East of Wallace's Line: Studies of past and present maritime cultures of the Indo–Pacific region.* Rotterdam: A.A. Balkema, pp. 197–222.

Romney, A.K. (1957). The genetic model and Uto–Aztecan time perspective. *Davidson Journal of Anthropology* 3:35–41.

Sahlins, M. (1957). Differentiation by adaptation in Polynesian societies. *Journal of the Polynesian Society* 66(3): 291–300.

Sahlins, M. (1958). *Social stratification in Polynesia.* Seattle: American Ethnological Society.

Sahlins, M. and E. Service (1960). *Evolution and Culture*. Ann Arbor: University of Michigan Press.

Sharp, A. (1956). *Ancient Voyagers in the Pacific. Polynesian Society Memoir 32*. Wellington: The Polynesian Society.

Sheppard, P., R. Walter and S. Aswani (2004). Oral tradition and the creation of late prehistory in Roviana Lagoon, Solomon Islands. *Records of the Australian Museum* Supplement 29: 123–32.

Sheppard, P., R. Walter and T. Nagaoka (2000). The archaeology of headhunting in Roviana Lagoon. *Journal of the Polynesian Society* 109(1): 9–37.

Somerville, B.T. (1897). Ethnographical notes in New Georgia, Solomon Islands. *Journal of the Royal Anthropological Institute* 26(4): 357–413.

Steward, J. (1955). *The Theory of Culture Change*. Urbana: University of Illinois Press.

Terrell, J. (1977). Geographic systems and human diversity in the northern Solomons. *World Archaeology* 9(1): 62–81.

Terrell, J. (1986). *Prehistory in the Pacific islands: a study of variation in languages, customs, and human biology*. Cambridge: Cambridge University Press.

Terrell, J. (1997). The postponed agenda: archaeology and human biogeography in the twenty-first century. *Human Ecology: An Interdisciplinary Journal* 25(3): 419–37.

Terrell, J.E., T. Hunt and C. Gosden. (1997). The dimensions of social life in the Pacific: human diversity and the myth of the primitive isolate. *Current Anthropology* 38: 155–95.

Vayda, A. and R. Rappaport (1963). Island cultures. In F. Fosberg (ed.), *Man's Place in the Island Ecosystem*, pp. 133–42. Honolulu: Bishop Museum Press.

Waite, D. (1990). Mon canoes of the western Solomon Islands. In A. Hanson and L. Hanson (eds), *Art and Identity in Oceania*, pp. 44-66. Honolulu: University of Hawaii Press.

Walter, R. and P. Sheppard (1996). The Ngati Tiare adze cache: further evidence of prehistoric contact between West Polynesia and the Southern Cook Islands. *Archaeology in Oceania* 31: 33–9.

Walter, R. and P. Sheppard (2001). Nusa Roviana: the archaeology of a Melanesian chiefdom. *Journal of Field Archaeology* 27(3): 295–318.

Weisler, M. (1993). Provenance studies of Polynesian basalt adze material: a review and suggestions for improving regional databases. *Asian Perspectives* 32: 61–83.

Weisler, M. (ed.) (1997). *Prehistoric Long-Distance Interaction in Oceania: An Interdisciplinary Approach*. Auckland: New Zealand Archaeological Association, Vol. Monograph 21.

White, L. (1949). *The Science of Culture*. New York: Farrar, Strauss and Young.

Woodford, C.M. (1909). The canoes of the British Solomon Islands. *Journal of the Royal Anthropological Institute* 39: 505–16.

Zelenietz, M. (1979). The end of head hunting in New Georgia. In M. Rodman and M. Cooper (eds), *The Pacification of Melanesia*, pp. 91–108. Ann Arbor: University of Michigan Press.

Chapter 12

Creating connections between Caribbean Islands: an archaeological perspective from northern Cuba

Jago Cooper, University College London

Introduction

The debate surrounding the perceived boundary created by the liminal edge where land meets water has been ongoing in island archaeology for many years. Arising from this debate is a discussion of whether marine islands constitute a viable unit of study separate from the traditional 'other' of mainland or continental archaeology (Rainbird 1999: 232). It has been forcefully argued that humans do not necessarily see the shoreline as a boundary (Ingold, 1993: 156) and, further, that the sea can be a positive space for communication and interaction (Gosden and Pavlides 1994: 162; Rainbird 2004). Whilst humans can and regularly do transcend the boundary between land and sea, it has also been argued that the boundary may have more relevance for studies that use the island as an isolate for flora, fauna and material resources (Terrell 2004). This environmental boundary between land and sea is far from clear-cut with complex transitional wetland environments blurring the boundary and numerous archaeological examples providing evidence of environmental translocation across it (Keegan and Mitchell 1986; White 2004). However, even given these complex issues of land-sea boundaries, many studies continue to perceive the island as a useful unit of analysis. Examples of such studies in the Caribbean are the provenance studies from the Lesser Antilles that have used islands as bounded units between which movement of materials can be identified and used to trace networks of interaction (Hofman 1993; Knippenberg and Gijn 1998; Rodríguez-Ramos 2002:16). This chapter will use initial results from archaeological fieldwork in northern Cuba to explore the detail with which inter-island interaction can be identified and interpreted in the Caribbean.

Caribbean island archaeologies

The insular Caribbean today, defined as an area bounded by the continental landmasses of North and South America, is a culturally diverse region within which many linguistic, political and ethnic boundaries can be studied. Archaeologists have been studying the different cultural boundaries of the Caribbean for many years and the early works of Rouse and others built upon culture historical frameworks popular at the time and have provided a long-lasting legacy for Caribbean Archaeology. Cultural groups were defined by lumping (or splitting) the distribution of associated material remains, with a heavy reliance on ceramic form and style. These cultures were often manifested in two dimensional maps that appear to show boundaries or frontiers that in many ways form hypothetical cultural islands (Rouse 1992: 8).

The widespread critiques of culture history in world archaeology have highlighted the complexity of culture and the lack of easily defined culturally homogenous groups around which valid boundaries can be created. This study of the boundary or limit of culture has been ongoing for many years through studies of trade, exchange and interaction. This has certainly been the case in the Caribbean where studies of local and regional interaction have appeared with regularity over the past fifty years (Curet 2004; Ulloa Hung 2000; Veloz Maggiolo 1974). These studies of interaction in the Caribbean could benefit from alternative studies that emphasize the flux, fluidity and fusion (Rainbird 2004: 245) of inter-island interaction that have been so eloquently discussed in other island theatres around the world (Broodbank 2000). Just as island archaeological studies have questioned the validity of the marine island boundary for humans and their movements so too does the study of interaction in an island framework help question the validity of the cultural boundary.

Cuban archaeology

Cuba is an island archipelago nation of over 4000 islands and islets that constitute more than 50% of the landmass of the insular Caribbean. Since Cuba gained independence from Spain in 1898, archaeology has provided an important influence on the social development of the nation. Harrington's book *Cuba before Columbus*, published in 1921, is a comprehensive view of Cuban prehistory that proved to be a turning point in archaeological approaches in Cuba. Harrington synthesized the work of previous Cuban archaeologists, such as Montané and Cosculluela by describing their findings and using their terminologies. Fortunately, Fernando Ortiz had access to this "importantísima contribución a la arqueología cubana" (Ortiz 1922: 10) and disseminated much of the information to a Spanish audience in his *Historia de Arqueologia Indocubana* published in Habana in 1922 (Ortiz 1922). Ortiz was a great patriot and to this day remains perhaps the most universally recognized Cuban researcher of the country's past; he is credited with identifying the past as a crucial building block for the construction of a Cuban identity in the present. While a full historiography of Cuban archaeology cannot be provided here, Irving Rouse provided an important subsequent contribution that must also be mentioned. Rouse worked in northern Cuba during the mid 1930s (Rouse 1942) and introduced a cultural framework established developed during his previous work in Haíti, based on ceramic typologies created from modes, types and styles (Rouse 1952). This classificatory framework was used to divide ceramic-age agriculturalist occupation of Cuba into Classic Taíno and Sub-Taíno, based on a distinction between Meillac and Chican ceramics. The same basic divisions based on ceramic typologies continue to underpin the cultural classification of ceramic period sites in Cuba today (Guarch 1978).

> The triumphant revolution of 1959 and the beginning of deep rooted social transformations on the island, permitted the rescue of our historical cultures as a profound expression of national unity... During these years the development of archaeological studies in Cuba allowed a better understanding of our aboriginal peoples. (Trincado Nelsa 2000: 105)

Revolutionary archaeology in Cuba has led to the development of a framework that is based on an economic model for stages of social development, influenced by Leninist, Marxist and Castroist theory (Tabio 1974). However, the national focus of research has meant that much of the work, with some minor modifications, has maintained the spatial and chronological cultural boundaries first established by Rouse in 1942 (Rouse 1942). An additional consequence of the revolution was a reduction in academic interaction with regional Caribbean neighbours. Although a number of Cuban archaeology publications have reported on the large amount of archaeological research conducted in Cuba and have received international distribution, their influence on regional Caribbean archaeology has nonetheless been limited, highlighted by the lack of Cuban archaeological literature cited in regional studies (Rouse 1992).

Since the collapse of the Soviet Union, the disruption of the Consejo de Ayuda Mutua Economica, and the start of the Periodo Especial in Cuba, a number of international collaborations have been established and opened up interest in improving regional interaction (Dawdy et al. 2005). The establishment of the journal *El Caribe Arqueológico*, with the help of Betty Meggers, has increased international collaboration and a number of international scholars have contributed to this Cuban journal. Therefore Cuban archaeology, with a wealth of existing data from over 100 years of archaeological research, is increasingly able to engage with, and contribute to, the development of regional interpretations.

A case study of interaction from Los Buchillones, northern Cuba

The site of Los Buchillones in many ways embodies the argument against the marine island isolate. The site itself spans the interface between land and sea by exhibiting dry and wet site occupation in the form of stilted houses (Valcárcel Rojas et al. 2006). The site is located on the north coast of central mainland Cuba in the Ciego de Avila province. Excavations at the site in the 1980s (Calvera Roses and Febles 1984), 1990s (Pendergast et al. 1999) and 2000s (Cooper et al. 2006; Cooper and Valcárcel Rojas 2004; Pendergast et al. 2003) have produced a large body of archaeological evidence with which to study this indigenous coastal community. The exceptional levels of organic preservation at the site provide a level of representational data that is not usually found at indigenous sites in the Caribbean. Comparative analysis of the material remains reveals a ceramic assemblage with broadly Meillac traditions typical of established 'Sub-Taíno' occupational models in Cuba (Guarch 1978; Rouse 1992). However, the organic artefacts and in particular the ornately carved wooden vessels, seats (*duhos*) and effigies reflect Chican elements that suggest wider 'Classic Taíno' influences. Such evidence has indicated that the spatial distribution of Meillac Sub-Taíno and Chican Classic Taíno in Cuba, based upon ceramic styles, might not be fully representative of the extent of cultural interaction. Our recent excavations at Los Buchillones produced archaeological material that can be used to study the wider terrestrial and marine interaction associated with the site's occupation. This paper focuses on results from my investigation of interaction from the site as a way of questioning whether or not boundaries of interaction can be identified.

Faunal assemblage Marine animals account for 80% of the assemblage at Los Buchillones (shell 58%, fish & marine mammals 18%, turtle 4%), terrestrial animals 13% and unidentified 7% (Rosario Pérez Iglesias et al. 2003). The wide variety of fish, shell and marine mammal found in this assemblage derive from habitats of mangrove, shallow water, turtle grass beds, reefs and pelagic waters (Newsom and Wing 2004: 24), not all of which could be found in the immediate environs of the site. The evidence for this exploitation of diverse resource zones also indicates the likelihood of efficient and effective transport technologies used to access different marine habitats away from Los Buchillones.

Tool assemblage Marine resources provided not only a subsistence base but also a frequently utilized technological resource. Shell tools and ornaments have been found in all excavations at the site and there is now an assemblage of 264 shell tools for comparative analysis. Excavations have not yet revealed large enough quantities of shell debitage, or objects discarded in the process of manufacture, to indicate a location for the manufacture of these shell tools. Tools found at Los Buchillones comprise scrapers, hammers, gouges (*gubias*), harpoon points, as well as a large selection of beads, pendants, shell carvings and evidence of musical instruments including a conch trumpet (Rodriguez Matamoros 1994). The majority of the beads and pendants are made from *Oliva reticularis*, a species not normally associated with dietary subsistence, suggesting that certain species were targeted specifically for the manufacture of adornments. A number of shell carvings have been found at the site, mainly in the form of teeth and eye inlays for wooden effigies. The lack of evidence for shell working at Los Buchillones suggests either that it was conducted in an as yet unexcavated area of the site, that shell was sourced, worked and transported from another location by the community of Los Buchillones, or that objects are being imported into the community through trade or exchange.

Marine Transport Ethno-historical sources discuss the presence of large sea-going canoes in the Caribbean that could carry up to 150 people (Rouse 1992: 16); however, a canoe of this size has yet to be excavated from an archaeological context in the Caribbean. At Los Buchillones, two canoe fragments have been recovered eroding from the shoreline within the area of the site but without stratified archaeological contexts. These fragments appear to come from the same canoe and join to form a hull 175 cm long and approximately 60 cm wide. The overall length of the canoe is indeterminable but it appears to have been a single-hulled canoe carved from a single tree trunk using petaloid axes and adzes. It is hoped that more evidence of marine transport will be uncovered in future excavations.

Island survey

An island survey was conducted over several seasons during 2004–2005 to obtain archaeological data to clarify whether there were definable boundaries of marine-based interaction around Los Buchillones. The survey was conducted on an archipelago of islands known as the Jardines del Rey or Sabana-Camaguey archipelago stretching out into the Bahama Channel from northern mainland Cuba (figure 12.1).

Full details of the survey methodology and archaeological discoveries found on eleven islands in the case study area are reported elsewhere (Cooper 2004; Cooper and Valcárcel Rojas 2004; Cooper et al. 2006). In 2005, excavations were conducted at several newly discovered sites with the primary objective of collecting an assemblage with which to make comparative studies. A secondary objective of the excavations was to collect samples for radiocarbon dating in order to establish a firmer chronology for the different archaeological sites in the islands.

One of the islands with the greatest concentrations of archaeological material is Cayo Hijo de Guillermo Este. This presently uninhabited island, 232 m long and 83 m wide, is located 25 km northeast of Los Buchillones as the crow flies

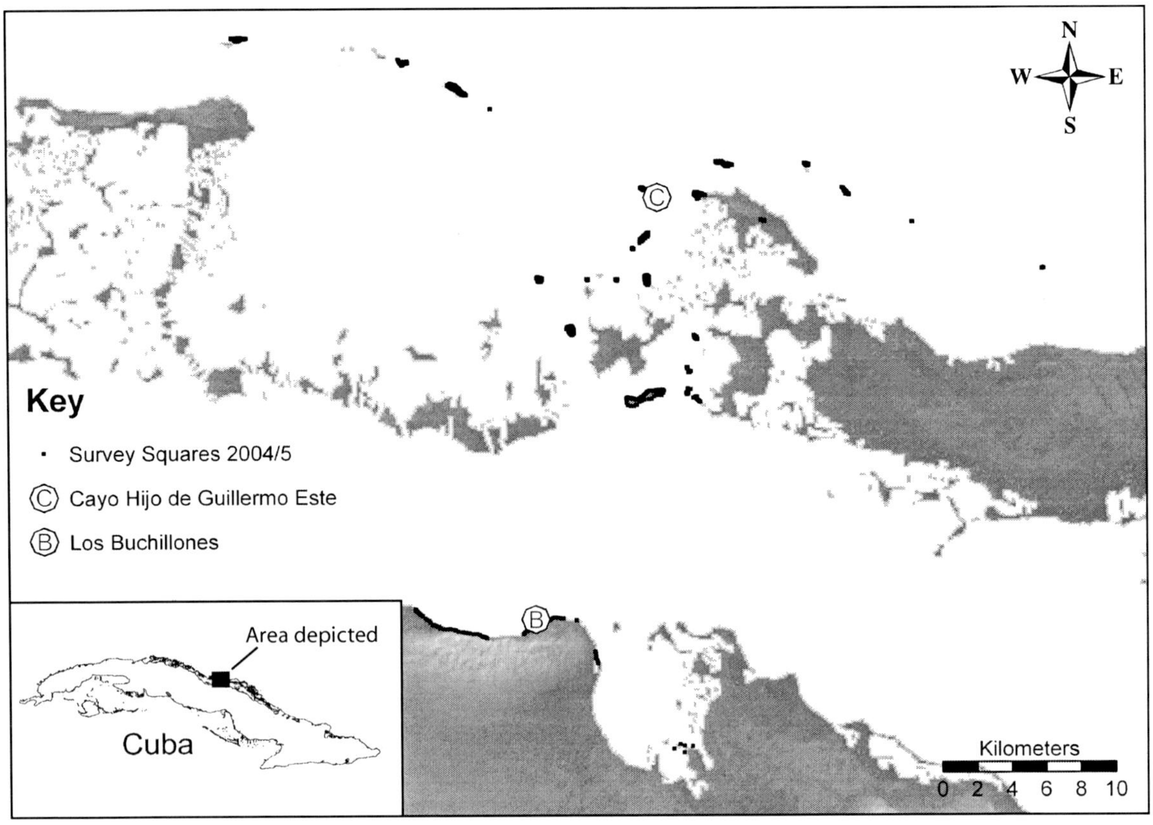

Figure 12.1: Case study area.

or 32 km and circa 6.5 hours by canoe. A number of different sites on the island were selected for excavation including two cave sites, two rock shelters and a 482 sq.m open area of the south-east quadrant of the island. Cave 1 on the island, with an area of 73 sq.m, contained large quantities of archaeological material that appeared to be comparatively undisturbed and provided a substantive body of archaeological material for comparative analysis. The cave was mapped and gridded into one metre squares and excavated (figure 12.2).

The cave environment is not conducive to the preservation of organic material and artefacts recovered in Layer 1 were limited to durable materials, animal (1380 Number of Individual Specimens), ceramics (21 NISP) and lithics (11 NISP).

Faunal assemblage Initial categories include: shell (77%), small mammal (*Capromys pilorides*, jutia) 10%, fish 5%, turtle 4%, crab (*Coenobita clypeatus,* hermit crab) 2%, birds 1%, unidentified 3%. The bone assemblage was then sent to Havana for specialist faunal analysis, the results of which are not available at the time of writing.

Shell constitutes 74% of all the material remains (by NISP) excavated from Layer 1 in Cave 1 and was removed for specialist analysis. The species in the shell assemblage are *Strombus* sp. 78%, *Cittarium pica* 6%, *Xancus angulatus* 4%, *Codakia orbicularis* 3%, *Oliva reticularis* 2%, *Murex brevifrons* 1%, *Fasciolaria tulipa* 1.1% with *Pinctada radiata, Nerita* sp., *Nerita peloronta, Strombus costatus, Arca zebra, Tellina radiata, Chiton* sp., *Diodora listeri* <1%. A taxonomic comparison between the shell assemblages from Cave 1 was made with the shell assemblage from Los Buchillones. There is a comparable range of species representation for both sites (table 12.1).

There are much denser concentrations of shell in Cave 1 and the proximity of the island to the reef suggests this might have been an initial processing station for the collection of marine resources. In addition to the species identification a number of shell tools have been identified using established Cuban shell tool typologies (Dacal Moure 1978) This tool assemblage can be compared to the shell tool assemblage from Los Buchillones (table 12.2).

A detailed study was then made of the worked shell artefacts. Whilst different shell tool densi-

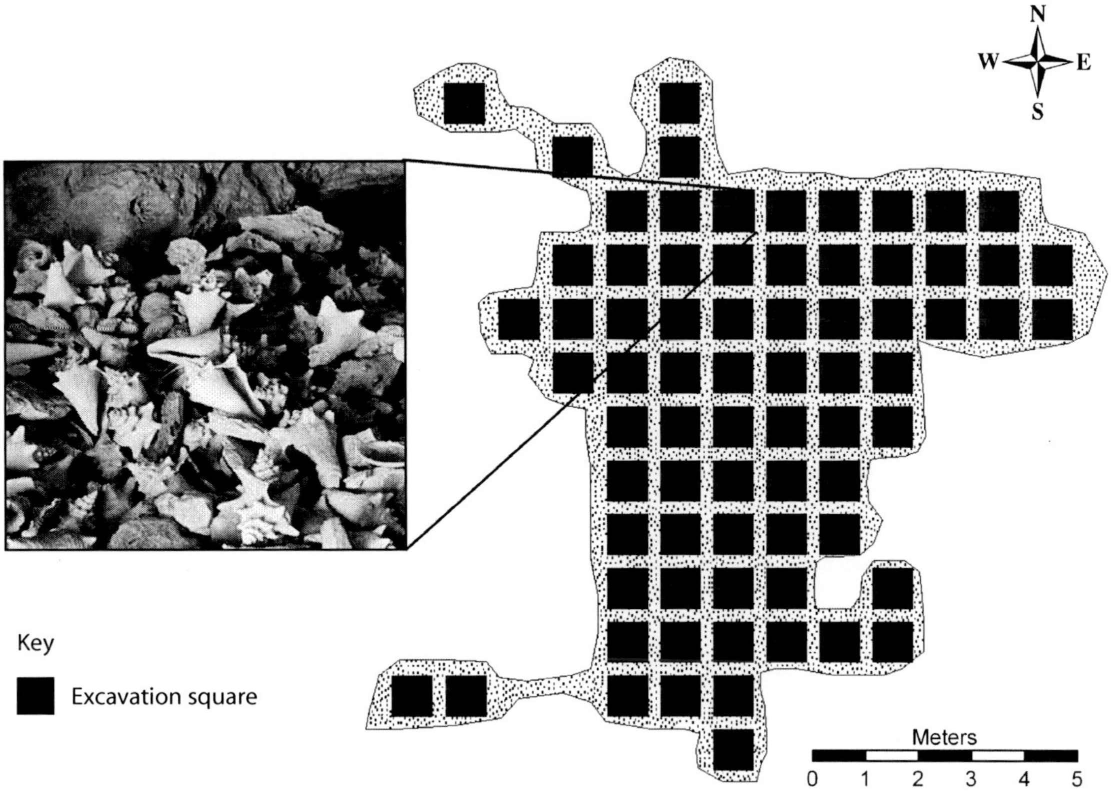

Figure 12.2: Cave 1 and shell debitage in one of the excavated squares.

Taxon	Cave 1	E-1/D2-1
Strombus gigas	477	20
Strombus sp.	179	5
Cittarium pica	48	1
Xancus angulatus (*Turbinella angulata*)	32	2
Codakia orbicularis	25	34
Oliva reticularis	15	8
Murex brevifrons (*Chicoreus brevifrons*)	12	2
Fasciolaria tulipa	8	1
Pinctada radiata	5	2
Nerita sp.	4	0
Nerita peloronta	2	0
Strombus costatus	2	5
Arca zebra	1	0
Tellina radiata	1	0
Chiton sp.	1	0
Diodora listeri	1	0
Lucina pectinatus	0	1
Menipe mercenaria	0	1

Table 12.1: NISP shell species excavated from Layer 1, Cave 1, Hijo de Guillermo Este and excavation units E-1, D2-1 at Los Buchillones.

Tool type	Cave 1	E1, E3, D2-1, D2-6
Hand pick	39	15
Point	33	45
Hammer	11	2
Bead	5	6
Scraper	7	188
Gouge	6	4
Vessel	6	2
Plate	6	3
Perforator	3	3
Spoon	3	2
Pendant	2	6
Knife	1	0
Axe-head	1	0

Table 12.2: Counts of tools excavated from Cave 1, Hijo de Guillermo Este and Excavation units E1, E3, D2-1 and D2-6, Los Buchillones.

ties might reflect different activities at the two sites, there is a clear similarity in the types of tools from both sites with most artefacts found in Cave 1 being comparable to those from Los Buchillones. The shell tools from Cave 1 and Los Buchillones also show similar species selection, manufacturing processes and use-wear patterns. However, there is one distinctive difference in the shell assemblage from Cayo Hijo de Guillermo Este and Los Buchillones, which is that the large quantity of shell debitage found on the island as compared to the mainland site. By using the ratio of NISP to MNI counts as a proxy indicator for shell working, the evidence of this activity in the islands becomes evident. An analysis of the manufacturing processes was also conducted and showed that a number of shell tools in Cave 1 are only partially worked or were abandoned in an unfinished state, often-following breakage during the manufacturing process. Analysis of the manufacturing process of these elements indicates that 29 of the identifiable tools remain unfinished, of these 24 are broken suggesting they were discarded during the process of manufacturing (table 12.3).

Lithic assemblage The geology of the case study area is entirely limestone and the weathering effects of the seawater, salt and strong winds made the exposed rocky islands very porous, exemplified by the numerous rock shelters and caves on Cayo Hijo de Guillermo del Este. This uniform geology means that any other lithic remains found on the islands were likely to be imported. Of the five lithic artefacts found in Layer 1, Cave 1, there were three unidentified flint fragments with no discernible modification and two limestone fragments with use-wear evidence of polishing, sharpening or filing. Use-wear analysis of the two limestone fragments is ongoing with an early hypothesis that they were used as wood files to smooth the wooden shafts hafted to the shell points for harpoon fishing.

Ceramic assemblage No viable sources of clay for ceramic production were found on the islands in the Jardines del Rey Archipelago within the case study area and it is likely that the ceramics found in the survey and subsequent excavations were imported. Macroscopic analysis of the ceramics from the caves suggests they were from griddles (*burens*) and closed globular vessels similar to examples found at Los Buchillones (Mesa González et al. 1994). Few ceramic remains were found on the islands, only 21 fragments were excavated from Layer 1, Cave 1 and one fragment was excavated from Layer 1, Cave 3 on the island. All of the fragments are <5 cm^2 with extensive surface erosion and therefore the potential for macroscopic comparative analysis with ceramic material from Los Buchillones is limited. Therefore, petrographic thin sections were made of 22 sherds from Cayo Hijo de Guillermo Este and 14 from Los Buchillones. Analyses of clay composition, firing temperatures, and inclusions revealed that the clay source and manufacturing processes for the ceramics from Cave 1 and Los Buchillones are likely to be the same. The closed globular vessels from both sites had poorly mixed clay fired at very low temperatures, with unsorted polycrystalline quartz and olivine and feldspar in-

Artefact type	Use wear	Broken	Unfinished	n=
Gouge	No	Yes	Yes	1
Gouge	No	No	Yes	3
Point	No	Yes	Yes	20
Point	No	No	Yes	2
Hand pick	No	Yes	Yes	1
Pendant	No	Yes	Yes	2

Table 12.3: Manufacturing process of unfinished shell tools excavated from Cave 1, Hijo de Guillermo Este.

clusions. The thin section from the griddle (*Buren*) sherds consisted of well-worked clay matrices fired at high temperatures with well-sorted and sub-rounded limestone and monocrystaline quartz inclusions. This indicated an association between the ceramics from Cave 1, Hijo de Guillermo Este and D2-6, Los Buchillones.

Chronology The chronology of indigenous sites in Cuba is not well established (Cooper, in press). This lack of temporal resolution arises from a paucity of radiocarbon dates in the country and the absence of a clearly sequenced ceramic typology. There are currently fewer than 200 recorded radiocarbon dates from archaeological sites in Cuba of which fewer than 80 come from Taíno sites circa 820–1650 AD (Pino 1995). Therefore, the presence of the ceramics in the caves only provides a broad time-span for the date of the site. Due to the wetland environment of Los Buchillones there are a number of organic artefacts that can be used for radiocarbon dating. Eighteen wood samples have been taken from house elements at Los Buchillones and have provided a potential occupation span for the site from 1220–1635 AD (Pendergast et al. 2002). In 2005, a further twelve samples were taken from D2-6 Los Buchillones, including four wood samples and eight shell samples, and sixteen shell samples were taken from ten of the newly discovered sites in the Jardines del Rey Archipelago. The shell dates were calibrated using relative marine offsets from regional data from Reimer's marine reservoir database (Reimer and Reimer 2001). Different δR values from sites in the region were evaluated using the wooden artefacts dated from the same archaeological contexts at Los Buchillones. The calibrated radiocarbon dates from D2-6 Los Buchillones provide a data range of 1264–1667 cal AD. The calibrated radiocarbon dates from Layer 1, Cave 1 provide a date range of 1232–1671 cal AD. These dates indicate that the archaeological material from Layer 1, Cave 1 in Cayo Hijo de Guillermo Este is contemporaneous with the occupation of Los Buchillones. Further discussion of site chronologies and radiocarbon calibrations can be found in Cooper 2007.

Discussion: interpreting inter-island interaction

Although only a 2000 sq.km transect of the Jardines del Rey archipelago has been surveyed, it appears that the majority of indigenous sites so far found have been on the islands furthest away from the nearest known coastal settlement of Los Buchillones. By plotting the location of the sites against bathymetric data within the case study area, a pattern emerges of site location adjacent to areas where the sea depth drops below 5 m, and this is where the reef habitat is located in the Bahama Channel. Cayo Hijo de Guillermo Este is the closest island to Los Buchillones that is also located adjacent to the reef. An initial examination of the faunal assemblage from Cave 1 on Cayo Hijo de Guillermo Este has illustrated intensive yet selective exploitation and processing of marine resources. It is possible that individuals from the coastal community of Los Buchillones could have travelled out from the mainland to offshore islands where they could establish fishing stations and shell processing workshops. The shell, fish and marine mammal species identified at Los Buchillones and Cave 1 on Cayo Hijo de Guillermo Este indicates exploitation of shallow, reef and pelagic water out in the Bahama Channel. There is evidence of specialized shell tool manufacturing and significant maritime skills and technology would have been needed to travel through the archipelago. These are only initial conclusions; however, there is enough evidence to develop a picture of a community whose members are regularly travelling distances of over 35 km to an island from which

Figure 12.3: Wickerwear basketry impressed pottery from Los Buchillones.

they undertake further journeys out to the reefs in the Bahama Channel.

The archaeological reality is that it is probably premature to identify wider interactions from Los Buchillones beyond the case study area because evidence for longer distance marine based travel can only be interpreted through indirect evidence of cultural transmission. The Chican style influences illustrated by the wooden effigies and elaborate wooden vessel styles indicate wider cultural influences from eastern Cuba, but their presence at Los Buchillones could merely be indicative of the unique preservation context of the site. Permission has not to date been granted to extend the case study area to include the Bahaman islands of Cayo Lobos and Cayo Guinchos located further off the Cuban coast in the Bahama Channel. The only means of identifying a wider interaction sphere that included the community of Los Buchillones is thus by comparing the archaeological assemblage with published material from outside Cuba. This comparison has produced initial evidence of Cuban-Bahaman contact: a number of the griddle (*buren*) fragments excavated at Los Buchillones have evidence of basketry impressions (figure 12.3).

Similar basketry impressed pottery has been found at Palmetto Grove and Pigeon Creek sites on San Salvador island in the Bahamas and detailed studies of this pottery have been conducted by Charlene Hutcheson (Berman and Hutcheson 2000; Hutcheson 2001). Hutcheson has reviewed photographs of the pottery from Los Buchillones and identified wickerwear basketry impressions. Similar wickerwear impressions have been found on 10.9% of the impressed sherds found at Pigeon Creek and 17.1% of the impressed sherds found at Palmetto Grove on San Salvador (Hutcheson 2001: 189). This analysis is in its early stages; however, when the possibility for interaction with the Bahamas is placed within the wider context of evidence for regular marine travel from Los Buchillones into and beyond the Jardines del Rey archipelago, it becomes possible to reconstruct the potential nature and form of inter-island interaction during this period.

References cited

Berman, M. J. and C. D. Hutcheson (2000). Impressions of a lost technology: A study of Lucayan-Taino Basketry. *Journal of Field Archaeology* Vol. 27 No. 4: 417–435.

Broodbank, C. (2000). *An Island Archaeology of the Early Cyclades.* Cambridge University Press: Cambridge.

Calvera Roses, J. and J. Febles (1984). *Evidencias Arqueológicas de Los Buchillones, Ciego de Avila: Carta Informativa. No. 55.* Departamento de Arqueología.

Cooper, J. (2004). Islas y Isleños en el Caribe: Interacción a través del paisaje. *Caribe Arqueologico* 8: 91–96.

Cooper, J. (2007). Island Interaction in the Prehistoric Caribbean: an archaeological case study from northern Cuba. PhD Thesis, University of London.

Cooper, J. (In Press). Pre-Columbian Archaeology of Cuba: a study of national site distribution patterns and radiocarbon chronologies. In S. M. Fitzpatrick and A. H. Ross (eds.), *Island Shores, Distant Pasts: Archaeological and Biological Approaches to the Pre-Columbian Settlement of the Caribbean.* University Press of Florida, Gainesville.

Cooper, J. and R. Valcárcel Rojas (2004). Pre-Hispanic settlements along the North Coast of Cuba: A pilot survey report from Los Buchillones. *Papers from the Institute of Archaeology* 15: 77-81.

Cooper, J., R. Valcárcel Rojas and P. Cruz Ramírez (2006). Gente en los cayos. Los Buchillones y sus vínculos marítimos. *Caribe Arqueologico* 9: 66–75.

Curet, L. A. (2004). Island archaeology and units of analysis in the study of ancient Caribbean societies. In *Voyages of Discovery: The Archaeology of Islands,* edited by S. Fitzpatrick, pp. 187–201. Praeger: Westport, Conn.

Dacal Moure, R. (1978). *Artefactos de Concha en Las Communidades Aborigenes Cubanas.* Museo Anthropológico Montané, Ciudad de la Habana.

Dawdy, S., A. Curet and G. La Rosa Corzo (2005). Introduction. In *Dialogues in Cuban Archaeology,* edited by A. L. Curet, S. L. Dawdy and G. La Rosa, pp. 1–25. The University of Alabama Press: Tuscaloosa, Alabama.

Gosden, C. and C. Pavlides (1994). Are islands insular? Landscape versus seascape in the case of the Arawe Islands, Papua

New Guinea. *Archaeology in Oceania* Vol. 29: 162–171.

Guarch, J. M. (1978). *El Taino de Cuba: Ensayo de reconstrucción etno-histórica.* Academia de Ciencias de Cuba: La Habana.

Hofman, C. L. (1993). Inferring inter-island relationships from ceramic style: a view from the Leeward Islands. *Proceedings of the 15th Congress for Caribbean Archaeology*: 1-6 and 1–17.

Hutcheson, C. (2001). Reweaving the strands: continued exploration into the basketry technology of prehistoric Bahamians. *Proceedings of the XVIIIth international congress for Caribbean Archaeology* No. 18: 185–198.

Ingold, T. (1993). The temporality of the landscape. *World Archaeology* 25: 152–174.

Keegan, F. W. and S. W. Mitchell (1986). Possible alochthonous Lucayan Arawak artifact distributions, Bahama Islands. *Journal of Field Archaeology* 13(3): 255–258.

Knippenberg, S. and A. v. Gijn (1998). Flint sources in the Lesser Antilles: implications for inter-island contact networks. Paper presented at the Proceedings of the 63th Annual Meeting of the Society for American Archaeology, Seattle.

Mesa González, I., J. Jardines Macías and J. Calvera Rosés (1994). Estudion preliminar de la cerámica del sitio arqueologico 'Los Buchillones', provincia de Ciego de Avila. In *Estudios Arqueologicos*, edited by J. Febles, L. Dominguez, F. Ortega, G. La Rosa, A. Martinez and A. Rives, pp. 66–79. Editorial Academia: La Habana.

Newsom, L. A. and E. S. Wing (2004). *On Land and Sea: Native American Uses of Biological Resources in the West Indies.* The University of Alabama Press: London.

Ortiz, F. (1922). *Historia de la Arqueologia Indocubana.* Teniente Rey 27: Habana.

Pendergast, D., J. Calvera, J. Jardines, E. Graham and O. Brito (2003). Construcciones de madera en el mar: Los Buchillones, Cuba. *El Caribe Arqueologico* 7: 24–32.

Pendergast, D., E. Graham, J. Calvera and J. Jardines (1999). Houses in the sea: excavation and preservation at Los Buchillones, Cuba. In *Enduring Records: The Environmental and Cultural Heritage of Wetlands*, edited by B. Purdy, pp. 71–82. University Presses of Florida: Gainesville.

Pendergast, D., E. Graham, J. Calvera and J. Jardines (2002). The houses in which they dwelt: the excavation and dating of Taino wooden structures at Los Buchillones, Cuba. *Journal of Wetland Archaeology* Vol 2: 61–75.

Pino, M. (1995). *Actualizacion de fechados radiocarbónicos de sitios arqueológicos de Cuba hasta diciembre de 1993.* Editorial Academia: La Habana.

Rainbird, P. (1999). Islands out of time: towards a critique of island archaeology. *Journal of Mediterranean Archeology* Vol. 12 No.2.: 216–234.

Rainbird, P. (2004). *The Archaeology of Micronesia.* Cambridge World Archaeology. Cambridge University Press: Cambridge.

Reimer, P. J. and R. W. Reimer (2001). A marine reservoir correction database and online interface. *Radiocarbon 43*: 461–463.

Rodriguez Matamoros, M. E. (1994). Reporte de nuevas evidencias artefactuales en el ajuar de concha de las comunidades aborigenes de la etapa de economia de apropriación. In *Estudios Arqueologicos*, edited by J. Febles, Dominguez, L., Ortega, F., La Rosa, G., Martinez, A. & Rives, A., pp. 13–27. Editorial Academia: La Habana.

Rodríguez Ramos, R. (2002) Dinámicas de Intercambio en el Puerto Rico Prehispánico. El Caribe Arqueológico (6): 16–22.

Rosario Pérez Iglesias, L., J. Jardines Macías and C. Rodríguez Arce (2003). *Estudio Arqueozoológico en Los Buchillones. Economía y Medio Ambiente.* Centro de Investigaciones y Servicios Ambientales y Tecnológicos. Ministerio de Ciencia Technología y Medio Ambiente. Submitted to Unpublished.

Rouse, I. (1942). Archaeology of the Maniabon HIlls, Cuba. *Publications in Anthropology* 24.

Rouse, I. (1952). Porto Rican prehistory: introduction; excavations in the west and north. *Scientific Survey of Porto Rico and the Virgin Islands* XVIII(3): 307–460.

Rouse, I. (1992). *The Tainos: Rise and decline of the people who greeted Columbus.* Yale University Press: New Haven.

Tabio, E. (1974). La Comunidad Primitiva. *Revolución y Cultura* Septiembre, 1: 7–14.

Terrell, J. E. (2004). Island models of reticulate evolution: The "ancient lagoons" hypothethesis. In *Voyages of Discovery: The Archaeology of Islands*, edited by S. Fitzpatrick, pp. 203–222. Praeger: Westport.

Trincado Nelsa, M. (2000). El Aborigen, La Historiografía y La Nacionalidad Cubanas. *El Caribe Arqueologico* 4: 100-105.

Ulloa Hung, J. (2000). Migraciones en el Caribe precolombino. *El Caribe Arqueologico* 4: 14–19.

Valcárcel Rojas, R., J. Cooper, J. Calvera Rosés, O. Brito and M. Labrada (2006). Postes en el Mar: Excavación de una estructura constructiva aborigen en Los Buchillones. *El Caribe Arqueologico* 9: 76–88.

Veloz Maggiolo, M. (1974). La Navegacion Prehistorica. In *Apuntes sobre Prehistoria de Santo Domingo*, edited by M. Veloz Maggiolo: Santo Domingo.

White, J. P. (2004). Where the wild things are: prehistoric animal translocation in the Circum New Guinea archipelago. In *Voyages of Discovery: The Archaeology of Islands*, edited by S. Fitzpatrick, pp. 147–164. Praeger: Westport.

Chapter 13

Size matters, but so does distance: autochthony and external influence in the cultural development of ancient Sardinia

Stephen L. Dyson, University at Buffalo, SUNY

Introduction

The Mediterranean (figure 13.1) is a world of islands. Hundreds of them dot the sea from the Turkish coast to the Straights of Gibraltar. Most are small places, uninhabited or sparsely inhabited throughout most of their history. However, even those, due to the long and complicated maritime history of the Mediterranean, have played larger roles, from a source of obsidian as in the case of Melos (Renfrew and Wagstaff 1982), to a watch post over a hostile shore like Malta. Four of the large islands, Crete, Cyprus, Sicily, and Sardinia, due to size, location, and resources were destined to a have special place roles in Mediterranean civilization. Each for reasons of location, settlement development, and place in the larger world of Mediterranean civilization had a different archaeological and political history (Patton 1996; Leighton 1999).

Sardinia is in land area the second largest of the Mediterranean islands with 24,000 sq.km. It has a diverse geography, and in spite of its sometimes forbidding landscape is capable of supporting a significant population engaged in agricultural and pastoral activities. That is demonstrated by the fact that for a certain period during the period of Roman occupation it was one of the chief suppliers of surplus grain to the capital (Rowland 1990). It has two mineral resources, obsidian and copper, that as far back as the Neolithic drew outsiders to its shores.

Sardinia is an interesting case study for those concerned with the comparative study of islands. Its strategic location and its resources have long made its control the goal of various imperial powers. Indeed the recent history of the island has focused on questions of colonialist dominance and resistance. However, the island landmass is sufficiently extensive, its topography sufficiently diverse, and its population sufficiently large, for it to develop autochthonous cultural systems. Moreover, for much of its early history Sardinia was at the margins of the eastern-central Mediterranean trading and colonial enterprises. Island history down to the early modern period can be interpreted in terms of a complex dynamic between internal dynamics and outside influences.

Sardinian studies and especially archaeological studies still have a strongly parochial quality. Most island archaeology is done by Sards educated at Sardinian universities. A limited number of scholars like Giovanni Lilliu of the University of Cagliari have exercised enormous influence and power (Lilliu 1988, 1995). An important ideological factor in Sardinian studies has been the resistance of the natives located in the interior to external powers that dominant the coast. The true Sardinia is that isolated, inward looking upland area, whose proud but parochial inhabitants preserved their own language, customs and cultural identity. They are portrayed well in the novels of the early twentieth century Nobel Prize winner Grazia Delleda. Such attitudes have often fed an autonomous, even nationalistic Sardinian political agenda. To the non-Sard the history of the island can often be dismissed as a case of 'involuted minicenters' with 'stagnant centers' (Broodbank 2000: 19; Braudel 1976: 150–1).

Relatively few non-Italian or even mainland Italian archaeologists have worked on the island. Those that have were often drawn there with very specific goals such as obsidian studies or the documentation of Punic colonization. This is to be regretted, for Sardinia has a very rich archaeological record, and possesses a whole series of important questions for students of islands. The

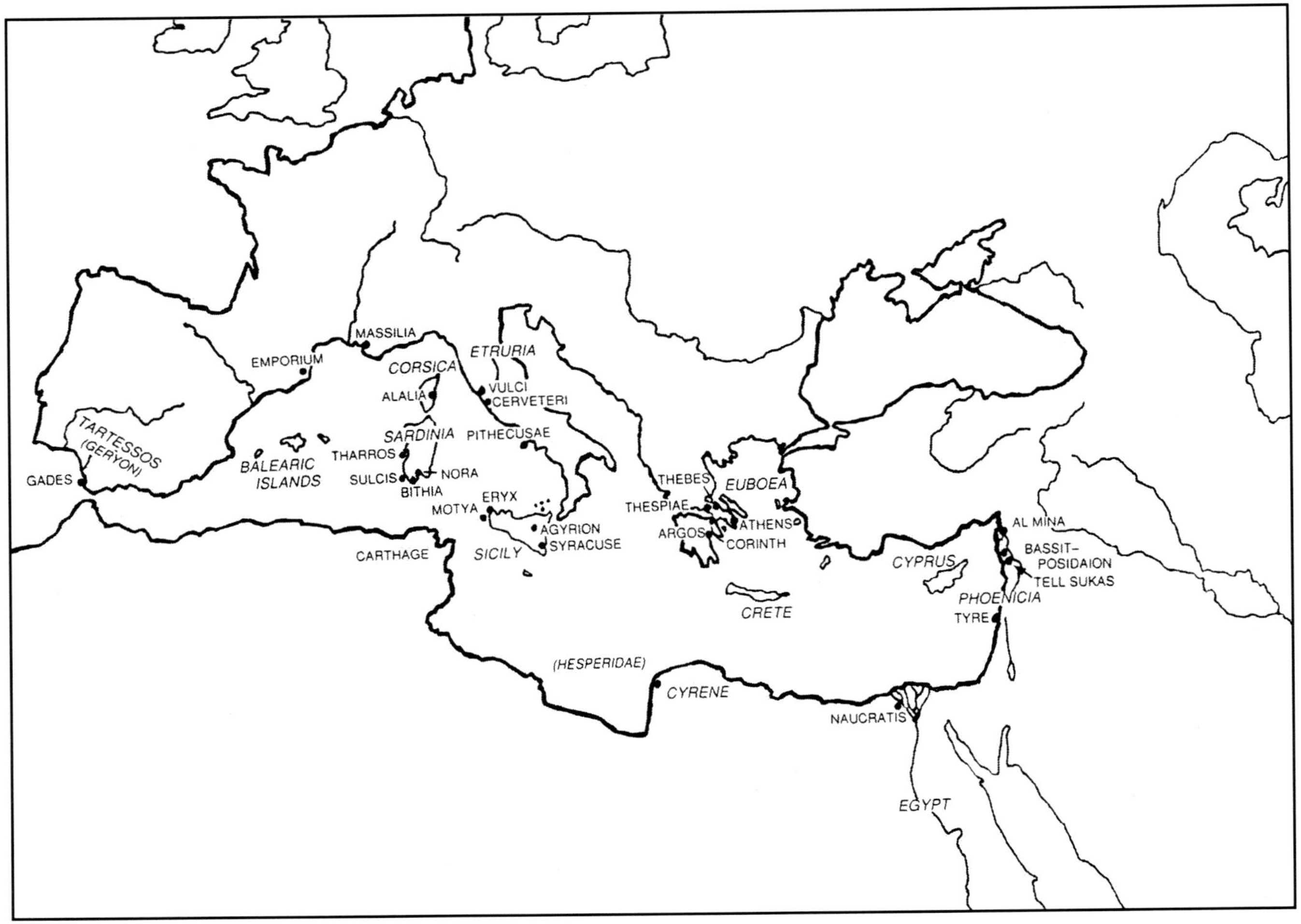

Figure 13.1: The ancient Mediterranean.

important research results of scholars like David Trump working on the Neolithic and Robert J. Rowland, Jr. writing on the Roman period bear that out.

Earliest occupation

The date of the first occupation of Sardinia is still a matter of dispute among prehistorians (Hofmeijer and Sondaar 1992; Cherry 1992). However, Sardinia was almost certainly inhabited by the Upper Palaeolithic and has been continuously occupied ever since. Initial settlement was made possible by the lowering of the sea level produced by the last glaciation. That significantly reduced the sailing distances from the mainland and made Corsica and Sardinia into a single land mass. With the rise of sea level produced by the ending of glaciation water, distances again increased. The islanders must have developed a relatively sophisticated fishing culture. However, due in part to the flooding of the earlier coastlines this has not yet been well documented in the island's archaeology. However, long distance fishing and trading have been documented in other parts of the Mediterranean as far back as the Mesolithic (Perlès 1992). Sardinia certainly remained in continuous contact with the wider Mediterranean. Both the Italian and North African land masses are relatively close, so some island-mainland interaction was possible.

However, small scale fishing and trading contact is very different from major movements of people. The Mediterranean is an unpredictable, treacherous sea. The Romans learned this to their cost, as several large fleets were destroyed by storms during the first Punic War (Goldsworthy 2003: 96–127). The distance from the Italian coast to the island is considerable (205 km), and there is not a significant archipelago of intermediate islands to help break the journey. Sardinia has a lower ratio of seaward visibility from any major external landmass of any of the large Mediterranean islands (Broodbank 2000: 40, fig. 4). The island could best be reached by using Elba and Corsica as stepping stones. However, the land of Corsica is much harsher than that of Sardinia, and would have held little attraction for Neolithic pioneers.

The east coast of Sardinia that faces Italy is mountainous and uninviting. It is the southwest of the island facing toward North Africa that is more suitable for settlement, but there is little archaeological evidence for extensive North African contacts before the Punic period. While Old World archaeologists have tended to use migration as a 'deus ex machina' to avoid analyzing indigenous cultural systemic practices, I would argue that no significant population movement into the island was likely until the well organized colonial settlements of the early first millennium BC.

The obsidian trade

The first cultural process that brought contact with the wider Mediterranean and reflected this complex dialogue between island initiative and external contact was the development of the obsidian trade. Sardinia is one of the major sources of obsidian in the west-central Mediterranean (Tykot 1996). The obsidian flows on Monte Arci in west-central Sardinia are massive and complex (Tykot 1997), and were known and exploited at least by the Early Neolithic. The presence of Sardinian obsidian has been documented in many areas of the Italian mainland and even up into southern France. However, the background to the development of the Sardinian obsidian trade is different from that of the other two Mediterranean island sources, Melos and the Lipari Islands—the latter were small islands, initially with small or non-existent populations. Outside sailors could come, locate and take the obsidian and move on (Torrence 1986; Perlès 1992).

Sardinia was a large land mass with a significant indigenous population when the trade began. Monte Arci is not located directly on the coast and hence would not have been easily discovered by casual voyagers. Detailed studies of the various Mont Arci flows by archaeologists like Robert Tykot have demonstrated complex patterns of trade both within and outside the island (Tykot 1996).

Certainly the development of the obsidian trade served both to bring the Sards in contact with each other and create the foundations of cultural interconnection that would make possible the later island cultures. It also served to connect Sardinia with the wider central Mediterranean. While some of the trade was probably in the hands of outsiders, there is no reason to doubt that much of it was carried out by Sardinians themselves. They knew the often isolated find places for obsidian and by this period they must have known how to traverse the sea. Unfortunately, the Sardinian Mesolithic is poorly defined, and our understanding of early coastal maritime adaptation limited.

The development of the obsidian trade coincides with the emergence of the Sardinian Neolithic (Rowland 2001: 10–23). The development of Neolithic culture on Sardinia represented a long and complex process with many of the phases still poorly documented (Trump 1984: Tykot 1999). Clearly the dynamics of the Sardinian Neolithic were greatly influenced by developments on the mainland, for the essential floral and faunal domesticates did not evolve on the island. On some of the larger Mediterranean islands such as Crete, the 'Neolithic package' was transported as a whole, integrated entity, suggesting the colonization by significant new populations that brought in an off-island culture (Broodbank and Strasser 1991). That was not the case of Sardinia. Rather the faunal and floral elements seem separate in their origins, filtered and integrated to form a distinctive Sardinian cultural system. The importation of the diverse elements was certainly related to the obsidian trade that regularly brought the Sardinians in contact with mainland culture. Sheep, cattle, and the knowledge of ceramic technology were the returns that the Sards got for the precious black stone. The Sardinian Neolithic reached its culmination in the fourth millennium BC with the so-called Ozieri culture that was represented by numerous, extensive villages, an elegantly decorated ceramic tradition and the beginning of monumental ceremonial centers. One of those at a place called Monte Accodi had a large earthen mount with access ramps that recalls in more primitive form the ceremonial centers at places like Cahokia in Illinois (Rowland 2001: 21–2; Emerson 1997). However, this movement toward centralization was short lived, and the transition to a more metal based society saw a return to localism (LoSchiavo 1988).

The Nuragic

It was in the middle-late second millennium BC that the most distinctive of Sardinian cultures, the Nuragic, came into being (Arslan et al. 1985). The name derives from the circular stone towers that are still dominant features in the Sardinian landscape. Over five thousands of these are known, and many more have certainly been destroyed. Sardinia has one of the densest prehistoric landscapes anywhere in the world. Indeed recent studies of the placement of the nuraghi in relation to each other and other monuments such as tombs have demonstrated the coherent, visual articulation of the Nuragic landscape (Blake 1998; 2001). The vast majority are single towered, dry-built stone structures. Their builders used the hard local basalt with great skill and employed such sophisticated techniques as corbelled arches. At most excavated nuraghi traces of a small village have been found adjacent to the tower.

A number of the Nuragic sites show traces of metal working, and small bronze objects are ubiquitous. The most characteristic artifacts of the Nuragic culture are small bronzes figurines, often referred to as 'bronzetti'. They were rendered in a semi-abstract style that has attracted the attention of modern art collectors and made them hot items on the clandestine antiquities market. Most are male figures, often warrior or shepherds. However, there are images of women, animals, ships, and even models of the Nuragic towers themselves. The imagery of the small bronzes that often mixes the clearly mythic with the realistic and their presence as ex-votos at indigenous shrines suggest religious functions.

The relatively simple forms of the nuraghi and their great number suggest a social structure based on extended families headed by patriarchs. The similarity of architectural design and construction techniques as well as the shared iconography of the Nuragic bronze figurines shows a high level of intercommunication among the Sardinian people. That was achieved in part through regional religious festivals held at sanctuaries centered on sacred wells. The importance of these common sanctuaries is demonstrated by the architectural sophistication of the larger well shrines that matches that of the best nuraghi.

It was relatively early in the Nuragic period that Sardinia first came in contact with the complex cultures of the Late Bronze Age Eastern Mediterranean (Rowland 2001: 55–7). A scattering of Mycenaean pottery has been found on the island, but no evidence for settlements and trading centers have been documented on Sicily (Leighton 1999: 147–186). It is clear that Sardinia was at the extreme end of the Mycenaean trading orbit, and the cultural impact of that Eastern Mediterranean developed culture was limited.

However, the evidence for the trade in copper indicates a more complex and nuanced system of cultural interchange between Sardinia and the eastern Mediterranean. The large number of Nuragic bronzes demonstrates the fact that Sardinia was in antiquity rich in copper ore. Cyprus was the other major copper/bronze metal working center in the Mediterranean. The two island were linked in a metal trade documented by large ox-hide shaped ingots found on both island and in wrecks on the sea bed in between. It has generally been assumed that this trade was structured on a classic core-periphery interchange model with the less developed Sardinians supplying raw materials to the Cypriotes in return for finished products. However, recent analyses of ingots found on Sardinia strongly suggest that their copper was of Cypriote origins (Stos-Gale and Gale 1992; Gale and Stos-Gale 1995; with more reservations Budd et al. 1995). Evidently the diverse and creative Sardinian bronze working craft industry was so active that it needed to draw upon external copper sources.

One expression of this growing cultural sophistication on the island was the appearance of 'complex nuraghi', multi-towered structures with curtain walls surrounded by villages of considerably greater size than those that normally developed around the single-tower nuraghi. We also have one example of a burial site at Monte Prama near Oristano with large-scale stone sculpture. These suggest the emergence of more complex chieftainships that possibly developed in response to the competition for internal and external resources and the desire to control access to the traders that came in from the sea.

The 'Sea Peoples' and later prehistory

The final evidence for the complexity of the Sardinian presence in the interconnected world of the central and eastern Mediterranean is the presence of ethnic groups identified by some scholars as Sardinian among the so-called 'Sea Peoples' who invaded Egypt at the end of the New Kingdom (Sandars 1978; Tykot 1994a: 62–67). Not only would their presence be a testimony to island military prowess, but also to active Sardinian nautical activity well beyond the shores of the island.

Sardinia did not experience the cultural hiatus that separates the Bronze and the Iron Ages in the eastern Mediterranean. The development of the Nuragic cultural system bridged the late second and early first millennium BC. The island's first contact with the new Iron Age societies that were emerging in the eastern Mediterranean was represented by the arrival of the first Phoenicians traders and settlers early in the first millennium BC (Aubet 1993). While they represented the first concentrated group of new settlers to arrive on the island, perhaps since the Upper Palaeolithic, their impact should not be exaggerated. Phoenician settlements were small enclaves that hugged the coast, finding shelter on small islands or protected headlands. Rarely were they able to project power into the interior, and they must have generally depended for their commercial interchange on good relations with the indigenous population. However, since Phoenician settlements extended from Sicily to North Africa and onto the Iberian peninsula Sardinia was now linked, however tentatively, to the wider world of the west-central Mediterranean.

The Classical and Roman period

Sardinia did not experience any of the Greek colonization that reshaped Sicily and southern Italy from the eighth century BC onwards. The Phoenicians and the Etruscans worked to keep the Greeks out of the Tyrhennian Sea. There was a short-lived Hellenic settlement on Corsica and a major colony at Massilia (modern Marseilles) on the Mediterranean coast of France (Boardman 1964: 218–222; Davidson 1992). The Greeks apparently did not try to compete with the Phoenicians on Sardinia. However, that does not mean that Greek goods have not been found on the island, for the Phoenicians were the great international carriers of the early first millennium B.C.

The power realities changed in the mid sixth centuries, when the Carthaginian state, which had developed from the Phoenician colony in North Africa, undertook a more thorough subjugation of Sardinia. The Carthaginians wanted the island for its resources, especially grain, and as a recruiting ground for soldiers for its mercenary armies. Resistance was intense, probably as much from the Sardinian Phoenicians as from the Nuragic natives. Certainly the Carthaginians gained relatively firm control of the southwest of the island. How much of the rest, especially the upland interior they really held, is a topic of much debate (Van Dommelen 1998). Sardinian scholars talk of frontier lines, but that is probably an anachronistic projection backward from the Roman imperial frontier system. What we do see is the gradual acculturation of the indigenous elite. That process had presumably started with the Phoenicians and intensified under the Carthaginians and the Romans. By the time the Romans arrived in the mid-third century BC the Carthaginians depended for their hegemony on Punicized Sards who had negotiated relations with the tribes of the interior. This is best attested by the first Sardinian hero Hampsicora, a Punicized Sard who led resistance against the Romans during the Second Punic War (Dyson 1985: 252–5)

The Romans seized Sardinia from the Carthaginians between the First and Second Punic Wars, partly for its resources and partly to prevent it from becoming a Punic base for attacks on the Italian mainland. The island was 'subjugated' in a series of brutal campaigns in

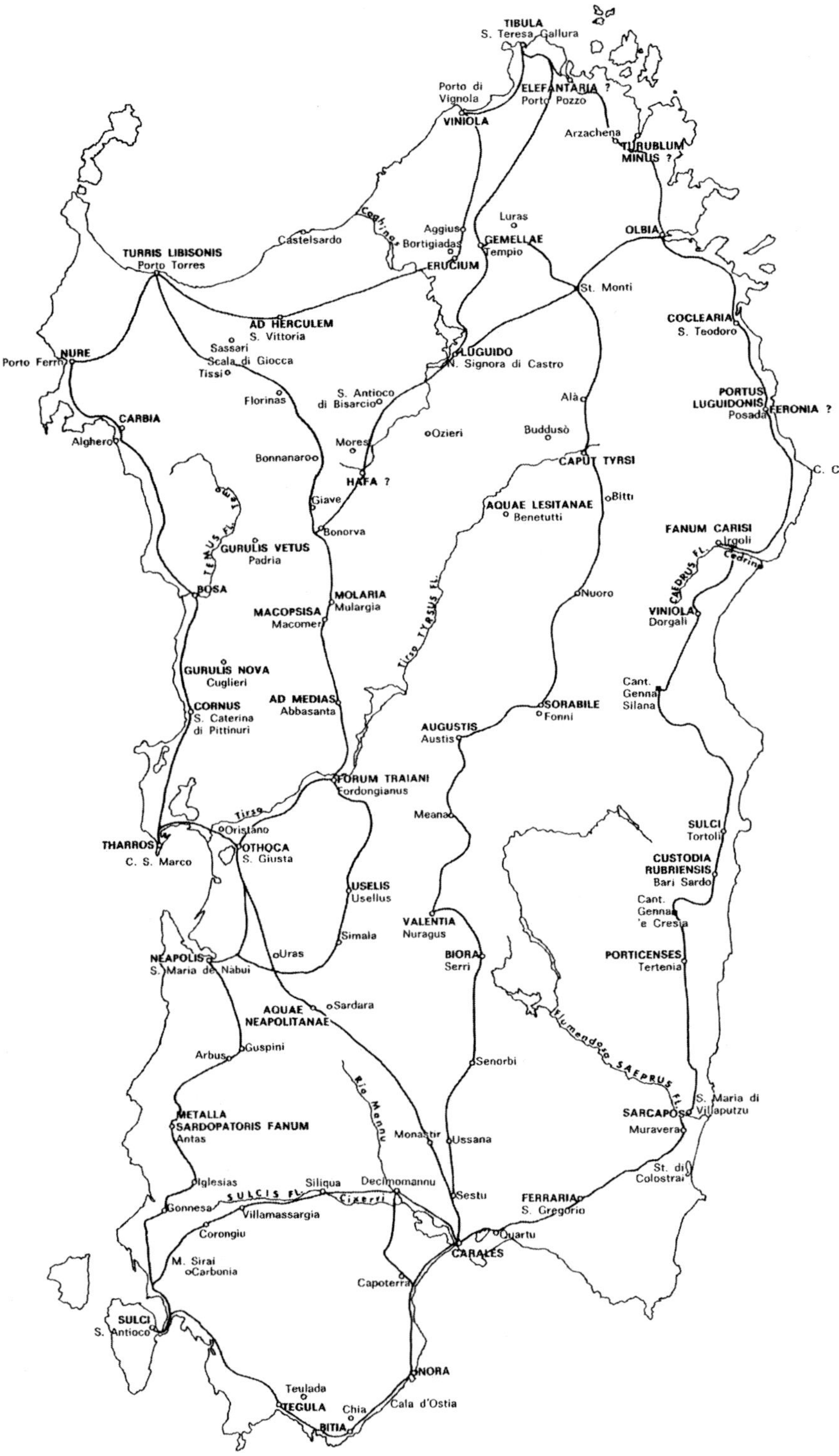

Figure 13.2: Roman roads and Roman place-names in Roman Sardina.

the late third and early second century BC, and the indigenous population decimated (Dyson 1978: 244–263). So abundant became Sardinians on the Roman slave markets that they became a watchword for an item of little market value. The Romans certainly invested heavily in an infra-structure of control. Roman Sardinia had as good a road network as any province, and the number of place names recorded along the roads would suggest a high level of interior urbanization (figure 13.2). The standard wisdom among Roman historians and archaeologists is that most of the nuraghi were abandoned, and their small settlements replaced by extensive slave operated estates. The surviving indigenous population was either pushed into the inaccessible reaches of the interior or resettled in the conquered lowlands. Roman Sardinia is fitted into a classic picture of imperial pacification duplicated at many different times and in many different places. Otherwise it became a secondary province of Rome, a source of some grain and a place of exile, known mainly for its unhealthy climate.

However, a fuller examination of the evidence suggests a picture that is more complicated and nuanced. Most of the larger Roman towns were located along the coast and developed out of pre-existing Phoenician and Carthaginian settlements. Research at interior sites mentioned on the Roman road itineraries has produced little evidence for well developed centers. We seem to be dealing mainly with posting stations and road-side markets rather than cities or even towns (Dyson 1992, 2000, 2003). Large villa sites, indicators of the presence of large estates, are rare outside of the coastal areas and the immediate hinterland of the big urban centers. This picture contrasts with that of most other western provinces, even the late conquered Roman Britain and suggests that the island did not undergo thorough Romanization.

Evidence has gradually accumulated that there was some continuity of occupation at many Nuragic sites well into the Roman period and even beyond (Webster and Teglund 1992). However, much of this information on the Roman presence at Nuragic sites has been derived from chance discoveries scattered thinly over many areas. Little concentrated landscape and survey archaeology on Roman period Sardinia has been undertaken.

Archaeological survey

For these reasons, my colleague Robert J. Rowland, Jr. of Loyola University, New Orleans and I decided in 1987 to start a systematic settlement survey in a single concentrated area of west-central Sardinia that was within the 'acculturated Roman frontier zone' (figure 13.3), had a large concentration of nuraghi, but also two major Roman settlements (Dyson and Rowland 1992). Those were the military, frontier control, and market center called Forum Traiani (modern Fordongianus) and Colonia Iulia Augusta Uselis (modern Usellus), a rare example of an attested Roman colonial settlement in Sardinia. The region was also linked to the coast by the Tirso, the largest river in Sardinia, and was in the hinterland of Tharros, one of the larger maritime trading centers on the island. Not only could trade objects be easily distributed throughout the region, but bulk agricultural products like grain could be readily transported to the coast and shipped to Rome. In other areas of the Roman Empire like mainland Italy and North Africa, emphasis on such cereal production transformed both the socio-economic structures and the settlement organization of the rural landscapes (Dyson 2003).

The first object of our research was the search for evidence of Roman presences at Nuragic sites, especially those in the hinterland of both Fordingianus and Usellus (figures 13.4 and 13.5). Survey at Nuragic sites is not easy. While the nuraghi are generally placed on the standard 1:25,000 topographic maps, finding them in the often thickly vegetated upland landscapes is often not easy. Since the sites are seldom cultivated and often covered with vegetation standard survey approaches with regular grids and transects were not easy applied. However, through intensive site walking we were able to recover evidence, mainly ceramic, for multi-period Roman occupation at many sites. It is not clear whether the towers themselves remained in use. One would think that the Roman military authorities would have discouraged that. However, the villages at their bases remained occupied, sometimes down to the end of Roman occupation of Sardinia. However, they seem to have undergone only marginal social or structural change. Except for the use of imported ceramics, glass, and lamps the sites show little signs of Romanization. Architectural forms and special organization seem little altered even after centuries of Roman rule.

We complemented these upland surveys with intensive fieldwork in the hinterlands of the major Roman settlements at Forum Traiani and Uselis. Using our experience on the Italian mainland, we attempted to maximize our finds by surveys focused on areas likely to yield Roman

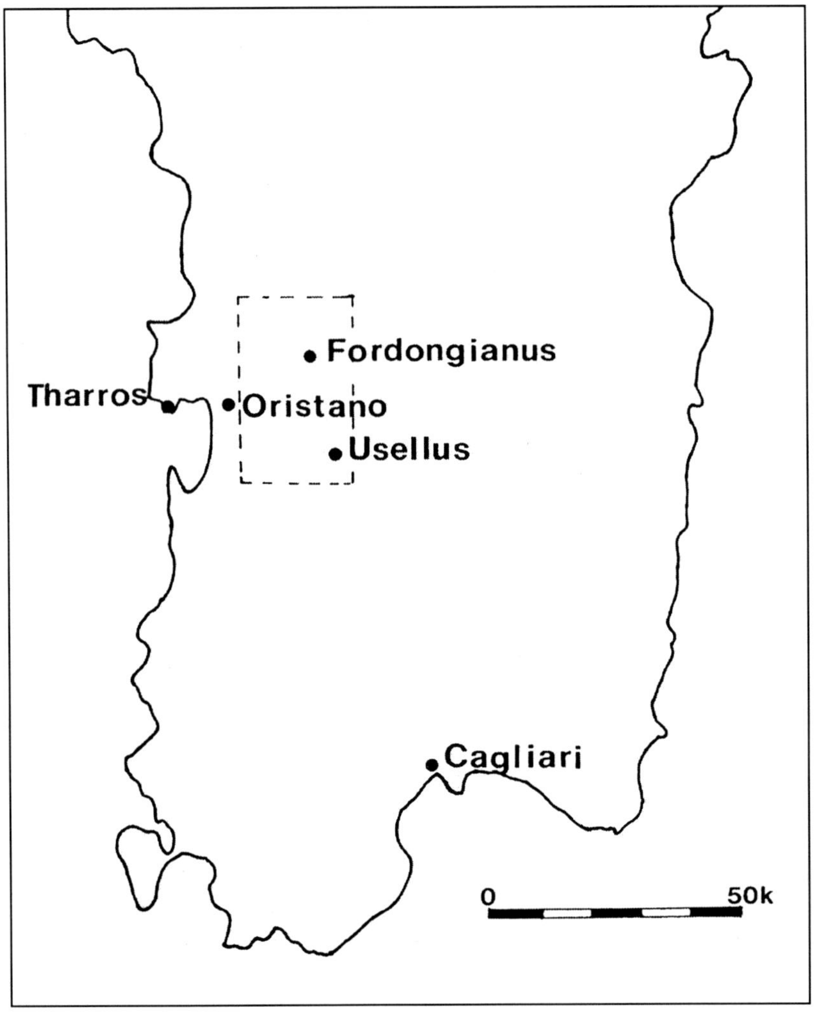

Figure 13.3: Area of the Dyson-Rowland settlement survey.

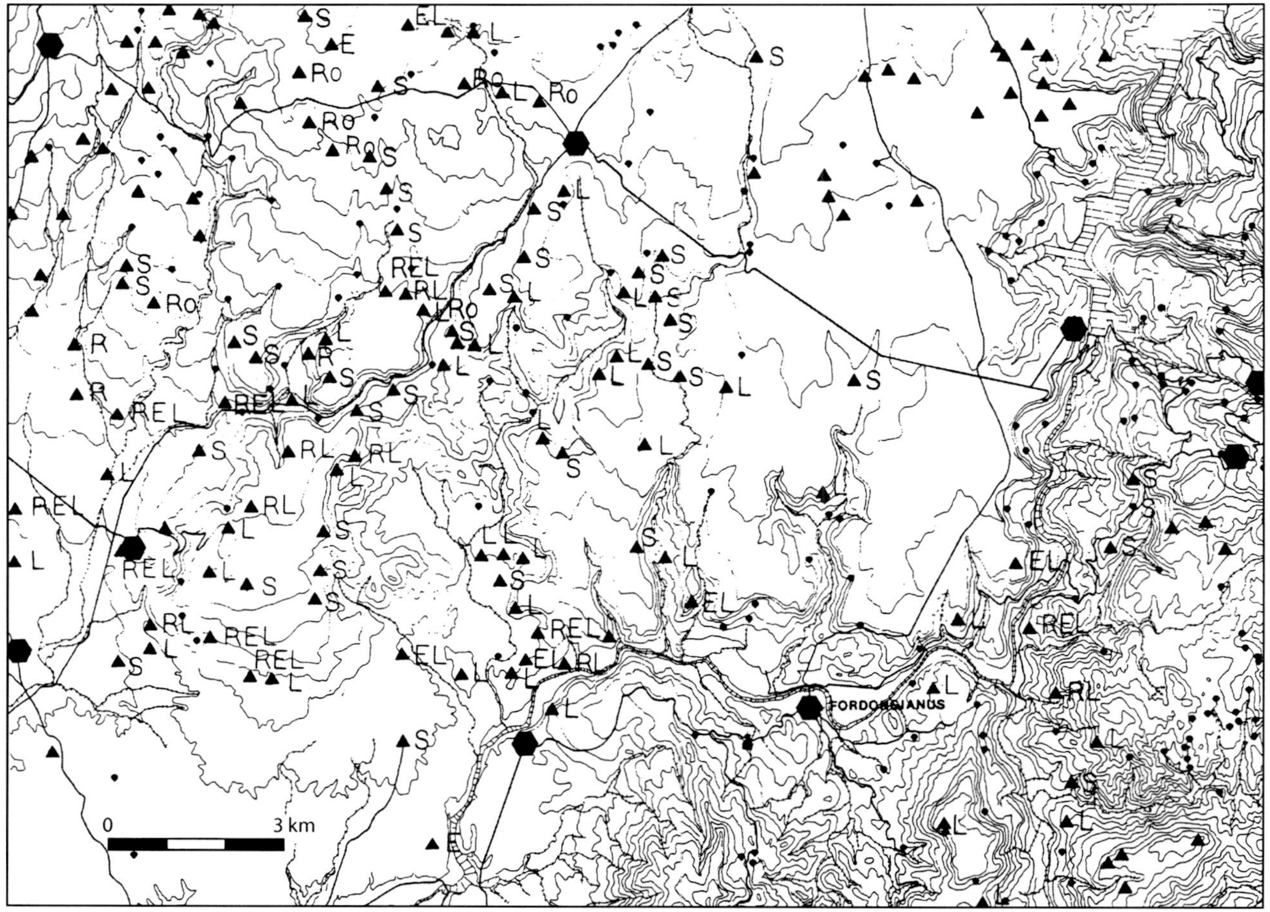

Figure 13.4: The results of the Fordongianus survey. S = Site surveyed with no Roman material found; R = Site occupied during the later Roman Republic (ca. 200–30 BC); E = Site occupied during the Early Empire (ca. 30 BC–AD 150); L = Site occupied during the Late Empire (AD 150–500); Ro - Site with Roman material that could not be more precisely dated.

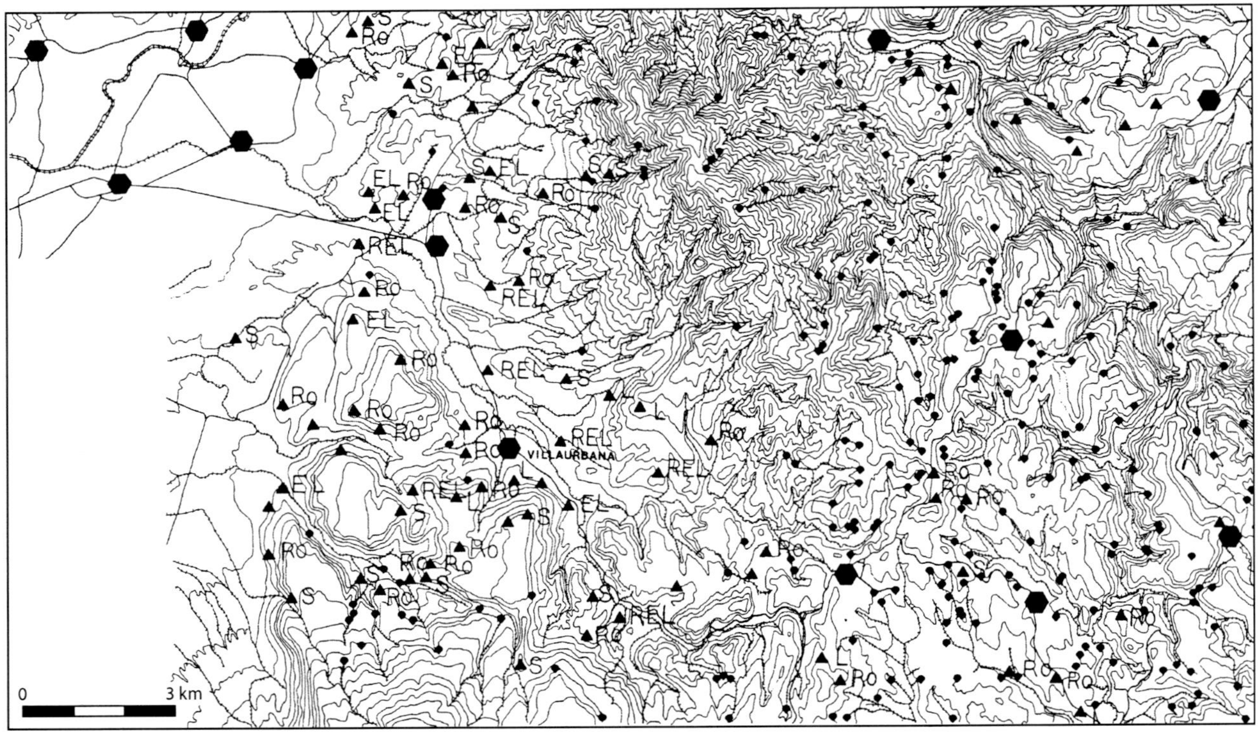

Figure 13.5: The results of the Usellus survey southwest of the ancient town. S = Site surveyed with no Roman material found; R = Site occupied during the later Roman Republic (ca. 200–30 BC); E = Site occupied during the Early Empire (ca. 30 BC–AD 150); L = Site occupied during the Late Empire (AD 150–500); Ro - Site with Roman material that could not be more precisely dated.

sites. However, few were discovered. In contrast to most parts of the Western Roman Empire, the hinterland of one of the most important Roman centers in Sardinia showed little sign of settlement development during centuries of Roman occupation.

Our evidence along with that from other research projects suggests a high level of indigenous continuity in the Carthaginian and Roman Sardinian countryside. It would be wrong to argue that nothing changed in Sardinia in the wake of Carthaginian and Roman imperialism. Clearly the detachment of the elites and their fuller integration with the imperial societies meant that indigenous Nuragic culture lost much creative vitality. We have nothing like the innovation in the arts and architecture that had characterized earlier periods in Sardinia's long history. Still, as scholars like James Scott have demonstrated, passive peasant resistance can be impressively effective in limiting the exercise of imperial power (Scott 1985). The Romans were pragmatists. They had a small bureaucracy, and for the size of their empire a small military that was stretched thin along extended frontier zones. Projecting imperial power was not easy. As long as the native Sards kept a reasonable level of peace and paid their taxes they were largely left alone.

Conclusion: contributions to a comparative island archaeology

What are the lessons of Sardinia as an island society. Clearly size mattered. It is a good example of a mini-continental island whose land area and diversity of landscape allowed the support of a considerable self-sustaining population. The Sardinians clearly had a long history of interaction with the sea. Special resources like obsidian and copper that drew outsiders to the island. However, the reality of distance, a landscape that was not easily mastered, limited settlement potential, and a marginal position in relation to the more sophisticated trading-colonial systems of the Mediterranean meant that Sardinian society developed within its own parameters. The long history of the Nuragic culture was a good testimony to that. In the Punic and Roman periods, outside impacts on the island did increase. However, unlike many other island experiences, this last round of ancient imperialism did not totally destroy the base of the indigenous society. Elite destruction and co-option may have sapped the vitality of the Sardinian cultural system, but the ways of the inland peasant remained relatively unchanged. These are some of the reasons that make Sardinia such a fascinating island society.

References cited

Arslan, E.A., F. Barecca, F. Lo Schiavo, eds. (1985). *Civiltà Nuragica.* Electa, Milano.

Aubet, M.E. (1993). *The Phoenicians and the West.* Cambridge University Press, Cambridge.

Blake, E. (1998). Sardinia's nuraghi: four millennia of becoming. *World Archaeology* 30:59–71.

Blake, E. (2001). Constructing a Nuragic locale: the spatial relationship between tombs and towers in Bronze Age Sardinia. *American Journal of Archaeology* 105: 145–162.

Boardman, J. (1964). *The Greeks Overseas.* Harmondsworth, London.

Braudel, F. (1976). *The Mediterranean and the Mediterranean World in the Age of Philip II: Volume 2.* Harper Colophon, New York.

Broodbank, C. (2000). *An Island Archaeology of the Early Cyclades.* Cambridge University Press, Cambridge.

Broodbank, C. and T.F. Strasser (1991). Migrant farmers and the Neolithic colonization of Crete. *Antiquity* 65: 233–245.

Budd, P., A.M. Pollard, B. Scaife and R.G. Thomas (1995). Oxhide ingots, recycling and the Mediterranean metals trade. *Journal of Mediterranean Archaeology* 8: 1–32.

Cherry, J. (1992). Palaeolithic Sardinians? Some questions of evidence and method. *Sardinia in the Mediterranean: A Footprint in the Sea,* edited by R.H. Tykot and T.K. Andrews. Monographs in Mediterranean Archaeology 3. Sheffield Academic Press, Sheffield. pp. 20–39.

Davidson, J. (1992). Greeks in Sardinia: myth and reality. *Sardinia in the Mediterranean: A Footprint in the Sea,* edited by R.H. Tykot and T.K. Andrews. Monographs in Mediterranean Archaeology 3. Sheffield Academic Press, Sheffield. pp. 384–393.

Dyson, S.L. (1985). *The Creation of the Roman Frontier.* Princeton University Press, Princeton.

Dyson, S.L. (1992). Roman Sardinia and Roman Britain. *Sardinia in the Mediterranean: A Footprint in the Sea*, edited by R.H. Tykot and T.K. Andrews. Monographs in Mediterranean Archaeology 3. Sheffield Academic Press, Sheffield. pp.484–492.

Dyson, S.L. (2000). The limited nature of Roman urbanism in Sardinia. *Romanization and the City: Creation, Transformations, and Failures*, edited by E. Fentress. Journal of Roman Archaeology Supplementary Series 38, 2000, Portsmouth. pp. 189–196.

Dyson, S.L. (2003). *The Roman Countryside.* Duckworth, London.

Dyson, S.L. and R.J. Rowland, Jr. (1992). Survey and settlement reconstruction in west central Sardinia. *American Journal of Archaeology* 96: 203–224.

Emerson, T.E. (1997). *Cahokia and the Archaeology of Power.* The University of Alabama Press, Tuscalosa, Alabama.

Gale, N.H. and Z. Stos-Gale (1995). Comments on 'Oxhide ingots, recycling, and the Mediterranean metals trade'. *Journal of Mediterranean Archaeology* 8: 33–41.

Goldsworthy, A. (2003). *The Fall of Carthage: The Punic Wars 265–146 BC.* Cassell Military Paperbacks, London.

Hofmeijer, G.K. and P.Y. Sondaar (1992). Pleistocene humans in the island environment of Sardinia. *Sardinia in the Mediterranean: A Footprint in the Sea*, edited by R.H. Tykot and T.K. Andrews, Monographs in Mediterranean Archaeology 3. Sheffield Academic Press, Sheffield. pp. 49–56.

Leighton, R. (1999). *Sicily Before History: An Archaeological Survey from the Palaeolithic to the Iron Age.* Cornell University Press, Ithaca, NY.

Lilliu, G. (1988). *La Civilta dei Sardi dal Paleolitico all'eta dei Nuraghi.* Il Maestrale, Turin.

Lilliu, G. (1995). *Cultura e culture. Storia e problemi della Sardegna negli scritti giornalistici di Giovanni Lilliu.* C. Delfino, Sassari.

LoSchiavo, F. (1988). Early metallurgy in Sardinia. *The Beginnings of the Use of Metals and Alloys*, edited by R. Maddin. The MIT Press, Cambridge, Mass. pp. 92–103.

Patton, M. (1996). *Islands in Time: Island Sociogeography and Mediterranean Prehistory.* Routledge: London,

Perlès, C. (1992). Systems of exchange and organization of production in Neolithic Greece. *Journal of Mediterranean Archaeology* 5: 115–164.

Renfrew, C. and J.M. Wagstaff, eds. (1982). *An Island Polity: The Archaeology of Exploitation on Melos.* Cambridge University Press, Cambridge.

Rowland, R.J., Jr. (1990). The production of grain in Roman Sardinia. *Mediterranean History Review* 5: 14–20.

Rowland, R.J., Jr. (2001). *The Periphery in the Center: Sardinia in the Ancient and Medieval Worlds.* BAR International Series 970, Oxford.

Sandars, N.K. (1978). *The Sea People Warriors of the Ancient Mediterranean 1250–1150 BC.* Thames and Hudson, London.

Skeates, R. (1993). Early metal-use in the Central Mediterranean region. *Accordia Research Papers* 4: 5–48.

Stos-Gale, Z.A. and N.H. Gale (1992). New light on the provenance of the copper oxhide ingots found on Sardinia. *Sardinia in the Mediterranean: A Footprint in the Sea*, edited by R.H. Tykot and T.K. Andrews. Monographs in Mediterranean Archaeology 3. Sheffield Academic Press, Sheffield. pp. 317–347.

Torrence, R. (1986). *Production and Exchange of Stone Tools.* Cambridge University Press, Cambridge.

Trump, D.H. (1984). The Bonu Ighinu Project and the Sardinian Neolithic. *Studies in Sardinian Archaeology* 1: 2–22.

Tykot, R. (1994a) Sea Peoples in Etruria? Italian Contacts with the Eastern Mediterranean in the Late Bronze Age. *Etruscan Studies* 1: 59–83.

Tykot, R. (1996). Obsidian procurement and distribution in the central and western Mediterranean. *Journal of Mediterranean Archaeology* 9: 39–82.

Tykot, R. (1997). Characterization of the Monte Arci (Sardinia) obsidian sources. *Journal of Archaeological Sciences* 24: 467–479.

Tykot, R. (1999). Islands in the stream: stone age cultural dynamics in Sardinia and Corsica. *Social Dynamics of the Prehistoric Central Mediterranean*, edited by R.H. Tykot, J. Morter, J.E. Robb. Accordia Specialist Studies on the Mediterranean 3, Accordia Research Institute, London. pp. 67–82

Van Dommelen, P. (1998). *On Colonial Ground: A Comparative Study of Colonialism and Rural Settlement in First Millennium BC West Central Sardinia.* Archaeologiscal Studies Leiden University 2, Leiden.

Webster, G.S and M. Teglund (1992). Toward the study of colonial-native relations in Sardinia from c. 1000 BC–AD 456. *Sardinia in the Mediterranean: A Footprint in the Sea*, edited by R.H. Tykot and T.K. Andrews. Monographs in Mediterranean Archaeology 3. Sheffield Academic Press, Sheffield: 448–473.

Chapter 14

'The isles afar off': taking a new look at Ireland's holy islands

Sharon A. Greene, University College Dublin

Introduction

Ireland is an island on the NW fringe of Europe with many smaller off-shore islands, most of which are situated off the western, Atlantic coast. Archaeological evidence tells us that some of these were inhabited or used by people from the Mesolithic (i.e., post ca. 9000BP) up to recent times and a small number are still occupied today. The period with which this paper is primarily concerned is known alternatively as the 'early medieval' or 'early Christian' period, ranging from about the fifth century AD to the twelfth century AD. As the latter term implies, this was the period that saw the arrival and spread of Christianity. It saw a number of other developments too, including the arrival of the written word, with the effect that it is the first period for which Ireland has indigenous historical records. The earliest of these were annals and religious texts, but from about the seventh century legal texts, hagiographical texts, myths, sagas and genealogies were written down. While the majority of surviving records date from about the seventh century onwards, some of these are considered to preserve older information. Even where they cannot be relied upon for historical fact with regard to politics and history, it is often the incidental information that provides the most interesting information for archaeologists—for example, saints' lives can tell us about daily life in the monastery. This historical information has been used at times to help explain archaeological evidence and vice versa (e.g., Murray 1979). In the case of island studies the historical records are useful when it comes to trying to understand attitudes to islands (see below) and also the nature of sea travel, for which there appears to be less archaeological evidence in this period, particularly for the west coast (Wooding 1996:6-7).

Viewing Irish islands over time

Ireland's off-shore islands first attracted archaeological, or antiquarian, attention in the nineteenth century—a period of extreme hardship for the people living in the islands, and in the west of Ireland in general. The land is still considered to be of very limited capability (Feehan 2003:228). The soils are generally poor and in the nineteenth century the land was seriously overpopulated. In the eighteenth and nineteenth centuries the population had expanded into marginal areas, including off-shore islands, which in many cases appear not to have been inhabited in the preceding few centuries. Those moving to the islands did not necessarily belong to a very strong maritime tradition. Although some larger craft were in use, the curragh, a hardy, skin-covered boat propelled by oar and sometimes a small sail, was the main water transport available. Many communities could not afford anything bigger. A discussion of fishing and boating traditions in Clare Island in the nineteenth and twentieth centuries shows that the inhabitants of this relatively large island with a primarily land-based economy were very late in adopting this type of boat. For example, there are records of two Clare islanders going to Inishkea around 1850 to see how the hide boats were made (Mac Cárthaig 1999:58).

Interested antiquarians in the nineteenth century were often frustrated in their efforts to get to off-shore islands by the weather (e.g., Macalister 1896:253). When they did manage to get to the islands it was usually a fleeting visit as overnighting was either not possible or not desirable, except with larger island communities like on Achill or the Aran Islands, as in many cases the islands were uninhabited. Therefore, the archaeological features visited were those most prominent in the landscape. These included large prehistoric stone forts and the ruined remains of small stone churches, often in cashels (stone-walled enclosures) and accompanied by inscribed stone crosses, beehive-shaped huts and burial grounds (figures 14.1 and 14.2).

Both local tradition and historical sources told them that these sites had been established by early saints. Many of the islands were still the destinations of pilgrimage (some continue to be so in the twenty-first century). Historical sources also told them that for a cleric in the early church, the highest form of worship was to become a hermit in some wild remote place, cutting himself off from society to better serve God (Stokes 1891:658). This seemed to provide the only reasonable explanation for the presence of churches in such remote and inaccessible places and all were duly given the label hermitage.

No attempt was made to investigate any potential secular use of the islands, however the modern inhabitants were considered by some antiquarians, and many scholars, artists and writers who followed them, to represent a rare remnant of 'real Ireland' (i.e., Ireland before the British Empire) in their undisturbed and peripheral existence on the edge of Europe (Robinson 1992:xv–xvii). In the late nineteenth and early twentieth century, various artists travelled to the islands to record this endangered way of life which some believed to have survived since prehistoric times.

In the nineteenth century, the early medieval period was considered the Golden Age of Ireland's past and many features of it were aspired to. These include the strong influence of the church at home and abroad, the strong rural character of society and its independence (Ireland had never been part of the Roman Empire and was not yet part of the British Empire). Therefore the Nationalist movement adopted a number of artefact and site-types from this period as symbols including ruined churches, high crosses and round towers. So, the islands with early Christian remains were implicitly marked out as 'sacred' to Irish identity in two ways—as storehouses of our ancient heritage both social and physical.

These nineteenth century attitudes had an immense effect on how twentieth century researchers approached island archaeology in Ireland. The assumption of remoteness is still deep-rooted. In the 1949 second edition of his widely read book *The Archaeology Of Ireland*, the only explanation R.A.S. Macalister could offer for the presence of the large stone forts (whose use we now know ranges from the Bronze Age to the early medieval period) on Aran and in other such "desolate" places on the west coast was that they were "the last shelter of expelled refugees, fleeing from some rapacious conqueror" who had driven them "step by step, back to the western coast and to the islands beyond" (1949:281). In an earlier edition of this book, he claimed that the island monasteries "afforded a retreat from the noisy world for those who wished to lead a life of contemplation, devotion, or study" (1928:244). These observations perhaps say more about the post-colonial Ireland Macalister was writing in than Iron Age or early medieval Ireland.

Relatively little research was carried out on the islands in the twentieth century, however significant contributions were made by Françoise Henry (1937, 1945, 1947, 1951, 1952) and more recently by Michael Herity (1977, 1983, 1989a, 1989b, 1990, 1995). In the past decade or so, publications covering archaeology and local history have also appeared and a small number of island sites have been excavated. However, the islands are still generally treated in a very singular fashion and the focus has remained strongly on the monastic sites (e.g., Marshall & Rourke 2000; Heraughty 1996) to the exclusion of any possible secular activity. The religious remains are still generally classed as hermitages, even where evidence has emerged of their growth into considerable monasteries (see below). This is despite the fact that the Archaeological Survey of Ireland has recorded ringforts and cashels (the typical secular settlement type of the period) on some of the larger islands and evidence is also emerging for unenclosed settlement in this period. At the beginning of the twenty-first century, the time is ripe to address the imbalances in this research and also to place these islands into a wider coastal landscape and economy for the first time.

The early medieval view—remote and distant isles?

Attempts to reconstruct the early medieval view of world geography using contemporary texts have shown the belief that the three continents of Europe, Asia and Africa were surrounded by a circular Ocean and that Britain and Ireland lay within this Ocean at the edge of Europe and thence at the edge of the civilized world (O'Loughlin 1997:14). The geographical works at this time were obviously strongly influenced by the Bible and other religious writings. Biblical references to the Ocean describe it as a dangerous place and the abode of demons (O'Loughlin 1997:15). Early Christian writings from the Mediterranean region said that an im-

Figure 14.1: Duvillaun viewed from the nearest mainland.

portant element of the life of a hermit or monk was to battle demons in remote places, such as deserts and it has thus been suggested that the island monks of Britain and Ireland could have envisaged themselves as being in a remote position with demons to battle in the Ocean (ibid:15). Thus there are tales of sea monsters being subdued, for example in the 'Life of Saint Columba' (ibid:22; Sharpe 1995, 197–8).

In AD 563 Saint Columba was forced to leave his native Ireland and established his monastery on the Scottish island of Iona (Sharpe 1995:12). This monastery thrived until the arrival of the Vikings in AD 802 (Ó Corráin 2001:17) and the cult of Saint Columba had widespread influence. Adomnán, an abbot of Iona wrote the 'Life of Saint Columba' a century later (Sharpe 1995:3) and was also responsible for the widely respected geographical text *De Locus Sanctis* written between AD 680 and 700 (O'Loughlin 1996:98). We know from his work that the earlier works of other European scholars available in Irish monasteries influenced their understanding of their geographical position in relation to the rest of the Christian world. While these and other available texts give us some information on the situation of Iona and of Ireland and Britain in the Ocean, there is little direct information relating to the western islands of Ireland. Nonetheless, indirect references, of which just a couple will be mentioned here, can be helpful in this respect.

The famous medieval tale 'Saint Brendan's Voyage' was first written down in the latter part of the eighth century AD (O'Loughlin 1999:1). On Brendan's voyage, which starts on the west coast of Ireland, various islands are encountered. O'Loughlin has noted that this tale is largely allegorical and there are two kinds of places encountered (1999:5). Some islands are like those that would be encountered as part of a normal ocean voyage. These islands of the 'real' world are referred to using concepts of time and distance that can be readily understood and are also the scenes of sinful behaviour. The last island encountered in this category is the 'Island of Delights', which is an island monastery of the kind that would have been familiar (through knowledge if not experience) to the writer's audience. It was described as an ideal monastery to which other monasteries should aspire (ibid:8). After leaving this they travel through a fog, losing sense of space and time and ultimately end up at the bright, heavenly 'Island of the Promised Land'. On their return journey they visit the 'Island of Delights' again (ibid:10).

Figure 14.2: Crucifixion slab on Duvillaun.

O'Loughlin has pointed out that an important lesson was being emphasized here—i.e., that leading a perfect monastic life, such as was happening on the 'Island of Delights', brought one closer to the gates of Heaven (ibid:17).

It could be argued that the insular setting of the 'Island of Delights' and 'Island of the Promised Land' tell us something about idealized attitudes to monastic islands of the time. However, while the monks on the island monasteries would have had an awareness of their position in the spiritual geography, they were also (necessarily) keenly aware of the very real nature of the Ocean. To the writers of the Bible the Ocean was a distant and dangerous thing but to the island monks of the Atlantic coastline it was much closer and had perhaps even more dangers. We know from the 'Life of Saint Columba' that they had excellent knowledge of winds, tides and the dangers of sea faring (O'Loughlin 1997:20) but this does not appear to have prevented them from carrying out numerous sea journeys for a variety of reasons. It is interesting in this regard that early Irish writers frequently used water imagery, and unlike their Mediterranean fathers, it was more often used in a positive sense. One Early Irish nature poet described the sea as "teeming with sea-life;/ home of seals,/ sporting, splendid" (Muhr 1999:193), emphasising the point that for coastal and island dwellers the sea was an important supplier of sustenance.

About AD 825, An Irish monk named Dicuill wrote the following: "Around our island Ireland there are islands, but some small and some minimal, Next to the island Britain [there are] many, some great, some small and some middling. There are some in the southern sea and some in the western, but they abound more in the part of the north west and north of that [island]. In some of these I have dwelt, some I have entered, some I have only seen, some I have read about" (Howlett 1999:129).

In the same document he relates information from "a certain religious priest" about islands situated two days voyage north of the northern islands of Britain in which "throughout almost one hundred years hermits from our [Ireland] making ships voyages have dwelt" (ibid:130). It is now generally accepted that the islands he refers to are the Faroe Islands (Arge 2000:155; see figure 14.3).

It appears that many monks went in search of "a place of retreat" (Sharpe 1995:127) in the Ocean and this probably explains the purpose or origin of some of the remains seen there today. However, there are also suggestions that we cannot assume all had this origin. Tírechán's *Collectanea*, written in about AD 680, mentions the fifth century Bishop Cethiacus, a Briton and disciple of Saint Patrick who appears to have had an establishment on an island off the Mayo coast (MacNeill 1964:130; Charles-Edwards 2000:48). If he was part of the mission to spread the word of Christ among the Irish, he is unlikely to have started with a hermitage. The position of his church is more likely to be explained as the donation of what was considered marginal land to a churchman by the local rulers, as was common at this time (Bitel 1986:17).

These few historical examples give an interesting impression of the relationship between early Irish churchmen and islands. In the earliest, it appears possible that an island, which is considered tribal property, is granted to an early ecclesiastic in the conversion period. By the time Brendan's Voyage was written down it appears the Irish clerics were already proficient at sea travel. Dicuill suggests that Irish monks were living on distant northern islands by about AD 725 (Howlett 1999:134) and in his own time, a century later, these islands were being visited by other clerics for relatively short periods of time (ibid). Dicuill himself was a well-travelled man familiar with the islands of both the Irish and British coasts (see quote above). To what degree the islands were considered special in a monastic context is still not clear. What is clear however is that most of them were within the sphere of travel and influence of others, suggesting that even those considered distant were not necessarily permanently isolated or remote.

Archaeological evidence

Early medieval texts tell us something of contact between island settlements and the wider world and archaeology can expand on this evidence. Excavations on the western Irish islands have been relatively rare and we are largely reliant on stray finds made over the years. Nonetheless, current research is beginning to suggest, and in some cases confirm, that the inhabitants of these islands were not culturally isolated.

Ardoileán or High Island is situated off the west Galway coast and is one of those islands that seems a place suitable only for a lonely hermitage (see figure 14.4). It is surrounded on all sides by high cliffs and, even at the recognized landing places, you literally take your life in your hands climbing to the island's surface, which is covered with rough, boggy grassland. The monastic site here was founded by St Fechín around the year AD 650 (Molloy et al. 2000:236). Apart from the well-preserved monastic remains

Figure 14.3: Principle places mentioned in the text, excluding the coasts of Mayo and Galway.

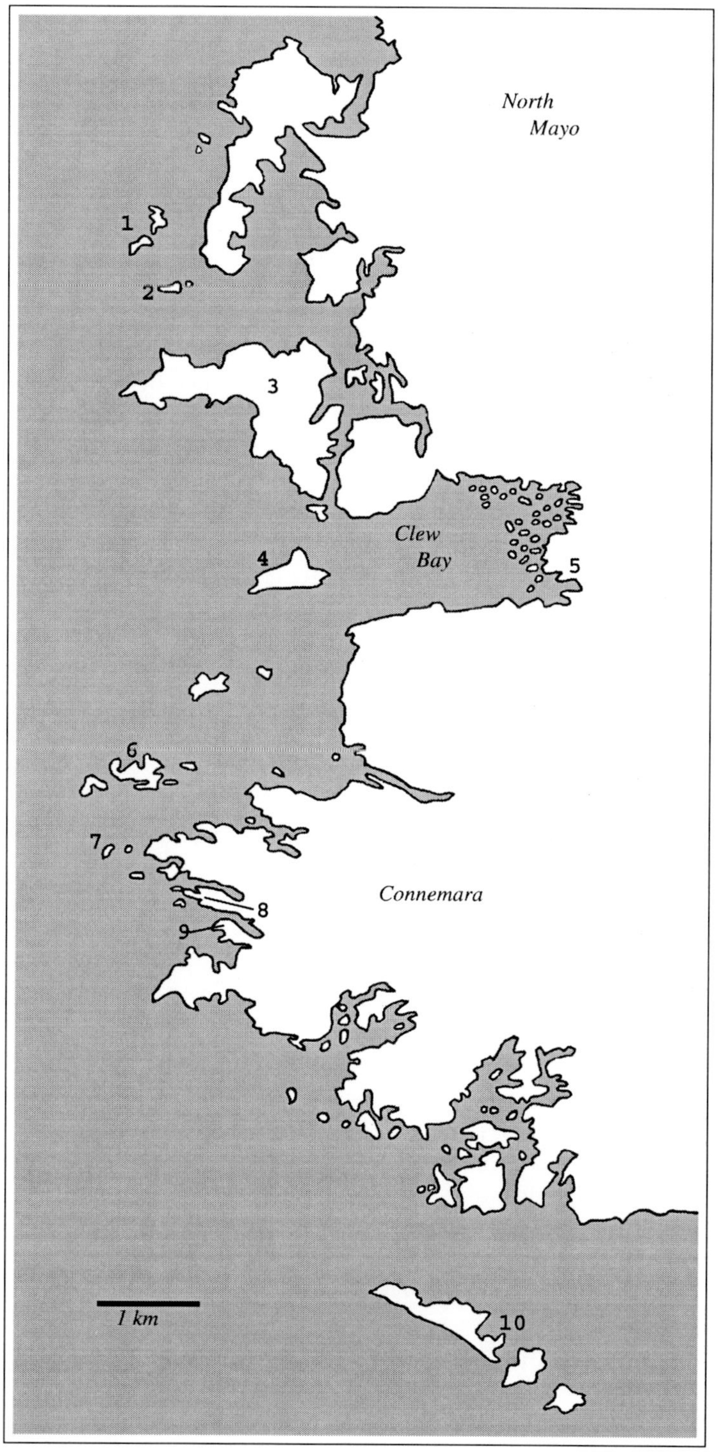

Figure 14.4: Principle places mentioned in Mayo and Galway: 1=Inishkea Islands, 2=Duvillaun, 3=Achill Island, 4=Clare Island, 5=Cushalugort, county Mayo, 6=Inishbofin, 7=High Island, 8=Eyrephort, county Galway, 9=Truska, 10=Aran Islands.

there are the remnants of a sophisticated early medieval watermill system that included the construction of dam, millrace, millpond and mill (figure 14.5).

Horizontal watermills were common in western Ireland in this period, but it has been estimated that to justify the construction of one it would have to be serving a population of 100 people or more (Rynne 2000:205). If this is the case, St Fechín's hermitage must have grown significantly by the time the mill was built in the late eighth/ninth century. The presence of the mill also suggests that this religious community was part of a network through which such technical innovations had travelled ultimately from mainland Europe and which also eventually brought knowledge of these horizontal water mills from the west of Ireland to the Western Isles and mainland of Scotland (Rynne 2000:198).

A similar strand of evidence with an agricultural theme is provided by rotary querns found on Inishkea North, off the north-west coast of Mayo. Caulfield noted the discovery of a quern of Scottish type in association with seventh/eighth century occupation and suggested that these 'Type D' querns may have originated in Scotland (1966:68). Also significant in the context of Rynne's suggestion that the High Island type of horizontal watermill influenced mainland Scottish watermills, is the fact that Caulfield's 'Type K' rotary quern, which has an unusual knob acting as a handle are only found in Mayo and mainland Scotland (1966:76). In this case he suggests the influence travelled in the opposite direction. Two of the four Irish examples are from Inishkea North island.

Christianity is often given the credit for the arrival and spread of new technology including changes in agricultural practice. This includes the general increase in arable farming in the early medieval period after the 'Late Iron Age Lull'. However, available palynological evidence from around Ireland, including a number of sites in the west is beginning to show that the increase in farming activity may actually predate the establishment of the monasteries (Hall 2000:368). This is even the case for one of the island monasteries. The increased pastoral and arable activity on Inishbofin appears to pre-date the historically well recorded establishment of a monastery there by Saint Colman in AD 665 (O'Connell & Ní Ghráinne 1994:82). There is a tradition that a chieftain called Guairim was a contemporary of Saint Colman's on Inishbofin (Concannon 1993:3) suggesting that it is a mistake to ignore secular island settlement.

Another interesting observation that can be made about agricultural remains, both micro- and macroscopic, is in regard to the kind of cereals used and even grown on the western islands. We know that barley, oats and rye were the most common cereals used in Ireland in this period. There is limited evidence for wheat and the impression from historical sources was that it was considered a luxury crop (Monk 1986:33). The climate and environment of the off-shore islands was best suited to barley (McCormick 1997:55), however on Inishbofin, wheat was the main crop grown. Rye was the main crop on High Island, but wheat and barley were also grown in small quantities (Molloy et al. 2000:240). Could the predominance of wheat be a cultural preference on the part of the clerics? If it is it does not go very well with the idea of self-denial often associated with hermits. Further research is required to allow us to comment on whether this is a wider phenomenon within Irish monasticism or if it is a specifically insular phenomenon. There is certainly evidence from other monastic sites of food types outside the expected range being consumed e.g., the high incidence of deer and seal bone on Iona (McCormick 1997:57–8).

To turn to more domestic artefacts, the discovery of pottery on an early medieval Irish excavation site is rare. Apart from the occasional discovery of imported wares, indigenous pottery, known as 'souterrain ware', has a localized distribution in the north east of the country (Ryan 1973:630). In recent years the distribution has not changed dramatically (Edwards 1990:74). However, various sherds of a plain, unglazed ware were found in excavations on Inishkea North by F. Henry, which have recently been identified as souterrain ware (Sandes, forthcoming). This suggests that either the pottery or the technique for making it somehow arrived on that island from the north-east. The fact that it does not occur in the intervening mainland areas between North Mayo and Ulster, suggests that it came there by sea, around the north coast.

The question of Scandinavians on the west coast

Assumptions about the remoteness of the west coast of Ireland are also perhaps partly responsible for the lack of research (until very recently) into the potential for archaeological evidence for the presence of Vikings on the western coastline. The Viking Age began when Scandinavian (mostly Norse) seamen and warriors crossed the North Sea to Britain and Ireland at the end of the 8th century, initially to pillage and plunder.

Figure 14.5: Stone-built monastic remains on High Island with millpond in background.

However, many ended up settling and in Ireland are given credit for the foundation of many of the coastal cities. Trade and travel routes in the Irish Sea and around the south coast as far as Limerick are well attested historically and archaeologically during the Viking Age. However, evidence on the west coast has been more limited. There are historical references to early raids on the west coast on the islands of Inishmurray and Inishbofin and at Roscam in Co Galway and slightly later references to battles between the 'foreigners' and local tribes at Connemara and Clew Bay (Ó Corráin 2001:17). Archaeology adds one Viking grave on the Connemara coast at Eyrephort (Sheehan 1988), a hoard of silver bracelets from Cushalogurt (Sheehan 2001:52) and more recently a small house accompanied by two burials and a Hiberno-Norse bone comb at Truska (Gibbons and Kelly 2003; see figure 14.4).

A series of stone buildings excavated on Inishkea North in 1946/50 may be offering evidence of either more permanent settlement or enduring influence here however (figure 14.6). Their structure, layout and internal features are very similar to complexes of Scandinavian stone buildings found elsewhere on the Atlantic coastline, particularly on the Western and Northern Isles of Scotland (e.g., Sharples & Parker Pearson 1999:53). Artefactual evidence to back up this interpretation is relatively slim, but includes a circular whalebone disc (Henry 1945:146), a large number of soapstone articles (mostly spindle whorls), and a bone comb of Dunleavy's Class G (1988:368). The comb is of a type found in Norway, Sweden, Greenland, Frisia, Scotland and England and Dunleavy dates its use in Ireland to between the ninth and thirteenth centuries AD (ibid). The use of whalebone is more common in Scotland than Ireland and Caulfield (1966:77) noted a parallel for this disc from an excavation in North Uist. Soapstone was a valuable commodity to the Norse and in the Scottish Isles is frequently used as an indicator of a Norse presence (Owen 1999:33). It occurs rarely on native Irish sites.

Thus evidence is slowly beginning to emerge that suggests connections between western Ireland and Scotland that may pre-date and include Scandinavian influence. Continuing research into the material culture of the inhabitants of these islands may reveal further evidence for Scandinavian presence and a Hiberno-Scandinavian culture as has been recognized elsewhere in the country leading to the acceptance of the presence of a sea route around the north and west coasts of Ireland in the Viking Age.

Figure 14.6: Complex of stone structures on Inishkea North (the low enclosing wall around the north and east sides is modern).

Conclusion

An interesting picture is beginning to emerge for the early medieval period in the islands of the west coast of Ireland. If a regularly used sea route along this coastline in the early medieval period can be recognized, even if only in the later part of the period, it forces us to look differently at the so-called remote holy islands and explains things like the presence of souterrain ware on Inishkea North. In the twenty-first century therefore, the islands and seascape are beginning to emerge as a zone of varied and complex potential, and activities and tales of idealized hermitages on the 'isles afar off' are slowly being replaced by the histories of ever-changing and developing island communities that can no longer be dismissed as insular, in the traditional sense of the word.

Acknowledgements

My doctoral research is funded by an Irish Research Council for the Humanities and Social Sciences Government of Ireland Scholarship. Sincere thanks to Matt Campbell and James Conolly (and all who took part) for a successful and stimulating conference; and to my supervisor, Dr. Aidan O'Sullivan for continued thought-provoking discussions.

References cited

Arge, S.V. (2000). Vikings in the Faeroe Islands' life. In W.W. Fitzhugh & E.I. Ward, (eds.) *Vikings—the North Atlantic Saga*, pp. 154–163. Washington: The Smithsonian Institute.

Bitel, L.M. (1986). Women's Ecclesiastic Enclosures in Early Ireland: A Study of Female Spirituality and Male Monastic Mentalities. *Journal of Medieval History* 12: 15–36.

Caulfield, J.J. (1966). *The Rotary Quern in Ireland.* Unpublished MA thesis, Department of Archaeology, University College Dublin.

Charles-Edwards, T. (2000). *Early Christian Ireland.* Cambridge: Cambridge University Press.

Concannon, K (ed.) (1993). *Inishbofin through time and tide.* Athlone: FÁS.

Dunleavy, M. (1988). A classification of early Irish combs. *Proceedings of the Royal Irish Academy* 88C: 341–422.

Edwards, N. (1990). *The Archaeology of Early Medieval Ireland.* London: B.T. Batsford Ltd.

Feehan, J. (2003). *Farming in Ireland—History, Heritage and Environment.* Faculty of Agriculture, University College, Dublin.

Gibbons, E.K. & Kelly, E.P. (2003). A Viking Age farmstead in Connemara. *Archaeology Ireland* 63: 28–32.

Hall, V. (2000). The documentary and pollen analytical records of the vegetational history of the Irish landscape AD 200-1650. *Peritia* 14: 342–371.

Henry, F. (1937). Early Christian slabs and pillar-stones in the West of Ireland. *Journal of the Royal Society of Antiquaries of Ireland* 67: 265–79.

Henry, F. (1945). Remains of the early Christian period on Inishkea North, Co Mayo. *Journal of the Royal Society of Antiquaries of Ireland* 75: 127–55.

Henry, F. (1947). The antiquities of Caher Island. *Journal of the Royal Society of Antiquaries of Ireland* 77: 23–38.

Henry, F. (1951). New monuments from Inishkea North, Co Mayo. *Journal of the Royal Society of Antiquaries of Ireland* 81: 65–88.

Henry, F. (1952). A wooden hut on Inishkea North, Co Mayo. *Journal of the Royal Society of Antiquaries of Ireland* 82: 163–78.

Heraughty, P. (1996). *Inishmurray. Ancient Monastic Island.* Dublin: O'Brien Press.

Herity, M. (1977). The High Island Hermitage. *Irish University Review* 7: 52–69.

Herity, M. (1983). The buildings and layout of Early Irish monasteries before the year 1000. reprinted in Herity, M. (1995). *Studies in the Layout, Buildings and Art in Stone of Early Irish Monasteries*, pp. 19-56. London: Pindar Press.

Herity, M. (1989a). Early Irish Hermitages in the Light of the *Lives* of Cuthbert. In G. Bonner, David Rollason, and Clare Stancliffe (eds.), *St Cuthbert, His Cult and His Community to AD 1200*, pp.45–64. Woodbridge: The Boydell Press.

Herity, M. (1989b). Cathair na Naomh and its cross-slabs. *Cathair na Mart* 9: 91–100.

Herity, M. (1990). The hermitage of Ardoileán, county Galway. *Journal of the Royal Society of Antiquaries of Ireland* 120: 65–101.

Herity, M. (1995). Two island hermitages in the Atlantic: Rathlin O'Birne, Donegal and Caher Island, Mayo. *Journal of the Royal Society of Antiquaries of Ireland* 125: 85–128.

Howlett, D. (1999). Dicuill on the islands of the North. *Peritia* 13: 127–134.

Mac Cárthaigh, C. (1999). Clare Island Folklife. In C. Mac Cárthaigh & K. Whelan (eds.), *New Survey of Clare Island. Volume I: History and Cultural Landscape*, pp.41–72. Dublin: Royal Irish Academy.

Macalister, R.A.S. (1896). Omey, Co Galway. *Journal of the Royal Society of Antiquaries of Ireland* 26:253.

Macalister, R.A.S. (1928). *The Archaeology of Ireland* (1st edition). London: Methuen & Co. Ltd.

Macalister, R.A.S. (1949). *The Archaeology of Ireland* (2nd edition). London: Methuen & Co. Ltd.

MacNeill, E. 1964 (J. Ryan, ed.) *Saint Patrick.* Dublin: Clonmore & Reynolds.

Marshall, J.W. & Rourke, G.D. (2000). *High Island, An Irish monastery in the Atlantic.* Dublin: Town House and Country House.

McCormick, F. (1997). Iona: the archaeology of the early monastery. In C. Bourke (ed.), *Studies in the Cult of Saint Columba*, pp. 45–68. Dublin: Four Courts Press.

Molloy, K., Fuller, J.L. & Conaghan, J. (2000). Vegetation and land-use history on High Island: the results of preliminary investigations. In J.W. Marshall, & G.D. Rourke, *High Island, An Irish monastery in the Atlantic*, pp. 233–243. Dublin: Townhouse.

Monk, M. (1986). Evidence from macroscopic plant remains for crop husbandry in prehistoric and early historic Ireland: a review. *Journal of Irish Archaeology* 3: 31–36.

Muhr, K. (1999). Water imagery in Early Irish. In F. Kelly & M. Ó Ríordáin (eds.), *Celtica Volume* 23, *Essays in Honour of James Patrick Carney*, pp. 193–210. School of Celtic Studies, Dublin Institute for Advanced Studies.

Murray, H. (1979). Documentary evidence for domestic buildings c.400-1200 in the light of archaeology. *Medieval Archaeology* 23: 81–98.

O'Connell, M. & Ní Ghráinne, E. (1994). Inishbofin: Palaeoecology. In P. Coxon & M. O'Connell (eds.), *Clare Island and Inishbofin*, pp. 61–101. Dublin: Irish Association for Quaternary Studies (IQUA).

Ó Corráin, D. (2001). The Vikings in Ireland. In A.-C. Larsen (ed.), *The Vikings in Ireland*, pp. 17–27. Roskilde: The Viking Ship Museum.

O'Loughlin, T. (1996). The view from Iona: Adomnán's mental maps. *Peritia* 10:98–122.

O'Loughlin, T. (1997). Living in the Ocean. In C. Bourke (ed.), *Studies in the Cult of Saint Columba*, pp. 11–23. Dublin: Four Courts Press.

O'Loughlin, T. (1999). Distant Islands: the topography of holiness in the *Nauigatio Sancti Brendani.* In M. Glasscoe (ed.), *The Medieval Mystical Tradition—England, Ireland and Wales*, pp. 1–20. Cambridge: D.S. Brewer.

Owen, O. (1999). *The Sea Road—A Viking Voyage Through Scotland.* Edinburgh: Canongate Books/ Historic Scotland.

Robinson, T. (1992). Place/Person/Book, Introduction to J.M. Synge, *The Aran Islands*, pp. vii–1. London: Penguin Books.

Ryan, M. (1973). Native pottery in early historic Ireland. *Proceedings of the Royal Irish Academy* 73C: 619–645.

Rynne, C. (2000). The early medieval monastic watermill. In J.W. Marshall, & G.D. Rourke, *High Island, An Irish monastery in the Atlantic*, pp. 185-214. Dublin: Townhouse.

Sandes, C. (forthcoming) The pottery. In S.A. Greene. *Françoise Henry's Excavations on Inishkea North, Co Mayo (1946 and 1950).* Seandálaíocht Monograph, Department of Archaeology, University College Dublin.

Sharpe, R. (trans.) (1995). *Adomnán of Iona: Life of St Columba.* London: Penguin Classics.

Sharples, N. & Parker Pearson, M. (1999). Norse settlement in the Outer Hebrides. *Norwegian Archaeological Review* 32: 41–62.

Sheehan, J. (1988). A reassessment of the Viking burial from Eyrephort, Co Galway. *Journal of the Galway Archaeological and Historical Society* 41: 60–72.

Sheehan, J. (2001). Ireland's Viking Age Hoards: Sources and Contacts. In A.-C. Larsen (ed.), *The Vikings in Ireland*, pp. 51–60. Roskilde: The Viking Ship Museum.

Stokes, G.T. (1891). The island monasteries of Wales and Ireland. *Journal of the Royal Society of Antiquaries of Ireland* 21: 658–664.

Wooding, J.M. (1996). *Communication and Commerce along the Western Sealanes AD 400-800*. Oxford: British Archaeological Reports, International Series 654.